BACKCASTING FOR A SUSTAINABLE FUTURE
the impact after 10 years

Jaco Quist

ISBN 978-90-5972-175-3

EBURON ACADEMIC PUBLISHERS
P/O Box 2867
2601 CW Delft
The Netherlands
www.eburon.nl / info@eburon.nl

Cover and graphic design: Annemarie van den Berg & Jonas Piet
Language editing: Gert Stronkhorst

BACKCASTING FOR A SUSTAINABLE FUTURE
the impact after 10 years

Proefschrift

ter verkrijging van de graad van doctor aan de Technische Universiteit Delft,
op gezag van de Rector Magnificus prof.dr.ir J.T. Fokkema,
voorzitter van het College voor Promoties,
in het openbaar te verdedigen op woensdag 11 april 2007 om 12.30 uur

door

Jacobus Nicolaas QUIST
landbouwkundig ingenieur
geboren te Oostburg

PREFACE

Dreaming about the futures we like and desire, who has not had that experience? At the same time we do not know the future in advance. Thus, thirteen years ago, when I started my quest on backcasting, sustainable futures and system innovations in various ways, I did not know that I would spend a major part of my working life until now on these subjects. But it worked out this way, after I was employed at the Sustainable Technology Development programme for a project on sustainable meat alternatives in 1994. There I became acquainted with backcasting for the first time. Afterwards, in 1997, I continued at the Technology Assessment Group of Delft University of Technology, where I conducted a backcasting experiment on sustainable nutrition and households. In addition to considerable teaching duties, I could also extend my understanding of innovation theory and technology assessment methods. All this merged together when I started writing this thesis in 2003 on the follow-up and spin-off of backcasting experiments.

While a thesis means conducting research and writing on your own, it is also a learning process that depends on and benefits from many other persons. To start with, I would like to thank all the interviewees. Without their willingness to talk about the backcasting experiments and their follow-up and spin-off, it would not have been possible to write this thesis. I would also like to thank my supervisors Wil Thissen and Philip Vergragt who both provided very useful, but complementary inputs and feedbacks during the course of this research. I would also like to thank Karel Mulder for providing the opportunity to write this thesis. I also thank him, as well as John Grin, for their occasional suggestions and useful discussions. Last but not least, I want to thank the members of my PhD committee for their willingness to be on the committee.

No researcher can do without colleagues; therefore, many thanks to my former and current academic and non-academic colleagues at what used to be the Technology Assessment Group and what is nowadays the Technology Dynamics and Sustainable Development Group at the faculty of Technology, Policy and Management of Delft University of Technology. When I got more research time, they had to do a larger share of the teaching. In particular, I would like to thank my colleagues Linda Kamp and Frida de Jong for reading various draft chapters. I am also pleased with the support of my paranimfs, who are my colleague Mariette Overschie and my brother Conno Quist. Finally, this book has benefited strongly from the language corrections by Gert Stronkhorst, as well as from the cover and graphic design by Jonas Piet and Annemarie van den Berg. All three did a great job.

In the final stage of writing I received support from various friends and family members of Annemarie Kamerbeek and myself, who took care of our daughter Maud during many weekends. Finally, although I have learned a lot from conducting this research as well as from writing this thesis, during this period I have learned the most through my partner Annemarie Kamerbeek, as well as from our daughter Maud. We went through some very hard times, as well as good times. That is why I dedicate this book to Annemarie.

Jaco Quist, March 2007

CONTENTS

Preface __ v

Contents __ vi

1. Introduction __ 9

1.1 Sustainable development and system innovations __ 9

1.2 Participatory backcasting __ 11

1.3 Research questions and relevance __ 12

1.4 Research approach __ 14

1.5 Outline of this thesis __ 15

Notes __ 16

2. An exploration of backcasting and related approaches __ 17

2.1 Introduction __ 17

2.2 Backcasting: a brief history __ 18

2.3 Comparing four backcasting approaches __ 24

2.4 A methodological framework for participatory backcasting __ 28

2.5 Conclusions __ 30

Notes __ 31

3. In search of useful theories and concepts __ 33

3.1 Visions __ 33

3.2 Learning __ 42

3.3 Stakeholder participation __ 47

3.4 System innovations and sustainability __ 51

3.5 System innovation theories __ 54

3.6 Networks __ 59

3.7 Conclusions __ 62

Notes __ 62

4. Conceptual framework and research methodology __ 65

4.1 Defining the research focus __ 65

4.2 Conceptual framework __ 68

4.3 Propositions __ 77

4.4 Research methodology __ 81

Notes __ 89

5. Backcasting for sustainable protein foods and its impact __ 91

5.1 Introduction __ 91

5.2 The NPF backcasting experiment __ 93

5.3 Analysis __ 97

5.4 Follow-up and spin-off __ 107
5.5 Further analysis __ 112
5.6 Conclusions __ 122
Notes __ 125

6. The impact of backcasting for sustainable household nutrition __ 129
6.1 Introduction __ 129
6.2 The SHN backcasting experiment __ 132
6.3 Analysis __ 138
6.4 Follow-up and spin-off __ 145
6.5 Further analysis __ 148
6.6 Conclusions __ 152
Notes __ 154

7. Backcasting for Multiple Sustainable Land-use in rural areas __ 157
7.1 Introduction __ 157
7.2 The MSL backcasting experiment __ 160
7.3 Analysis __ 165
7.4 Follow-up and spin-off __ 174
7.5 Further analysis __ 180
7.6 Conclusions __ 189
Notes __ 192

8. Comparing cases and testing propositions __ 195
8.1 Introduction __ 195
8.2 Comparing the backcasting experiments __ 195
8.3 Comparing follow-up and spin-off __ 201
8.4 Evaluating the propositions __ 207
8.5 Internal and external factors __ 212
8.6 Evaluating research methodology and indicators __ 214

9. Conclusions, reflections and recommendations __ 217
9.1 Conclusions __ 217
9.2 Reflections __ 221
9.3 Backcasting and methodological reflections __ 231
9.4 Recommendations __ 239

Epilogue: an evocation of an ideal backcasting experiment __ 242
References __ 244
Appendix A Interview checklist __ 265
Appendix B List of interviews and contacts __ 266
Abbreviations __ 268
Summary __ 271
Samenvatting __ 277
Curriculum Vitae __ 284

1

INTRODUCTION

This chapter introduces the need for system innovations towards sustainability (1.1), before introducing the possible contribution of participatory backcasting to such system innovations (1.2). It also proposes several research questions how to study this contribution (1.3) and describes the approach to this research (1.4), as well as the outline of this thesis (1.5).

1.1 Sustainable development and system innovations

Contemporary societies face the challenge of realising sustainable development and have to deal with underlying persistent and complex sustainability problems. In the coming decades industrialised countries like the Netherlands need to reduce their environmental burden enormously, especially when taking into account that the world population will increase, global wealth will grow considerably, and a fair distribution of the wealth growth between industrialised countries and developing countries is highly needed. Some, like Von Weizsäcker *et al.* (1997), argue that it is necessary to improve the environmental efficiency in the industrialised world by a factor 4, which would enable us to double current wealth, while halving the environmental burden. Others advocate that a tenfold reduction of materials flow per unit of service, including related production, is required over the next 30-50 years (Schmidt-Bleek 1994), or that it is necessary to meet societal needs even twenty times more efficiently in environmental terms by 2050, thus with a factor 20[1] (Weaver *et al.* 2000).

Sustainable development has been defined as *"a development in which the needs of the present generation are fulfilled in such a way that future generations will be able to meet their needs too"* (WCED 1987: 43). In themselves, sustainability and sustainable development are complex, ambiguous and explicitly normative concepts. In addition to involving long time frames, these concepts cover multiple aspects, multiple levels, multiple interpretations, potentially conflicting interests and numerous actors. Although it is not known what sustainability exactly is, sustainable development is widely considered as a promising and desirable direction, leading to a highly reduced environmental burden, a major reduction in the use of resources and a fairer distribution of wealth. This should lead to a much better balance between ecological, economic and social aspects than the one that exists today.

Obviously, sustainable development represents a very complex challenge. For instance, the pursuit of environmental improvement with a factor 10 to 20 requires technological, cultural, organisational and institutional changes at the level of socio-technical systems. The different types of changes are related to one another and develop in a co-evolutionary way, which enhances the complexity involved. Changing socio-technical systems also requires using a time frame of several decades and involving many actors. Examples of such socio-technical systems include specific industrial sectors, households, agriculture, the transportation of goods and the mobility of persons.

Several terms are currently used to describe this type of system change for the pursuit of sustainable development, such as system innovations towards sustainability (Quist and Vergragt 2004, Elzen *et al.* 2004), transformations of socio-technical systems or industrial transformations (Olsthoorn and Wieckzorek 2005), transitions towards sustainability (Rotmans *et al.* 2001, Elzen *et al.* 2004), and shifts to sustainable production and consumption systems. Geels (2005) refers to technological transitions, which he describes as the change process from one socio-technical system to another. Essentially, all these terms and concepts cover roughly similar concepts, although at the same time they are also slightly different when it comes to less decisive aspects. Differences depend, for instance, on where the boundaries are put and how exactly the definitions are phrased. In this thesis the term system innovation towards sustainability is used; it is defined as *"the transformation process from one socio-technical system to another, sustainable one"*. Despite the limited differences between the various concepts and terms, in this thesis I assume that several system innovations towards sustainability together make up a transition. Consequently, the transition to sustainable agriculture in the Netherlands can be seen as consisting of several system innovations towards sustainability, such as related to arable land farming, greenhouse and arable land horticulture, as well as to livestock production and dairy farming.

Addressing changes at the level of socio-technical systems includes dealing with uncertainty and complexity. Uncertainty is inherent to long term futures. Complexity is determined by the large number of variables involved, the large number of actors and their interests at stake, the combination of technological and non-technological changes, as well as the complicated social processes taking place during their development and implementation in society. Thus, complex sustainability problems and the system transformations necessary for dealing with them are multi-actor, multi-level and multi-aspect; this includes both technical and social aspects (e.g. Rotmans *et al.* 2001, Elzen *et al.* 2004).

Transformation of socio-technical systems[2] is further complicated as current institutions, structures and rule systems guide existing practices within existing socio-technical systems. As a consequence, existing rule systems are not equipped to solve new complex and persistent problems[3] like sustainability problems (Grin 2006, Grin *et al.* 2004), or to deal with disruptive and rapid changes. Instead, existing structures and institutions gradually emerge from particular practices and develop gradually. The resulting structures and institutions support specific practices, activities and patterns of activities by actors, and make these recursive and resilient to changes, while constraining others[4]. However, at the same time the actions and interactions of actors are the seeds of change. As a result, adaptations in existing practices and the emergence of new practices may eventually lead to new institutions and structures, and to new practices and adjusted socio-technical systems. It is exactly this mechanism that allows system innovations towards sustainability to develop.

There is an interesting relationship between sustainable development and system innovations towards sustainability and what Beck (e.g. 1997, 2006) has called reflexive modernisation. Reflexive modernisation is the response to regular or simple modernisation that has resulted in the present socio-economic system with its wealth in the developed countries, its unequalled use of resources, serious environmental problems and global inequity. Reflexive modernisation aims at preventing unwanted social side-effects, such as the unfair distribution of risks; it requires adopting an integral viewpoint on new technologies and taking into account various actor perspectives as well as a range of social aspects[5]. Reflexive modernisation is thus a way of development that deals more adequately and timely with side-effects or, even better, prevents them beforehand.

In the view of Beck (2006: 33), reflexive modernisation requires a fundamental transformation of industrialised societies, in which the quest for sustainable development is an important driver. Sustainable development can thus be seen as a way of dealing with Beck's plea for reflexive modernisation (Grin 2006), although it is certainly more ambitious in its social and environmental goals (e.g. a fair distribution of wealth and resources). As a consequence, pleas are heard not only for system innovations towards sustainability, but also for reflexive governance of sustainable development and system innovations towards sustainability (Voß *et al.* 2006)[6].

1.2 Participatory backcasting

Addressing complex sustainability problems by system innovations towards sustainability requires participatory approaches. Such approaches should have a long-term system orientation and take a broad notion of sustainability into account, as well as the social dynamics of complex social change processes. Stakeholder involvement is crucial. Not only are their stakes affected, stakeholders also have essential knowledge and necessary resources.

Participatory backcasting is one such approach . It has considerable potential to explore and evaluate possible system innovations towards sustainability (Weaver *et al.* 2000, Quist and Vergragt 2006). Backcasting means literally looking back from the future. It can be seen as the opposite of forecasting, which looks from the present to the future in a prospective way. In backcasting the desirable future is envisaged first, before it is analysed how it could be achieved by looking back from this future and identifying what steps need to be taken to bring about that future. In addition, it is also possible to look back from an undesirable future and to determine what to do to avoid this (Robinson 1990). While most existing scenario and foresighting approaches focus on likely or possible futures, the major distinction with backcasting is its explicit normative nature, based on setting normative goals and constructing normative desirable futures. Dreborg (1996), for instance, has argued that backcasting is particularly interesting in the case of complex and persistent problems, where there is a need for major change, dominant trends are part of the problem, when there are externalities that cannot be satisfactorily solved in markets, and in case of sufficiently long time horizons alternatives are allowed that need long development times.

The origin of backcasting goes back to energy studies in the 1970s, and before that to what was called normative forecasting (Jantsch 1967). In the early 1990s the focus shifted towards exploring sustainable futures, stakeholder involvement and achieving and shaping stakeholder support, follow-up and implementation (Quist and Vergragt 2006). In the Netherlands, participatory backcasting was introduced in 1992 at the Sustainable Technology Development programme, which focused on future technologies that make it possible to meet future societal needs in a sustainable way (Weaver *et al.* 2000, Vergragt 2005). Another example of participatory backcasting is the international 'Strategies towards the Sustainable Household (SusHouse)' project, which dealt with making household functions like nutrition, shelter and clothing care sustainable (Quist *et al.* 2001a, Green and Vergragt 2002, Vergragt 2005). In the Netherlands, participatory backcasting has also been applied to generating and debating options for responding to climate change (Van de Kerkhof 2004) and to exploring sustainable industrial paint chains (Partidario 2002). Abroad, participatory backcasting has been applied in Canada (e.g. Robinson 2003) and Sweden (e.g. Holmberg 1998).

The above-mentioned indicates the potential of participatory backcasting in exploring and initiating system innovations towards sustainability. The generated images of desirable sustainable future are a key element in this research and are called future visions. Participatory backcasting may thus be of great help in exploring new solutions to complex sustainability problems. As existing institutions and rule systems are not equipped for this, future visions allow testing and playing around with new rules that are part of the envisioned future (Grin 2006, Grin *et al.* 2004, Quist *et al.* 2005), as well as developing new and creative ideas and support for follow-up steps.

Participatory backcasting studies can be seen as social experiments. In this research the term backcasting experiment is used to refer to participatory backcasting studies. Nevertheless, there are considerable differences in the way the approach has been applied, in the results that have been achieved, in the type and degree of stakeholder involvement and in the degree of follow-up and diffusion after the backcasting study is completed (Quist *et al.* 2005). The analysis of the follow-up and impact of backcasting studies has been a relatively neglected area, as well as the investigation of the underlying mechanisms. An interesting research topic emerges here, which could shed more light on these particular aspects of participatory backcasting.

1.3 Research questions and relevance

Generally speaking, reports on participatory backcasting experiments focus on the results with regard to a particular topic, or on the way a particular approach has been applied (Weaver *et al.* 2000, Green and Vergragt 2002, Partidario 2002). Reports occasionally pay attention to the process aspects of stakeholder participation, or to learning effects among stakeholders. For instance, Van de Kerkhof (2004) has extensively evaluated a participatory backcasting experiment, concerning the generation and discussion of options to mitigate climate change. She focused especially on the dialogue and learning processes among participating stakeholders. Loeber (2004), who has carried out a profound process-oriented evaluation of a particular backcasting experiment at the STD programme, has also looked at the learning effects and to what extent they resulted in follow-up in the short term. Available studies are either limited to testing and evaluating a particular elaboration of backcasting in a single project (Partidario 2002, Van de Kerkhof 2004), focus on methodology and results (Weaver *et al.* 2000, Green and Vergragt 2002, Quist *et al.* 2001), or have in-vestigated process-related and learning aspects shortly after the backcasting study was completed (Van de Kerkhof 2004, Loeber 2004). Coenen (2000) has looked at the impact of backcasting experiments that were part of the STD programme three years after they were completed. None of these studies is aimed at evaluating the impact after five years or more, although they have been relevant to the development of participatory backcasting and the evaluation of its immediate outcomes and short-term effects.

Although little is known about the longer-term impact of backcasting experiments, there are considerable differences in the degree in which follow-up and diffusion have been achieved after five or ten years (Coenen 2000, Quist *et al.* 2005, Quist and Vergragt 2006). It would, therefore, be very interesting to increase the knowledge on the relationships between (participatory) backcasting experiments – mostly organised as a project bounded in time and capacity – and their longer-term follow-up and broader effects after five to ten years. This requires an evaluation study, which includes both backcasting experiments and their follow-up and spin-off after five to ten years. This may also shed light on what factors determine follow-up and spin-off, assuming that such factors may have both a positive and a negative influence. Looking at the follow-up and spin-off of backcasting experiments, as well the factors influencing these effects was exactly what the purpose of this study was at the outset. It was also my aim to analyse if the impact of backcasting experiments can be related to system innovation theories, as well as to take a closer look at theories on visions, higher order learning and stakeholder participation; in this thesis I report on the research and its results.

Research questions

Based on the research topic discussed above, two main questions have been developed to guide this research.

A. *What factors determine the impact of backcasting experiments after five to ten years?*

B. *How should participatory backcasting be applied for exploring and shaping system innovations towards sustainability?*

The first main question focuses on what factors determine and explain the follow-up, spin-off and broader impact of participatory backcasting experiments. It is a 'what' question, looking for a theoretical explanation. The second main question is a practical 'how' question and focuses on how backcasting practices can be improved, especially with respect to system innovations towards sustainability. It is intended to improve the guidelines for applying backcasting in various contexts and under various conditions. The main questions have been elaborated in a set of more detailed research questions:

1. How has backcasting evolved and how does it relate to other approaches?
2. How can backcasting experiments for system innovations towards sustainability and their impact after ten years be related to various theories including system innovation theories?
3. How can backcasting experiments and their follow-up, spin-off and broader effects five to ten years after completion be conceptualised and evaluated?
4. How has backcasting been applied, and what results have been achieved in backcasting experiments?
5. What has been the impact of backcasting experiments after several years in the domains of government, business, public and research, in terms of follow-up activities, spin-off, and wider effects?
6. What factors may influence the emergence of follow-up and spin-off of backcasting experiments?
7. In what ways can the application of backcasting for system innovations towards sustainability be improved?
8. What are possible recommendations for various actors that are involved in backcasting experiments for system innovations towards sustainability?

Briefly, research questions 1-3 refer to literature research, theorising and conceptualising. Research questions 4-6 refer to the empirical part of this research. Research questions 7-8 focus on reflections and recommendations. In this study we thus look at the critical factors that explain spin-off, as well as at the causal relationships between the spin-off and impact, and the original backcasting experiments.

Potential relevance

It is expected that the results of this research will improve our understanding of backcasting for system innovations towards sustainability in a multi-actor setting, involving companies, public interest groups, government and knowledge institutes. It aims at enhancing the understanding of factors enabling implementation, follow-up and broader effects and how they relate to changing structures and institutions. This study may reveal some of the underlying social mechanisms involved in achieving follow-up, and shed light on what the functions of the normative future can be in these processes, for instance when new actors enter the scene or join particular follow-up activities. It may also improve our understanding of the impact of specific methods and the way they can be applied in particular settings. In this way it may enable practitioners to improve methods and practices in participatory backcasting considerably and make these more effective in terms of follow-up, implementation and learning. This study may also shed light on whether and how the impact of participatory backcasting after five to ten years can be related to (system) innovation theory.

Finally, it is expected that this study sheds some light on how existing theories on visions, learning and participation relate to visions, learning and participation in backcasting experiments.

1.4 Research approach

The focus of this research is on analysing the follow-up and spin-off of participatory backcasting experiments after five to ten years and on determining what factors affect the emergence of follow-up and other effects (see Figure 1.1). This requires a research approach in which both the backcasting experiment itself (the first box on the left in Figure 1.1) and its impact and follow-up (depicted by the second box in Figure 1.1, which is one in the middle) are studied in such a way that the factors that exert influence can be identified. As the figure indicates, a distinction is made between internal and external factors. While the former originate from the backcasting experiment and the follow-up, the latter originate from the surrounding socio-technical system or its context.

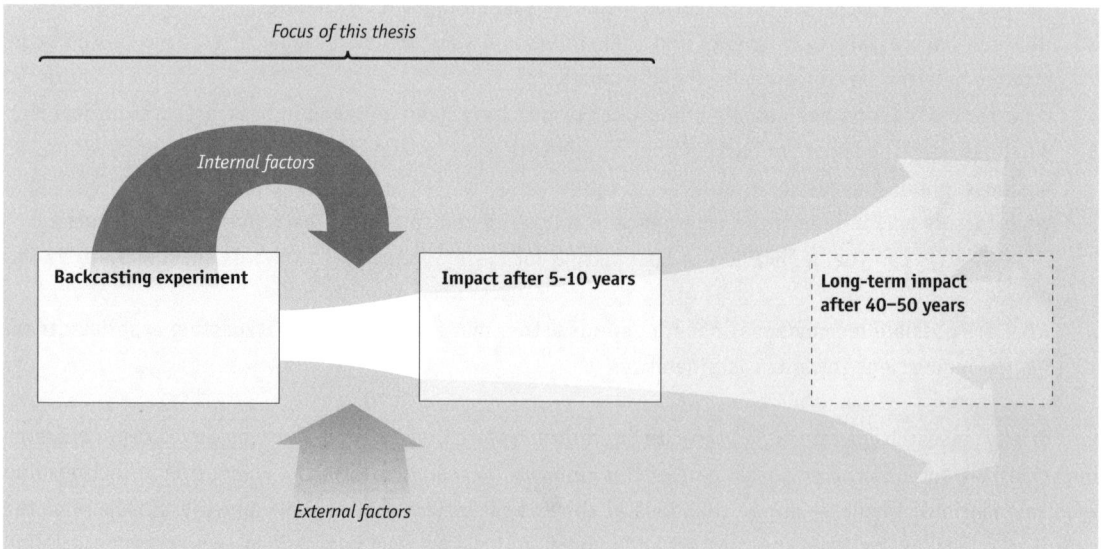

Figure 1.1 *The backcasting experiment, the impact after five to ten years, and the possible long-term impact after 40-50 years*

This research does not investigate whether future visions have been realised, as such an evaluation can only be carried out in the longer term (see the third box on the right in Figure 1.1), usually after 30 to 50 years. In addition, scenario literature has shown that scenarios or future visions are never realised exactly as they are planned. Nevertheless, a basic assumption of this research is that future visions may guide and facilitate certain developments. If this is a valid assumption, then it may be possible after five to ten years to analyse to what extent the system innovation towards the vision is 'on track'. This should include studying the influence of the future vision both during the backcasting experiment and when establishing follow-up activities. Furthermore, on a trajectory towards a sustainable future, future visions will likely be re-evaluated and adjusted, which will influence activities and decision-making processes, and result in phenomena that have become visible after five to ten years.

This research calls for an empirical ex-post case study approach. Since establishing follow-up or diffusion requires time, it is important to select backcasting experiments that were completed some time ago (preferably more than five years). From the viewpoint of this research it is interesting to include backcasting experiments with substantial follow-up as well as backcasting experiments with little or no follow-up.

This research has been organised as follows (see Table 1.1). First, an exploration of the backcasting literature and related approaches has been conducted. Based on this, a methodological framework for participatory backcasting is proposed. As a next step, various bodies of literature have been studied in search of useful building blocks for a conceptual framework. The theoretical exploration also looks at theories of system change in the field of innovation studies. This is followed by developing propositions and a research methodology. Next, three cases are evaluated using a qualitative research approach; data are collected mainly through semi-structured interviews and document analysis. Next, the cases are compared and the propositions evaluated. Finally, conclusions are presented and discussed, followed by providing recommendations.

Table 1.1 *Overview of research steps and content of this thesis*

Activities	Content
Defining research topic and proposing research approach (Step 1)	Chapter 1 Introduction
Methodological exploration of backcasting and related approaches and identification of relevant concepts (Step 2)	Chapter 2 (research question 1) Backcasting
Theoretical exploration of relevant concepts for this research (Step 3)	Chapter 3 (research question 2) Theory on visions, learning, stakeholder participation, system innovations and networks
Development of conceptual framework, propositions and case study research methodology (Step 4)	Chapter 4 (research question 3)
Case studies (Step 5) Case I: Novel Protein Foods (NPF) Case II: Sustainable Household Nutrition (SHN) Case III: Multiple Sustainable Land-use (MSL)	Chapters 5-7 (research questions 4-5)
Testing propositions and comparing cases (Step 6)	Chapter 8 (research question 6)
Drawing conclusions and providing recommendations (Step 7)	Chapter 9 (research question 7-8) Conclusions, reflections and recommendations

1.5 Outline of this thesis

This thesis is organised as follows (see also Table 1.1). The introduction in Chapter 1 is followed by an exploration of backcasting and some related approaches in Chapter 2. Chapter 3 contains a theoretical exploration of visions, learning, stakeholder participation and system innovation theory. In Chapter 4 a conceptual framework is developed using the results of the preceding chapters. It also develops a set of propositions and a research methodology, and it deals with case selection. In Chapter 5-7 three backcasting experiments and their effects after ten years are evaluated. Chapter 5 deals with Novel Protein Foods (NPF) as a sustainable alternative to the consumption and supply of meat that began as part of the STD programme. Chapter 6 evaluates Sustainable Household Nutrition (SHN), a backcasting experiment that was part of the Sustainable Households (SusHouse) project. Chapter 7 looks at Multiple Sustainable Land-use (MSL) a backcasting experiment that also was conducted at the STD programme. Chapter 8 compares the three cases and evaluates the propositions developed in Chapter 4. Finally, in Chapter 9 conclusions are drawn and recommendations are provided to various groups of actors that can be distinguished in backcasting experiments.

Notes

1   Factor 20 by 2050 is derived from the so-called IPAT formula and is based on a doubling of the world population combined with a fivefold increase of wealth per capita on average worldwide, while halving the total global environmental burden. At present the popular version of the IPAT equation is $I=P \times A \times T$: the environmental impact (I) equals the product of population size (P), the degree of affluence (A) per capita, and the environmental impact from production and consumption used to produce one unit of affluence (T). There is presently an academic debate taking place about what factor would be required, how to translate this factor in environmental terms and what indicators and methodology should be used (see, for instance, Reijnders 1998).

2   It is emphasised here that it is impossible to change socio-technical systems in a simple hierarchical way, as they are the result of constructive interaction processes among actors involved. The basic assumption here is that change processes are the result of social shaping processes and that it is possible to influence – and to a certain extent guide – such processes.

3   Other major problems existing institutions are unable to cope with adequately, are for instance mobility and congestion, water management, health care, and socio-economic issues both in the main cities and in rural areas.

4   A similar pattern has been found when developing and studying the concept of technological regimes that also enable certain patterns of technology development, while constraining others (e.g. Rip and Kemp 1998).

5   Preventing side-effects in current technology development practices has always been a major aim of technology assessment, even before the concept of reflexive modernisation emerged; both are rooted in similar social concerns. Since technology assessment evolved into constructive technology assessment reflexivity has been a major concern (e.g. Schot 2001).

6   For instance, Voß et al. (2006) argue that 'reflexive governance' is needed for sustainable development. A distinction can be made between first and second order reflexivity. First order reflexivity relates to reconsidering and adapting actors' and individuals' own actions and behaviour. By contrast, second order reflexivity is more far-reaching and has to do with reconsidering and adjusting structures and institutions.

7   Another emerging and presently widely discussed approach is transition management (Rotmans et al. 2001, Rotmans 2003). It is discussed and related to backcasting in Chapter 2.

2

AN EXPLORATION OF BACKCASTING AND RELATED APPROACHES

This[1] chapter introduces various types of future studies (2.1), before reporting on a literature survey of backcasting and related approaches (2.2). Next, this chapter compares four selected backcasting approaches (2.3), before developing a methodological framework for participatory backcasting (2.4) and presenting conclusions (2.5).

2.1 Introduction

Before taking a look at backcasting and related approaches, I briefly discuss various types of futures that are distinguished in future studies. Building on Dunn[2] (1994: 195) I distinguish between likely futures, possible futures and desirable futures. Each type of futures can be related to a group of approaches in future studies[3].

The vast majority of forecasting approaches focuses on likely futures and are projective in nature, using, for instance, trend extrapolation and quantitative historical data. The drawback here is that traditional forecasting is only reliable in the case of well-defined and relatively stable systems like existing markets and in the short term (e.g. De Laat 1998).

Most foresighting and scenario[4] approaches focus on possible futures. Well-known examples are the context scenario approach (e.g. Van der Heijden 1997, Thissen 1999, Enserink 2000) and global system scenario approaches, such as the one by the IPCC (ww.ipcc.org) and the one by Meadows and colleagues (Meadows 1972, Meadows *et al.* 1992). It is also possible to combine elements from possible futures and normative futures. For instance, the Netherlands Council for Government Policy, which has been a major foresighting organisation in the Netherlands for several decades, has used normative views like varying political visions and different action perspectives for defining different policy goals, as well different scenarios (e.g. WRR 1980, WRR 1983, WRR 1992, WRR 1994). I will return to this approach in 9.3.

Approaches focusing on normative or (un)desirable futures are the least widely applied. These approaches focus on desirable, yet attainable futures. Backcasting is a well-known example of such an approach. Before the emergence of backcasting reference was made to normative forecasting (e.g. Jantsch 1967), while in France desired futures are referred to in what has been called 'la prospective' (Godet 2000)[5].

As approaches using desirable or normative futures are highly important from the viewpoint of sustainable development, this has resulted in an increasing interest in this type of future studies approaches in general and in backcasting in particular. Backcasting and related approaches are explored in the next section.

2.2 Backcasting: a brief history

2.2.1 Backcasting in energy studies: soft energy paths

The origin of backcasting dates back to the 1970s, when Lovins (1976, 1977) proposed backcasting as an alternative planning methodology for electricity supply and demand (Robinson 1982, Anderson 2001). He called this method 'backwards-looking analysis', while Robinson (1982) proposed the term 'energy backcasting'. Assuming that future energy demand is mainly a function of current policy decisions, Lovins argued that it would be beneficial to describe a desirable future or a range of desirable futures and to assess how such futures could be achieved, instead of focusing only on likely futures and projective forecasts. The assumption was that, after identifying the strategic objective(s) in a particular future, it would be possible to work back to determine what policy measures should be implemented to guide the energy industry in its transformation towards that future. In the 1960s, Jantsch (1967) had already dealt with normative forecasting[6], which can be seen as a predecessor of backcasting.

At that time energy studies using backcasting were especially concerned with so-called soft energy (policy) paths, which took a low energy demand society and the development of renewable energy technologies as a starting point. These studies were a response to regular energy forecasting practice. This was based on trend extrapolation and projected rapidly increasing energy consumption and focused strongly on large-scale fossil fuel and nuclear technologies to deal with the estimated growth. The response led to numerous studies on soft energy paths (e.g. Lovins 1977, Robinson 1982), comparing them to regular ones (e.g. Lönnroth et al. 1980, Johansson and Steen 1980, Goldemberg et al. 1985). Interestingly, backcasting has been applied regularly in energy studies since then (e.g. Mulder 1995, Mulder and Biesiot 1998, Anderson 2001, MacFarlane 2001, Hennicke 2004), sometimes under the header energy end-use analysis.

Whereas the focus of energy backcasting was on analysis and on developing policy goals, the backcasts of different alternative energy futures were also meant to reveal the relative implications of different policy goals (Robinson 1982: 337-338), and to determine the possibilities and opportunities for policy-making. Robinson has always emphasised that the purpose of backcasting was not to produce blueprints, but to indicate the relative feasibility and different social, environmental and political implications of different energy futures (Robinson 1990: 823). Robinson (1982) also worked out the principles defined by Lovins into a sequential six-step methodology for energy and electricity futures. The central step was developing an outline of the future economy by constructing a model of the economy in a final future state, followed by an energy demand scenario that corresponded to the results of the model. Recently, Anderson (2001) adapted the energy backcasting approach, with the aim of reconciling the electricity industry with sustainable development. He takes into account wider environmental and social responsibilities, as well as non-expert knowledge, and includes the development of supporting policies within his methodology.

In summary, the early focus in backcasting was on exploring and assessing energy futures and on their potential for policy analysis in the traditional sense of supporting policy and policy-makers, usually adopting a government-oriented perspective.

2.2.2. Backcasting for sustainability

It was realised that the backcasting approach could potentially be applied to a much wider range of subjects, due to its characteristics and normative nature. For instance, Robinson (1988) discussed the wider conceptual and methodological issues of backcasting; this includes the role of learning and unlearning with respect to existing dominant views about the future, the issue of broadening the process to a larger group of potential users and how to alter the hegemony of existing dominant perspectives. Elsewhere, Robinson

(1990: 822) mentioned that backcasting is not only about how desirable futures can be attained, but also about how undesirable futures can be avoided or anticipated.

Robinson's (1990) paper also marked the move towards the application of backcasting to sustainability and illustrates the interest in Sweden, as the paper reported on a study supported by the Swedish Energy Research Council. In Sweden, a strategic interest in alternative energy futures had developed (Johansson and Steen 1980, Lönnroth et al. 1980), which was followed by substantial efforts in conceptual development (e.g. Dreborg 1996, Holmberg 1998, Höjer and Mattsson 2000). Dreborg (1996) has argued that traditional forecasting is based on dominant trends and is therefore unlikely to generate solutions based on breaking trends. Due to their normative and problem-solving character, backcasting approaches are much better suited to address long-term problems and sustainability solutions. Dreborg also emphasises that our perception of what is possible or reasonable may be a major obstacle to real change – which is in line with earlier remarks by Robinson (1988) about (un)learning and the dominance of existing perspectives. Scenarios or future visions of a backcasting project should, therefore, broaden the scope of solutions to be considered by describing new options and different futures. According to Dreborg, backcasting is particularly useful when applied to complex and persistent problems, when dominant trends are part of the problem, when externalities are at play, when there is a need for major change and when time horizon and scope allow development of radical alternative options. Sustainability problems obviously combine all these characteristics (Dreborg 1996).

Dreborg (1996) also focuses on the conceptual level beyond Robinson's stepwise method and relates backcasting to Constructive Technology Assessment (CTA). The purpose of CTA is to broaden the technology development processes and the debate about technology with environmental and social aspects, as well as to enhance the participation of social actors like public interest groups, in addition to the traditional participants in such processes (see 2.2.4 for a further explanation on CTA). A distinction can be drawn between the analytical side and the constructive and process oriented side of backcasting (Dreborg 1996). With respect to the analytical side, the main result of backcasting studies are alternative images of the future, thoroughly analysed in terms of their feasibility and consequences. With respect to the constructive-oriented side, backcasting studies should provide an input to a policy developing process in which relevant actors should be involved. Results of backcasting studies should therefore be addressed to many actors, including political parties, government authorities, municipalities, organisations, enterprises and a general public that needs to be well informed.

Höjer and Mattsson (2000) have suggested that backcasting and regular forecasting are complementary rather than conflicting opposites. They favour backcasting in cases where existing trends are leading towards an unfavourable state. Although this is in line with Dreborg's (1996) argument, they add a forecasting step to their backcasting approach in which forecasts and the desired vision are compared. If the vision is unlikely to be reached based on the most reliable forecasts, model calculations and other estimates, backcasting studies should be used to generate images of the future that fulfil the targets. Höjer and Mattsson (2000: 630) also emphasise the importance of scrutinising how to attain the desirable future by working back from the desirable future to check the physical and social feasibility of the route or pathway towards that future. This requires not only identifying the necessary measures and actions for bringing about that future, but also using models and regular forecasting tools to quantify the consequences of different measures.

Backcasting for sustainability has been applied in Sweden on a range of topics, for instance sustainable transportation systems (Höjer 1998, Höjer and Mattsson 2000, Roth and Kaberger 2002, Åkerman and Höjer 2006), for air transport (Åkerman 2005) and to explore futures for a region like the Baltic Sea (Dreborg et al.1999). Despite the plea (Dreborg 1996) in Sweden to broaden backcasting with a range of social actors and participants, the backcasting studies referred to above are not participatory and have a strong analytical focus. Expert involvement in backcasting studies has occasionally been reported (Höjer 1998, Banister *et al.* 2000a). Although the Natural Step backcasting methodology, which aims at sustainable companies, can be seen as participatory, it focuses on internal stakeholders and employees (Holmberg 1998, Holmberg and Robèrt 2000).

The interest in backcasting for sustainability is still growing. Studies have been conducted in many countries, albeit especially in Europe. In addition to the Swedish examples mentioned above, there have been various backcasting studies on water, mobility and mobility technologies (Falkenmark 1998. Geurs and Van Wee 2000, Banister *et al.* 2000a, Banister *et al.* 2000b, Marchau and Van der Heijden 2003, Geurs and Van Wee 2004). Attempts have also been made to combine backcasting with other approaches (Höjer 1998, List 2003, Robèrt 2005, Dortmans 2005, Marchau and Van der Heijden 2003, MacDonald 2005), while recently different backcasts and visions for the future of hydrogen were discussed (McDowell and Eames 2006).

2.2.3 The shift to participatory backcasting

The shift towards participatory backcasting using broad stakeholder involvement started in the Netherlands in the early 1990s. Participatory backcasting has been applied in the Netherlands since that time, first at the government programme for Sustainable Technology Development (STD) that ran from 1993-2001 (Vergragt and Jansen 1993, Weaver *et al.* 2000) and later in its EU funded spin-off, the research project 'Strategies towards the Sustainable Household (SusHouse)', which ran from 1998 to 2000 (Green and Vergragt 2002, Quist *et al.* 2001a). Both initiatives focused on achieving sustainable need fulfilment in the distant future, using a backcasting approach that included broad stakeholder participation, future visions or normative scenarios, and the use of creativity to reach beyond existing mind sets and paradigms.

Inspired by the Swedish practice, Vergragt and Jansen (1993) mentioned backcasting as part of the philosophy of the STD programme. They described the basic idea (1993: 136) as *"to create a robust picture of the future situation as a starting point, and start to think about which (technical and other) means are necessary to reach this state of affairs. Such a view of reality is not a scenario or a product of forecasting, but should be seen as a solid picture that can be accepted by the technological spokesmen right now."* Like Dreborg (1996), Vergragt and Jansen (1993) emphasised the link with Constructive Technology Assessment, including a broadening of technology development processes with sustainability aspects and the participation of social actors like public interest groups in such processes, in addition to the traditional participants. Elsewhere, Vergragt and Van der Wel also emphasise the importance of implementation and follow-up (Vergragt and Van der Wel 1998: 173). *"Future visions alone are not enough: Backcasting implies an operational plan for the present that is designed to move toward anticipated future states. Backcasting, then, is not based on the extrapolation of the present into the future – rather, it involves the extrapolation of desired or inevitable futures back into the present. Such a plan should be built around processes characterised as interactive and iterative."* This implies that many stakeholders need to be involved and that there is continuous feedback between future visions and present actions.

Weaver *et al.* (2000: 74), reporting on the approach and results of the STD programme, describe backcasting as a possible tool for establishing shared visions of desirable future system states and for securing a 'systems' perspective. They also emphasise that backcasting can be used to define feasible short-term actions that can lead to trend-breaking change, in other words putting vision into action (Jansen 2005). Weaver *et al.* (2000: 72-78) refer to backcasting not only as an overall methodology, but also as a concept and as a specific step in the full methodology. This may cause some confusion, as these authors relate the term backcasting to different aspects. Furthermore, different tools and methods can be applied as part of the overall backcasting methodology (Aarts 2000, Weaver *et al.* 2000).

While the focus of the STD programme was on sustainable technologies, the SusHouse project, which was an international project initiated from the Netherlands, aimed at developing and testing strategies for sustainable households in the future. The backcasting methodology used stakeholder workshops, creativity methods, normative scenarios, scenario assessments and backcasting analysis (Vergragt 2000, Quist *et al.* 2001a, Young *et al.* 2001, Green and Vergragt 2002). In the SusHouse project it was originally assumed that all backcasting activities could be concentrated in a single workshop. However, it turned out that backcasting activities took place during a large part of the project; not only during the stakeholder workshops, but also during the scenario elaboration and scenario analysis activities of the research teams (Quist *et al.* 2000). Quist *et al.* (2000: 8-16) also mention the link with CTA, the connections with the field of Creative Problem Solving (e.g. Isaksen 2000) and the importance of (conceptual) learning by the stakeholders and researchers involved. Tassoult (1998) has reported on the integration of a particular creativity method entitled 'Future Perfect' as part of a backcasting project on sustainable washing; the final results of the complete project have been reported by Vergragt and Van der Wel (1998). Green and Vergragt (2002), reporting on the results of the SusHouse project, conclude that stakeholders should not only be involved in constructing normative scenarios, but also in the economic and environmental assessment of the normative scenarios. Elsewhere, Vergragt (2005) emphasises that, although it is important that stakeholders share a future vision, in itself that is not enough to achieve implementation and follow-up; he argues that it is also important to understand the culture and interests of stakeholders and their reasons for participating both in the backcasting study and in follow-up activities.

Since it was applied in the STD programme, participatory backcasting has become a well-known and widely applied approach in the Netherlands (see also Quist 2006). For instance, backcasting and normative future visions have been applied in strategic research programmes at DLO, the main Dutch research organisation for agriculture and rural areas (e.g. Grin *et al.* 2004, Poot 2004). A participatory backcasting approach has been applied to the subject of climate change, involving stakeholders in a debate on different futures meeting Kyoto targets (Van de Kerkhof *et al.* 2003, Van de Kerkhof 2004). Partidario has elaborated and applied a backcasting approach similar to the SusHouse methodology to study sustainable futures for industrial paint chains in the Netherlands and Portugal (Partidario and Vergragt 2002, Partidario 2002). Rotmans *et al.* (2001: 23-24), working on transition management, also refer to backcasting from the future as part of their transition management approach. Jansen (2003) has paid attention to backcasting in national foresighting programmes and has compared these to backcasting in the STD programme. The STD approach has also been elaborated in a backcasting methodology for vision development and the integration of spatial functions (e.g. agriculture, leisure, nature, landscape, etc) in rural areas (De Kuijer and De Graaf 2001). This methodology has been applied in various regions in the Netherlands (De Graaf *et al.* 2003, De Graaf and De Kuijer 2004a, De Graaf and De Kuijer 2004b). There is also an interest in developing more quantified ways of backcasting, which has been applied to the design of transition strategies for sustainable transportation chains (Suurs *et al.* 2004).

The shift towards participatory backcasting has also taken place in other countries. For instance, in Sweden, 'The Natural Step' (TNS) methodology, as reported by Holmberg (1998), Holmberg and Robèrt (2000), is a backcasting methodology focusing on strategic planning for sustainability in companies (see also www.thenaturalstep.org). It has been applied successfully within corporations like Ikea, carpet producer Interface and Scandic hotels (Holmberg 1998; for a detailed account, see Nattrass and Altomare 1999). After assuring commitment from top level management, as many employees as possible are involved and consultation takes place at all levels of the organisation to generate ideas about how to become a sustainable corporation; in several cases external stakeholders were involved as well. This example shows that it is possible to apply backcasting at the level of particular organisations.

In Canada, Robinson, elaborating upon his extensive experience in backcasting (e.g. Robinson 1982, 1988, 1990), has also developed backcasting further and has included participation. Robinson (2003) emphasises the importance of social learning, interactive social research, and the engagement of non-expert users in backcasting studies, and he has called this 'second generation backcasting'. This form of backcasting has been applied to the Georgia river basin in West Canada and has been related to participatory integrated assessment (Tansey *et al.* 2002), using a modelling tool based on the QUEST approach, which enables residents to engage in interactive construction of future images for the river basin; users are also asked to evaluate the scenario outputs in terms of their desirability and to match them with personal preferences. As it is possible to iterate by adjusting inputs, it enables users to continue towards future visions that provide a better match with their preferences. This contributes to learning among users (Robinson 2003).

Other examples of participatory backcasting have taken place in Sweden (Carlsson-Kanyama *et al.* 2003a) and Belgium (Keune and Goorden 2002, 2003). The former was part of an international project on sustainable cities in five European countries (Carlsson-Kanyama *et al.* 2003b) including the Netherlands (Falkena *et al.* 2003). Finally, local scenarios and participatory backcasting have been combined with continental scenarios (Kok *et al.* 2006a, Kok *et al.* 2006b).

2.2.4 Related approaches

This section briefly deals with two groups of approaches and methodologies that may be relevant to participatory backcasting[7]:

(1) Participatory approaches that develop and use normative or desirable futures, but have not been labelled as backcasting.
(2) Participatory approaches in Constructive Technology Assessment (CTA) and Participatory Technology Assessment (PTA).

Participatory approaches using normative futures

Transition Management (TM) is a strongly emerging approach in the Netherlands aiming at transitions towards sustainability (Rotmans *et al.* 2000, Rotmans *et al.* 2001, Kemp and Rotmans 2004, Rotmans 2003). TM uses normative future visions as a core element, in addition to long-term thinking as a framework for short-term policy, a focus on learning and thinking in multiple aspects, multiple domains and multiple levels. The relationship with backcasting is sometimes referred too (Rotmans *et al.* 2001), although differences with backcasting have also been emphasised (Rotmans 2003). TM has a strong policy orientation and emphasises the need to keep various options open by working towards different visions at the same time. It has been adopted by the Dutch government and is currently being applied to future energy supply systems in the Netherlands (www.energietransitie.nl). It has also been advocated that TM can be applied to any complex societal problem and offers a management strategy for the government for dealing with such societal problems through transitions (Rotmans *et al.* 2001). I will return to the Transition Management approach in 9.3.

Elsewhere, Rotmans *et al.* (2000) and Van Asselt *et al.* (2005) have reported on normative scenarios for sustainable areas in Europe in which stakeholders were involved. Street (1997) has described the use of scenario workshops as a participatory approach to sustainable urban living, involving citizens and other local stakeholders. Mayer (1997) has reported on the so-called Awareness Workshop Methodology, which also involves local stakeholders in scenario development on local environmental improvement. Grin *et al.* (1997) have proposed Interactive Technology Assessment (ITA) in which both broad stakeholder involvement and the construction of normative future visions by stakeholders are core elements. Developing desirable futures has also been combined with context scenarios for the Scheldt river basin, though in an educational setting (Ruijgh-van der Ploeg and Verhallen 2002). Raskin *et al.* (2002) have reported on efforts by the Global Scenario Group of the Tellus Institute, which has resulted in a set of global futures as well as strategies for achieving a sustainable future.

In 't Veld (2001) has proposed a participatory foresight approach entitled TO3[8] after an evaluation of future studies for spatial planning policymaking in the Netherlands (Stuurgroep T&O 2001). This approach (In 't Veld 2001: 27-37) combines participation, orientation, design and analysis in the generation of normative future images. It assumes trans-disciplinary research, connecting heterogeneous knowledge, combining creativity and reflexivity, taking into account ambiguity and uncertainties, and using experts and other stakeholders to connect and relate heterogeneous knowledge. The approach also assumes process demands (e.g. type and range of participation, or the transparency of the process) and knowledge demands (e.g. quality of the future images).

Constructive and Participatory Technology Assessment

Other approaches that are relevant to backcasting are the participatory approaches like the ones practiced in what has been referred to as Constructive Technology Assessment (CTA, see Schot and Rip 1996), Participatory Technology Assessment (PTA, see Schot 2001). In general, the focus of these types of approaches is not on generating normative or desirable futures, but on broadening the process of technology development and related decision making with societal aspects and societal actors. The purpose is that technology driven futures of technology developers are adjusted to societal aspects and concerns, as well as to broaden the public debate on technology. This group of approaches is interesting, as they focus on broadening innovation and technology processes, though in general not on a system level. These approaches do not use sustainability as the only target, but treat sustainability aspects as part of a set of relevant aspects. CTA/PTA approaches may use normative visions, as shown by Grin *et al.* (1997), but not necessarily. Mayer (1997) has given an overview of participatory methods that can be used in CTA and PTA.

Another CTA methodology is Strategic Niche Management (SNM, e.g. Hoogma *et al.* 2002), while others have referred to a similar methodology as Bounded Socio-Technical Experiment (BSTE, see Brown *et al.* 2003). Strategic Niche Management refers to experimenting with new technological options in a space protected from regular market pressures, to enable stakeholders (both producers, users, regulators) to learn about a new technology, articulation of user demands and the embedding in its context. While market niches can survive under market conditions, so-called technological niches cannot, and as a result they need protection, which can be provided by governments through subsidies, taxes or levying, but it also requires support from other actors. Rather than looking at the longer term, this approach focuses on concrete technologies and artefacts that are already available and can be tested in a pilot. SNM may be part of a long-term strategy, which includes implementation of new technologies. However, it is still unclear how the removal of the protection mechanisms (as suggested in SNM) could take place in a balanced way, without 'killing' the technology in an early phase and without offering protection for too long.

Fonk (1994) has developed a CTA methodology called Future Images for Consumers[9]. This methodology focuses on dialogue and discussion between major actor groups, like technologists and scientists, producers, as well as consumer groups and public interest groups. It starts with developing actor group specific future images assuming that these are rather homogeneous, while emphasising the role of consumers or end-users. These images are not necessarily images of a desirable future, but rather images based on the expectations and assumptions of relatively homogeneous groups of actors. Next, the images serve as a starting point for discussion between these groups, with the aim of identifying issues of consensus and disagreement.

Another emerging methodology refers to socio-technical scenarios (Elzen *et al.* 2002, Elzen *et al.* 2004, Hofman *et al.* 2004, Hofman 2005), building on the same insights as CTA. These authors argue that this type of scenarios makes it possible to take into account the complexity and multi-level nature of transitions or system innovations. The approach focuses strongly on niches that can become stepping stones for transitions and technological regime change, as well as on the possible pathways to realise them. This methodology is expert and analysis driven, rather than stakeholder driven. Its explicit aim is to take into account the social aspects of technology development and system innovations, as well as the mutual influence between society and technology.

2.3 Comparing four backcasting approaches

In the literature survey presented in the previous section a number of backcasting studies and experiments were discussed, in which a range of methods were applied. It has been argued that reports on backcasting methods are hard to find and that therefore conventional methods have to be combined in backcasting approaches (Marchau and Van der Heijden 2003: 266). This calls for a more detailed exploration of various backcasting approaches, which is the topic of this section. In the literature both the term backcasting approach and the term backcasting methodology are used. In my view the term backcasting approach should be used to describe (backcasting) approaches in general and more abstract terms, whereas I use the term (backcasting) methodology for more elaborated varieties, for instance when applied in concrete cases.

In this section four different backcasting approaches are more extensively discussed, using selected papers from the literature review with a comprehensive description of a particular backcasting approach or methodology. The four backcasting approaches are:
-   The backcasting approach as proposed by Robinson (1990);
-   The Natural Step backcasting approach (Holmberg 1998, Nattrass and Altomare 1999, Holmberg and Robèrt 2000);
-   The STD backcasting approach as described by Weaver *et al.* (2000) and Aarts (2000);
-   The backcasting approach as applied in the SusHouse project, based on Vergragt (2000, 2005), Quist *et al.* (2001a), and Green and Vergragt (2002).

The four approaches are summarised in Table 2.1, which shows key assumptions, proposed steps and examples of methods. Descriptions of methods can be found in the references.

Robinson's (1990) backcasting approach
Robinson (1990: 823) characterises his approach as explicitly normative and design-oriented, aimed at exploring the implications of alternative development paths as well as the underlying values. It starts by defining future goals, objectives and constraints for both the defined system and its external context, followed by the construction of future scenarios, which can be based on criteria set externally to the analysis. Next, the scenarios must be evaluated in terms of socio-economic, technological and physical feasibility and of policy implications. Iteration of scenario construction is needed to avoid physical inconsistencies

Table 2.1 *Four backcasting approaches*

	Robinson's backcasting approach	TNS backcasting approach	STD backcasting approach	SusHouse backcasting approach
Key assumptions	> *Criteria for social and environmental desirability are set externally to the analysis* > *Goal-oriented* > *Policy-oriented* > *Design-oriented* > *System oriented*	> *Decreasing resource usage* > *Diminishing emissions* > *Safeguarding biodiversity and ecosystems* > *Fair and efficient usage of resources in line with the equity principle*	> *Sustainable future need fulfilment* > *Factor 20* > *Time horizon of 40-50 years* > *Co-evolution of technology & society* > *Stakeholder participation* > *Focus on realising follow-up*	> *Stakeholder participation* > *Factor 20* > *Sustainable households in 2040* > *Social and technological changes are needed* > *Achieving follow-up is relevant*
Methodology (steps)	*(1) Determine objectives* *(2) Specify goals, constraints and targets & describe present system and specify exogenous variables* *(3) Describe present system and its material flows* *(4) Specify exogenous variables and inputs* *(5) Undertake scenario construction;* *(6) Undertake (scenario) impact analysis*	*(1) Define a framework and criteria for sustainability* *(2) Describe the current situation in relation to that framework* *(3) Envisage a future sustainable situation* *(4) Find strategies for sustainability*	*(1) Strategic problem orientation* *(2) Develop sustainable future vision* *(3) Backcasting – set out alternative solutions* *(4) Explore options and identify bottlenecks* *(5) Select among options & set up an action plan* *(6) Set up cooperation agreements* *(7) Implement research agenda*	*(1) Problem orientation and function definition* *(2) Stakeholder analysis and involvement* *(3) Stakeholder creativity workshop* *(4) Scenario construction* *(5) Scenario assessments* *(6) Stakeholder backcasting and strategy workshop* *(7) Realisation follow-up and implementation*
Examples of methods	> *Social impact analysis* > *Economic impact analysis* > *Environmental analysis* > *Scenario construction methodologies* > *System analysis & modelling* > *Material flow analysis and modelling*	> *Creativity techniques* > *Strategy development* > *Employee involvement* > *Employee training*	> *Stakeholder analysis* > *Stakeholder workshops* > *Problem analysis* > *External communication* > *Technology analysis* > *Construction of future visions* > *System design & analysis*	> *Stakeholder analysis* > *Function & system analysis* > *Backcasting analysis* > *Stakeholder workshops* > *Scenario constructio* > *Scenario evaluation (consumer acceptance, environmental, economic)*

and to mitigate or avoid adverse impacts.

This resulted in the six step approach summarised in Table 2.1. The approach does not specify who will set the criteria and the future goals and how this will be done. Stakeholder participation is also not included. The focus is on analysis and policy recommendations. It is acknowledged that the analysis must be connected to the policy process, which can be done by involving relevant government agencies as well as the wider public.

No reference is made to particular methods, but various groups of methods are mentioned, such as different types of scenario impact analyses, modelling and scenario approaches. The approach combines analysis and design, supported by modelling.

The Natural Step (TNS) backcasting approach

Holmberg (1998) describes a backcasting approach for strategic sustainability planning in companies and other organisations that consists of four steps. The first step is to define relevant sustainability criteria for the organisation under study, based on four principles that are listed as key assumptions in Table 2.1. The second step consists of an analysis of the present situation, the present activities and competences of the organisation and the supply and consumption chain of which the organisation is a part. This makes it possible to identify sustainability bottlenecks. In the third step future options and future visions are en-visaged with the help of employee involvement, for which creativity techniques can be applied. The future options and visions for the organisation need to be widely discussed within the organisation and can imply new activities. Finally, in the fourth step strategies are developed to move from the present towards the desired situation.

Although Holmberg (1998) does not elaborate on particular methods, he refers to employee involve-ment, discussing the results widely within organisations, creativity techniques, developing relevant sustain-ability criteria, strategy development, training and consulting employees, and translating the outcomes into the organisation's activities and policies.

STD backcasting approach

Weaver *et al.* (2000) have described the backcasting approach of seven steps as depicted in Table 2.1. According to Weaver *et al.* (2000: 76), Steps 1-3 are designed to develop a long-term vision based on a stra-tegic review of how a need might be met in the future in a sustainable way and backwards analysis is used to set out alternative solutions for sustainable need fulfilment. Step 4 and step 5 are meant to clarify the short-term actions that are needed to realise that future, and can be seen as a joint action, R&D and policy agenda. Steps 6 and 7 deal with implementation, facilitating stakeholder cooperation and realising the ac-tion agenda. The idea is that stakeholders who are involved in the backcasting projects set up cooperation enabling the implementation of research and follow-up agendas. The approach allows iteration and moving forwards and backwards between two steps.

Basic assumptions include the factor 20 environmental improvement by 2040, high-level stakeholder involvement, a focus on the sustainable fulfilment of societal needs, a focus on follow-up and agenda-set-ting and a focus on technological options, while acknowledging that technology development is bounded by cultural and structural conditions.

Aarts (2000) has made an overview of the methods that have been applied as part of the STD backcasting approach. She mentions several methods for the development of future visions, like essays by experts, the TvC methodology, stakeholder interviews and creativity workshops, the Delphi method, while she also refers to the relevance of visualisation and communication. In addition, she refers to problem analysis, stakeholder analysis and stakeholder involvement, as well as to methods for organising and managing projects on options that originate from the backcasting analysis and the future vision. Finally, she emphasises the relevance of methods for the transfer and dissemination of outcomes and follow-up agendas.

SusHouse backcasting approach

The SusHouse project used a backcasting approach to develop strategies for sustainable households (Vergragt 2000). Basic assumptions include (i) a factor 20 environmental improvement (ii) broad stakeholder involvement, (ii) development of normative future scenarios, and (iv) taking into account follow-up and implementation.

The approach was divided into seven steps, as shown in Table 2.1 (see also Quist *et al.* 2001a, Green and Vergragt 2002). For each household function a stakeholder analysis was performed, covering stakeholders on the demand side, the supply side, research bodies, government and public interest groups. Selected stakeholders participated in stakeholder creativity workshops with the aim of identifying sustainable ways of future function fulfilment. The results were used for construction of normative, so-called design-oriented scenarios. These scenarios were assessed in terms of environmental gain, consumer acceptance and economic credibility. Scenarios and assessment results were discussed in a second set of stakeholder workshops, which also focused on follow-up proposals, research agendas and policy recommendations in line with the scenarios. In both series of workshops backcasting techniques were applied, for which a set of guiding questions has been developed (Quist *et al.* 2000a). Backcasting analysis was also carried out during scenario construction by the research teams involved. Thus backcasting techniques, participatory methods, analytical methods, design methods and management and communication methods were all applied as part of this backcasting approach. Finally, the approach allows for iterative cycles. After each round of assessments, scenarios can be adjusted after which the scenario assessments can be conducted again.

Comparison

There are various differences and similarities between the different backcasting approaches. With regard to the differences, both Robinson's backcasting approach and the TNS approach do not contain a separate backcasting step. They reserve the term backcasting for the overall approach. By contrast, the SusHouse approach and the STD approach contain a separate backcasting step. In all four examples the overall backcasting approach provides a framework consisting of steps in which various types of methods can be applied. While the SusHouse methodology and the STD approach both contain an explicit backcasting step, no reference is made to methods or tools for this step. This would suggest that the backcasting step is underdeveloped in methodological terms. Sometimes, but not always reference is made of iteration or iterative cycles within the approach. Sometimes the use of modelling is emphasised (e.g. Robinson 1990, Robinson 2003), whereas in the STD approach and the SusHouse approach modelling is not explicitly referred to, or is considered part of further elaboration after the vision has been developed.

All four approaches contain analytical methods and design methods. Participatory methods can be found in three approaches; the exception is Robinson's backcasting approach, which was not intended to be participatory in nature. All four approaches contain steps in which future visions or (normative) scenarios are constructed, as well as steps in which the current situation or present system is analysed. In addition, all four approaches deal with basic assumptions and starting points that can be part of the first step or set external to the backcasting study. All four approaches include activities like elaboration and (scenario) analysis. Also, the TNS approach, the STD approach and the SusHouse approach contain steps dealing with operational aspects of implementation and follow-up, strategies and agenda setting. In the STD approach the importance of operational management and coordination has been emphasised. Finally, learning is important in all four approaches.

2.4 A methodological framework for participatory backcasting

2.4.1 Developing a methodological framework
The purpose of this section is to develop a generic methodological framework for participatory backcasting that has the potential to cover the full range of participatory backcasting approaches found in the literature. For this I build on the following findings of the previous sections.
- Backcasting approaches function as methodological frameworks in which different types of methods can be applied. Backcasting approaches do not prescribe specific methods, but allow combining various methods.
- In the four approaches that were analysed, reference has been made to design methods and analytical methods. More recent approaches include participatory or interactive methods. These approaches also emphasise agenda setting, strategy development and follow-up and implementation and have separate steps corresponding to these elements.
- All four approaches include the development of desirable future visions or normative scenarios. They also include an analysis of the present situation or the problem under study, as well as analysis of generated future visions or normative scenarios.
- Two recent approaches include a separate backcasting step in which looking back from the desirable future is explicated, although methods in this step are not clearly articulated.
- In one case (Aarts 2000), explicit reference has been made to practical organisation, communication and project management, which refers to another group of relevant methods.
- The four approaches show different sets of basic starting points and assumptions, which are partially normative. These assumptions can either be set before the backcasting study or in the first step of a backcasting study.
- From another foresighting approach described by In 't Veld (2001) I take the starting point that in participatory backcasting both process and knowledge demands can be set. Because of the normative nature of backcasting I add a third type of demands, which I call normative demands.
- Goals should not only relate to the content of the future vision, as emphasised in earlier backcasting approaches, but need to cover all aspects of the backcasting experiment and may be process related or refer to realising follow-up and implementation.

2.4.2 Participatory backcasting in five steps
Despite the fact that different backcasting approaches show differences in methods applied, ways of stakeholder involvement and number of steps (Robinson 1990, Holmberg 1998, Weaver *et al.* 2000, Quist *et al.* 2001a), it is possible to develop a methodological framework for participatory backcasting consisting of five steps:

STEP 1: Strategic problem orientation
STEP 2: Develop future vision
STEP 3: Backcasting analysis
STEP 4: Elaborate future alternative & define follow-up agenda

STEP 5:   Embed results and agenda & stimulate follow-up

It is assumed that setting the normative assumptions and goals is part of the first step, as is achieving agreement on the normative assumptions among stakeholders involved. However, sometimes the normative assumptions are set before the problem orientation starts, for instance as part of an overall structure or programme. This was the case in the Netherlands with the STD programme, where the time horizon of 40 years, the factor 20 and the focus on sustainable need fulfilment were set as general assumptions, and were approved by key persons at the participating ministries and some leading industries in the Netherlands before specific backcasting studies were started.

In addition, if more than five steps are identified in a particular backcasting approach, generally speaking it is possible to place specific steps parting one of the suggested five steps. Occasionally, (e.g. Holmberg 1998) the fourth and the fifth step have been combined into one. In the SusHouse project embedding outcomes and realising follow-up were considered very difficult, and these were therefore left out of the methodology (Quist *et al.* 2001a). However, because embedding and initiating follow-up and implementation are of crucial importance, I argue that is justified to distinguish them as a separate step in the approach.

It must be emphasised that, although the approach is depicted in a linear way, it definitely is not. Iteration cycles are possible, while there is also a mutual influence between consecutive steps. Although it may be interesting to conceptualise the approach as a set of activities that all need to be carried out, rather than in a linear sense, in practical applications it remains necessary to use a transparent time frame that can be communicated to the stakeholders involved and to other external parties.

In a backcasting experiment the process has a dynamic nature, which means that stakeholders may leave the process, while new stakeholders may join it. Stakeholders are important, not only because of their context-specific knowledge, but also because they help endorse results and realise the proposed action agenda and specific follow-up. Four major societal groups can be distinguished: companies, research bodies, government and public interest groups and the general public.

2.4.3 A toolkit of four categories of methods

A wide range of methods and tools is necessary in a participatory backcasting framework. Four groups of methods and tools can be distinguished that together form the outline of a toolkit. Three groups of tools and methods relate to (1) stakeholder participation, (2) design and development, and (3) analysis. The fourth group involves tools and methods for process and stakeholder management. It must be noted that each step of the backcasting approach generally requires tools and methods from all four categories. It is possible for different steps to involve different tools and methods from the same group.

Participatory tools and methods make up the first group, which includes all the tools and methods that are useful for involving stakeholders and generating and guiding interactivity among them. It includes specific workshop tools, tools to generate stakeholder creativity, tools that help stakeholders in specific backcasting activities and tools for participatory vision and scenario construction. Mayer (1997) has provided an interesting overview of participatory tools and methods, while Slocun (2003) has provided a toolkit.

Secondly, there are design tools and methods. These are not only meant for scenario construction, but also for elaboration and detailing systems as well as process design.

Thirdly, backcasting involves analytical tools and methods. These relate not only to the assessment of scenarios and designs, like consumer acceptance studies, environmental assessments, economic analyses, but also include methods for process analysis and evaluation, stakeholder identification and stakeholder analysis.

Fourthly, backcasting also requires management, coordination and communication tools and methods. This includes methods for communication, to shape and maintain stakeholder networks that originate from the backcasting study and for process management (e.g. De Bruijn *et al.* 2002).

2.4.3 Possible goals in backcasting

In the literature regarding energy backcasting, the goal orientation reflects the desirable future states. Here, backcasting is considered as an approach applied in a backcasting experiment involving stakeholders. As a consequence, goals should not only reflect the desirable futures, but also the process side. Then, possible goals for backcasting experiments may include:
- Generation of future visions and analysing these;
- Putting visions and options on the agenda of relevant arenas;
- Developing a follow-up agenda with activities for various groups of stakeholders in line with the envisioned desirable future;
- Participation of a wide range stakeholders;
- Awareness and learning among the stakeholders involved with respect to the future vision, the consequences, the agenda and the views and perspectives of others;
- Realising follow-up and stakeholder cooperation.

It must be noted that specific goals can also be more or less relevant in a specific backcasting experiment or can be achieved in a particular step. It is also important to realise that for a backcasting experiment different demands can be made. As mentioned in 2.4.1, a distinction can be drawn between normative demands, process demands and knowledge demands.

Finally, backcasting as proposed here is both inter-disciplinary and trans-disciplinary. It is inter-disciplinary in the sense that it brings together and integrates methods and knowledge from various disciplines. It is trans-disciplinary because it involves stakeholders, stakeholder knowledge and stakeholder values.

2.5 Conclusion

A distinction has been drawn between likely futures, possible futures and normative or desirable futures and the associated types of future studies. Backcasting belongs to the third type of future studies. This type of future studies has been less widely applied, but it is becoming more popular because of applicability in sustainable development. Backcasting is not the only approach that uses normative or desirable future visions; it belongs to a family of participatory approaches that all use normative futures or normative scenarios.

Backcasting originated in the 1970s and was originally developed as an alternative to traditional energy forecasting and planning. It has evolved into a participatory approach involving a wide range of stakeholders, with the aim of not only identifying and analysing radical sustainable alternatives, but in many case also of setting agendas, spreading information, realising follow-up activities and achieving other effects. Backcasting has been applied to a range of topics and levels, like regions, companies and various sociotechnical systems, like the mobility system. Stakeholder participation also varies. This ranges from expert and employee involvement, to broad involvement covering a wide range of stakeholders.

The term backcasting can refer to a concept, a study, an approach, a methodology or an interaction process among participating stakeholders. Backcasting can also refer to backwards-looking analysis or backcasting analysis, which is the specific step of looking back from the desired future. In other words, different people use different definitions, which is why one should always specify what is exactly meant when using the term backcasting.

In this chapter I have proposed a methodological framework for participatory backcasting, based on the analysis of four backcasting approaches. This framework consists of five steps and the outline of a toolkit containing four groups of methods and tools: design tools, participatory tools, analytical tools and management, coordination and communication tools. The backcasting approach is not only inter-disciplinary (combining and integrating tools, methods and results from different disciplines), but also trans-disciplinary (through the involvement of stakeholders). The framework distinguishes three types of demands: normative demands, process demands and knowledge demands. In addition, different goals are distinguished that relate to process aspects, content aspects or both.

Finally, three key concepts can be identified in participatory backcasting: (1) desirable futures, also called future visions, (2) stakeholder participation, and (3) learning by stakeholders. These are the starting point for the theoretical exploration presented in the next chapter.

Notes

1   Parts of this chapter are based on Quist and Vergragt (2006), but updated here; these parts are 2.2 (but not 2.2.4) and 2.4.

2   In fact Dunn (1994:195) uses the terms plausible futures, potential futures and normative futures, but these are similar to the terms likely futures, possible futures and desirable futures, respectively. Recently, Börjeson et al. (2006) use the terms predictive scenarios (what will happen), explorative scenarios (what can happen) and normative scenarios (how can a predefined target be achieved), thus making similar distinctions.

3   In addition to the basic distinction used here, numerous other typologies for scenario and future studies are possible and proposed in the literature (e.g. Van Notten et al. 2003).

4   Though most scenario approaches focus on possible futures, in the literature reference is also made to business as usual scenarios that depict likely futures and to normative scenarios that resemble desirable futures,

5   There is a French scenario tradition, which is referred to as 'futuribles' or the 'strategic prospective (la prospective' (Godet 2000). Within this tradition has both been dealt with possible futures and desirable futures and combining these in strategic scenario planning (e.g. Godet 2000). This French tradition has not further been covered in this study.

6   Jantsch distinguished between explorative forecasting, which is forward-looking, and normative forecasting that should be used for setting goals in technology development. He used the Manhattan project and various other defence related programmes as examples.

7   As a consequence, I do not focus on other groups of methodologies and approaches like participatory policy analysis, participatory integrated assessment or so-called soft systems methodologies.

8   In Dutch TO3 stands for 'Toekomst-Orientatie, Ontwerp en Onderzoek'.

9   In Dutch abbreviated as TvC (Toekomstbeelden voor Consumenten). It has been applied as part of several backcasting projects within the STD programme (Weaver et al. 2000, Aarts 2000: 24-25). It may be used to develop desirable future images; it was not designed for that, but to incorporate consumer aspects early in the R&D process.

3

IN SEARCH OF USEFUL THEORIES AND CONCEPTS

Chapter 3 presents a survey of various relevant bodies of literature. It explores visions in technology development and system innovations (3.1), actor learning (3.2), stakeholder participation (3.3), system innovation theories (3.4-3.5), and network theories (3.6), before drawing conclusions (3.7). The purpose of the chapter is to identify useful theories and concepts that can be used to conceptualise backcasting experiments as well as their follow-up and spin-off.

3.1 Visions

3.1.1 Visions of the future

In Chapter 2 it was indicated that (shared) future visions and (shared) future vision development are crucial in a participatory backcasting experiment. The basic assumption is that future visions can be seen as shared multi-actor constructions that may have the potential to guide actor behaviour, especially if generated in a participatory or collective process (Grin and Grunwald 2000, Quist *et al.* 2001a). Visions may thus have the potential to provide an orientation for joint action, especially in situations where existing rule sets and institutions are not effective or valid (Dierkes 1996, Grin 2004, Grin *et al.* 2004, Quist *et al.* 2005).

Numerous examples of future visions can be identified, such as the mechanisation of agriculture in the previous century in many countries, the modernisation and electrification of the kitchen in the Netherlands (Van Otterloo 1990), the vision guiding the Delta waterworks[1] in the Dutch river estuary in 1950s – 1970s, the broad orientation in the 1950s with regard to rebuilding the Netherlands after the 2nd World War, the vision of the salmon returning to the river Rhine (e.g. De Groot 2002) and putting a man on the moon (and back to earth). Dierkes *et al.* (1992, 1996) have studied visions in technology development and found them in the early telecom system, the development of the typewriter and the emergence of the diesel engine. Other examples include Ford's vision of mass-manufacturing of cars, reducing costs and making cars accessible to a larger group of people in the beginning of the 20th century and the vision guiding the development of the Brazilian city of Curitiba. In the latter example the vision was developed and put into practice by an urban planner who became mayor of Curitiba and inspired an entire city to realise the vision (e.g. Rabinovitch 1996, Meurs 1998, Buis *et al.* 2002). Mission statements of large companies can also be considered (future) visions that guide their strategy and operations. Finally, the concept of sustainable development, as defined in 'Our common future' (WCED 1987), can also be seen as a broad future vision providing direction and guidance.

Future visions are not necessarily always desirable images of the future. For instance, Rachel Carson's book Silent Spring (1963) presented a gloomy future of a world full of pesticides. Other examples include the pessimistic prospect of the first report to the Club of Rome (Meadows 1972), the vision[2] on the depletion of the ozone layer in the 1980s, or more recently the climate change scenarios of the IPCC (www.ipcc.org). Such undesirable, catastrophic visions can also influence technology development and policy-making. For instance, wars, trade embargos, or natural disasters like global flue epidemics, earth quakes, tsunamis, tropical storms, flooding, or volcanic eruptions, can all be put in catastrophic visions that occasionally become reality and to which people can respond. However, most of these catastrophes are often only responded to after they have taken place; examples include the tsunami warning system for South and South East Asia and the Delta water works in the Netherlands after the 1953 flooding. Another example is the increase in energy efficiency since the 1970s, which has been strongly influenced by the first report to the Club of Rome (Meadows 1972), as well as by the oil crises in the same decade. The depletion of the ozone layer has also been responded to after confirmation by empirical evidence (see, for instance, Mulder 2005). Finally, the IPPC scenarios on climate change can also be seen as undesirable visions that may guide global actions aimed at mitigation. In that light the Kyoto protocol is just an early step of a response that has just begun.

Visions can also be or become controversial and contested (e.g. Brown *et al.* 2000, Grin and Grunwald 2000, Smits 2002a). This occurred with the highly disputed vision of 'atoms for peace' on nuclear energy, as well as in the ongoing debate on genetic modification of foods. Such controversies spark debates among actors with opposing worldviews and different visions on the desired future of society (Grin 2000). In fact, the vision on the depletion of the ozone layer was at first also contested, before it was accepted and responded to. The scenarios and views on climate change have also been heavily contested and debated, although at present support is growing and resistance is decreasing. Interestingly, in early backcasting papers it was proposed that undesirable future visions can be used to analyse how they can be avoided (Robinson 1990). In Technology Assessment the importance of visions has also been referred to (e.g. Smits[3] *et al.* 1995: 283, Vig and Paschen 2000), while the importance of assessing alternative visions in technology development (Dierkes *et al.* 1996, Grin and Grunwald 2000) has also been stressed.

A related issue is the threat of how utopian social visions turned out to be human and social disasters, when put in practice through totalitarian rules (De Geus 1996, Achterhuis 1998, Grin 2000: 10). Achterhuis (1998) has explored utopian thinking since the early Renaissance and has concluded that he would reject living in all societies described in utopian visions. Achterhuis criticises especially the limited personal freedom and limited plurality in utopian societies, while he also argues that realising utopias is impossible without totalitarian rule. However, De Geus (1996, 2002) takes an opposite view. He has argued not to discredit utopians alone and has pleaded in favour of investigating ecological utopias for ideas and directions for the desirable development of our society and use them as a navigational compass. Both De Geus (1996) and Grin (2000) argue that most utopias were created in response to socio-economic problems in the era that they were written. This allows us to learn from them, without taking them too literally. Avoiding side-effects and the totalitarian risk, however, requires a democratic framework and public legitimacy (Grin 2000).

In conclusion, visions seem widespread in society, as well as in system innovations and technology development. Different types of visions have been distinguished, such as desirable (positive) visions and undesirable (negative) visions that should be avoided. Both desirable and undesirable visions can be or become contested, while they also can 'gain' acceptance. In the following sections I look into more detail in various concepts of visions, followed by a comparison (in 3.1.6).

3.1.2 Leitbilder and vision assessment

German studies on technology development have resulted in the concept of Leitbild, which can be translated into English as vision or guiding image (Dierkes *et al.* 1992, Dierkes *et al.* 1996, Mambrey *et al.* 1994, Mambrey and Tepper 2000). The two key elements of the vision concept presented by Dierkes *et al.* (1996: 54-55) are: (1) The vision is shared, making is possible to unite people from different backgrounds, for instance different knowledge fields or disciplines that are needed to develop the technology, and; (2) The vision guides the behaviour and actions of the people who share or support the vision. Visions are shaped in a decentralised bottom-up type of social interaction processes among numerous actors. Visions influence these processes and the outcomes at the same time. This happens in a co-evolutionary way similar to the co-evolutionary development of technology and society. As a consequence, the visions cannot be enforced top-down in a hierarchical way.

According to Dierkes *et al.* (1996: 29-30) a vision is at the same time a vehicle for interaction, communication, explanation and discussion among both technologists and non-experts, and also a vehicle for broader reflection on normative choices and effects. The vision provides meaning and coordination to those spread over different locations and contributes to developing a particular technology. They are not very exactly expressed, especially not in the early stages of a new technology. Nevertheless, they are capable of establishing and structuring connections between different and mostly autonomous social subsystems and actor networks (Mambrey and Tepper 2000: 36-37). The concept of Dierkes *et al.* thus concerns shared visions for emerging technologies and technological artefacts even if they do not exist already; a successful vision mobilises those needed to develop a technology or an artefact.

The characteristics of Leitbilder mentioned above can be found in the two main functions of Leitbilder that Dierkes *et al.* (1992, 1996) distinguish: (A) guidance (referring to the German word Leit) and (B) image (referring to the German word Bild). Both are split into three sub-functions, which are depicted in Table 3.1. Together, according to Dierkes et al, the functions provide strength, attractiveness and stability. This includes connecting and motivating different people from different social backgrounds and knowledge disciplines to cooperate on a shared set of goals to be achieved in the future.

Table 3.1 *Functions of visions in technology development (based on: Dierkes et al. 1996)*

A. Guidance	
A1. Collective projection	Creating an overall shared, yet attainable, goal of technology development or innovation
A2. Synchronous pre-adaptation	Providing orientation towards the long-term overall goal through alignment and synchronisation of interaction and learning processes among actors from different knowledge fields and networks
A3. Functional equivalent	Replacing existing systems of rules and types of decision-making logic and making space for developing new rules and sets of rules facilitating the emerging network
B. Image	
B1. Cognitive activator	Activating different branches of sciences and technology for generating knowledge for realising the vision
B2. Individual mobiliser	Providing attractiveness and opportunities to individual actors and mobilising individual actors
B3. Interpersonal stabiliser	Binding heterogeneous actors and providing every day decentralised coordination at different locations

Guidance has to do with the direction and generation of shared goals, the linking and synchronisation of interaction and learning processes among individuals and actors from various social subsystems, such as different knowledge disciplines. Guidance also includes providing a space in which the existing and constraining rules and institutions can be ignored and alternative rules can be tried. According to Dierkes *et al.* (1996) the shared goal must combine both desirability and attainability at an abstract level, although adjustment must be possible if shared views change (sub-function A1). Radical innovations require the linking of many different actors and different knowledge fields and networks, and also the integration of knowledge from different fields and the use of new knowledge from these fields that may still have to be produced. In other words, radical technological innovations or new technologies often emerge on the border of two or several knowledge disciplines, where interaction can take place and result in rearrangement of knowledge and expectations. This requires synchronisation and alignment (sub-function A2). Furthermore, more radical innovations also require adjusting existing rules and rule systems or even new ones, because elements and knowledge need to be integrated and combined from several disciplines, having different rules and rule systems. Such a space and alignment, when provided by the vision, was defined as the functional equivalent (sub-function A3).

The image part of the vision consists also of three sub-functions. Firstly, the vision provides the cognitive challenge and activation leading to generation of relevant knowledge (sub-function B1). It also provides attractiveness including opportunities to a range of relevant actors that leads to mobilisation of actors and resources (sub-function B2). Finally, it provides coordination and stability among the actors and individuals involved and their daily activities at different locations (sub-function B3). The latter includes practical coordination in interactions, relationships and dependencies among different actors, as well as the overall network and ensuring that all work in line with the vision.

Three additional issues can be mentioned. Firstly, Dierkes *et al.* (1992, 1996) emphasise the presence of a vision champion, which is an individual or a group of individuals who coordinate different activities and necessary interactions at the same time. Dierkes *et al.* refer to vision champions like the engineer-entrepreneur Diesel in case of the Diesel engine and Edison in case of telephones and early wireless telephone systems. Secondly, there seems to be a balance needed between centralised and de-centralised coordination, but this issue is not dealt with. Thirdly, the Leitbild guides the emergence of a network and the activities in the network, while the network influences the vision as well (see also Grin 2000: 20). Dierkes *et al.* (1996) illustrate this mutual influence in the Diesel engine case, where manufacturers in the network decided to put a working machine on the market as soon as possible to become a player in mining and freight transportation and thus adjusted the vision. At the same time Diesel was still aiming to develop the perfect engine from a theoretical thermodynamic viewpoint.

The concept of Leitbilder has both been narrowed and broadened. Kuusi and Meyer (2002) have connected the concept of Leitbilder in technology development to foresighting, using nanotechnology as an illustration. They narrow the Leitbilder concept to rather small-scale technological applications, re-conceptualising Leitbilder as emerging technological paradigms in line with Dosi (1982). Elsewhere, (Mambrey and Tepper 2000, Mambrey *et al.* 1994) the idea of metaphors has been proposed; metaphors act within a broader Leitbild at the level of teams, projects and products.

Broadening the concept of Dierkes *et al.* (1996), Grin and Grunwald (2000) have dealt with changing guiding visions as a prerequisite to changing existing socio-technical systems. They write (Grin and Grunwald 2000: 1) *"one way to shape socio-technological systems is through the visions that guide their development… the assumption is that these visions exist already in most societal sectors, that these visions tend to reproduce the ways in which these sectors have developed hitherto, and that a critical discussion of these visions is a prerequisite for changing the course of development"*.

These authors address two new issues. Firstly, visions are widely present and guide developments in socio-technical systems, but they tend to reproduce themselves. This means that visions tend to facilitate certain directions of development, while constraining other directions or radical changes. This can be illustrated by the existing vision in many industries that focuses on reducing labour costs, while the increase of resource productivity is largely neglected. Another example is the vision in agriculture in the Netherlands focusing on increasing production volumes, while neglecting ecological and social aspects on a system level. Secondly, discussing such visions is a condition for change in socio-technical systems. This calls for broad participation and actor learning at the level of values and mental frameworks, and achieving some kind of agreement or congruence. The proposed mechanism to achieve change then is that mental change or value change at a group or subgroup level results in changes in the overall vision, which eventually leads to changes in the entire socio-technical system.

Such visions can also be found at the level of society as a whole, where opposing visions on a technology co-exist and compete. Opposing visions are rooted in totally different beliefs, values and mental frameworks and are shared by large groups in society, resulting in fierce debates, such as on nuclear energy and GMO in foods. Elaborating upon Fresco (1998) Grin has characterised three visions[4] on food and GMO that compete with each other and fuel the ongoing debate on this issue. These three visions show at least three different 'shared' visions of the future at the level of major social groups, thus at the level of society.

Visions are thus on the one hand seeds for change, shaped by actors, and on the other hand they are rooted in existing structures and carried on by the actors in these structures. Grin connects future visions and actors within a certain structure, which both constrains and enables the expectations that actors use. He writes (Grin 2000: 10-11) *"...the ways in which technology and its social context are being shaped along with each other, on the basis of a set of common or at least mutually congruent expectations. These expectations[5] concerning technology, its social context and the relations between the two, are projections for the future rooted in an actor's assessment of past experiences. Both evaluations of the past and expectations for the future reflect the values, worldviews, and deep preferences of those who hold them. ...they represent the bounded rationality of their 'possessor'. Thus expectations are not arbitrary but remain within a particular 'horizon of expectations'... These horizons delimit the range of people's attainable futures."* As a consequence, expectations partly determine or influence what futures and future visions can become attainable. The quote suggests that expectations are an important resource of actors when constructing and evaluating visions.

3.1.3 Technological expectations and promises

While in Germany the focus has been on Leitbilder as guiding images for (radical) technology development, a scholarly interest has also developed in dynamics of technological expectations and promises, especially in the Netherlands and the UK (Vergragt *et al.* 1989, Van Lente 1993, Geels and Smit 2000, Brown *et al.* 2000, Brown and Michael 2003, Van Merkerk and Van Lente 2005). In this line of research there is: (1) a strong emphasis on sharing and shared expectations and promises surrounding technology development, though often supported by a generic positive belief in technology as found among so-called technology supporters; (2) a focus on the rhetorical component[6] of technological expectations and promises; and, (3) the nested character of expectations. Expectations can be articulated at a local (niche) level, an intermediate level (of sectors) and a macro (national and international) level. When expectations at different levels are related, this is referred to as the nested character of expectations. In addition, expectation and promises thus need to be shared by individuals and individual actors to get meaning.

Van Lente (1993), who was one of the first to study technological expectations and technological promises, investigated cases at different levels: the local (micro) level, the meso-level and the national level. He has proposed a matrix of expectations that can be positioned at different levels. He also distinguishes between (1) search expectations guiding researchers and engineering, which relate to searching routines in Dosi's technological paradigm (Dosi 1982) and the concept of technological regimes of Nelson and Winter (1977, 1982), and; (2) niche expectations that provide protection to the niche and the search processes in the niche, for instance by senior managers or executive board members[7], (3) expectations at the level of societal trends with longer time horizon, which back niche expectations. Here, the nested character of expectations and promises is also at play. For instance, a promise at an intermediate (meso) level can facilitate or stimulate research projects at a niche level, while a successful niche project can support a promise at the meso-level of a (sub)-discipline. Van Lente also distinguishes three dimensions of expectations, as described in Table 3.2. The three levels, the three types of expectations and the three types of dimensions form together the matrix of expectations for an emerging technology or an emerging technological discipline.

Table 3.2 *Dimensions in the matrix of expectations (based on Van Lente 1993: 51)*

(1) *Cognitive dimension*	This concerns the possibilities and difficulties of a technological development. It may be based on an extrapolation of a theory or a scientific hypothesis.
(2) *Guiding dimension*	Expectations may guide the actions of actors. It shows routes to follow and options to choose, while some directions are labelled as the more successful or promising ones. It influences decisions about R&D strategy, heuristics to follow and how to reduce uncertainty.
(3) *Legitimating dimension*	Actors may use stabilised shared expectations to legitimise their actions and decisions inside and outside the organisation, for instance when mobilising resources. Doing otherwise needs to be accounted for and explained.

There are two issues that would warrant closer inspection. Firstly, what could be a possible distinction between promises and expectations, and, secondly, how do emerging promises and expectations fit in the concept of technological regimes (see 3.5), and the hierarchy of rules in the technological regime as proposed by Van de Poel (1998, 2002).

With regard to the first issue, Van Lente (1993) uses both the terms promises and expectations, but without distinction. However, Van Lente's (1993) appears to suggest implicitly that promises are located at a higher level and are more generic than expectations. For instance, he refers to the promise of technology at the macro level, to the promise of membrane science and technology at the meso-level and to the promise at the niche level of research projects in companies. With respect to expectations he consequently refers to the complex matrix of expectations suggesting there are a considerable number of them. He also proposes a promise-requirement cycle in which promises must be made true.

My proposal is to distinguish between broad technological promises (see also Table 3.3), which are backed by a set of supporting expectations that can be related to each other as explained above. In addition, a promise as proposed here agrees quite well the Leitbild concept discussed above. This can be a first step in relating expectations and promises to guiding visions and Leitbilder, and, possibly, facilitating further theory development. However, normative visions as generated in backcasting experiments are unlike Leitbilder and technological promises based on a mixture of normative principles and expectations.

With regard to the second issue, in this line of research expectations and promises are not considered to be located at the same level as visions or guiding images (Van de Poel 1998, 2000, 2003), but at a slightly lower level from a hierarchical viewpoint. For instance, in technological regimes (see later in this chapter in 3.5 for an explanation) the guiding principle provides guidance and orientation to the entire technological regime on an overall level. Van de Poel (1998, 2002) has positioned expectations and promises on a separate level just below the top level of the guiding principle, providing guidance to innovation and novelty, but within the boundaries of the technological regime.

Van Lente (1993) has discussed promises and expectations for both emerging technologies and scientific fields. This concerns similar phenomena as described by Dierkes *et al.* when dealing with emerging technologies guided by Leitbilder. Van de Poel views upon promises and expectations within the existing technological regime. My proposal is therefore to distinguish between promises and expectations fitting in the existing regime on the one hand and, on the other hand, the promises and expectations supporting emergent technologies or radical innovations that conflict with the existing regime and break with its constitutive rules.

Table 3.3 *Refining relationships between promises, expectations and regimes*

Existing literature	Proposed adjustments
No clear distinction between technological promises and expectations. Expectations can be related and nested (Van Lente 1993).	Technological promises are supported and constituted by sets of interrelated and nested expectations.
Promises and expectations are emergent, while context influences are not taken into account.	Emerging technological promises and expectations face competition from alternative emergent promises and existing technological regimes.
Promises and expectations on emerging technologies and disciplines are not related to promises and expectations in existing technological regimes (Van Lente 1993, Van de Poel 1998).	Promises and expectations are bounded by a technological regime and its guiding principle, while emerging promises and expectations may compete and conflict with the technological regime.
In technological regimes are guiding principles higher in the hierarchy than promises and expectations (Van de Poel 1998).	Technological promises on emergent technologies backed by expectations can, through a process of maturing, become part of a new guiding principle.

3.1.4 Visions in transitions and system innovations towards sustainability

The relevance of sustainable future visions providing guidance has been strongly advocated in sustainable technology development (Vergragt and Jansen 1993, Weaver *et al.* 2000), in system innovations towards sustainability (Quist and Vergragt 2000, Quist and Vergragt 2004) and in transition management (Rotmans *et al.* 2001, Kemp and Rotmans 2004, Loorbach and Rotmans 2006, Kemp *et al.* 2005, Berkhout *et al.* 2004, Wiek *et al.* 2006).

In transition management visions are referred to as *"a framework for formulating short-term objectives and evaluating existing policy... these visions must be appealing and imaginative and be supported by a broad range of actors."* (Rotmans *et al.* 2001: 23). In sustainable technology development (Weaver *et al.* 2000: 71-74) reference has been made to the development of shared visions on desirable future systems or societal need fulfilment, which encompasses new technological trajectories related to key areas of societal needs. This requires the commitment of stakeholders in shared problem definition, in shared vision development and in the innovation trajectories and innovation processes that can be derived from these shared visions.

So far, the different functions of visions have received limited attention in transition management and system innovations towards sustainability. The emphasis has largely been on overall approaches, end-states, overall mechanisms of system change and on generating visions. However, some recent papers have dealt with this issue (Wiek *et al.* 2006, Smith *et al.* 2005). Smith *et al.* (2005) propose five functions of future visions for system innovations and transitions, which are given in Table 3.4. These authors assume that different actors exert varying degrees of influence on constructing these visions and may deploy their resources in a strategic way.

Briefly, Smith *et al.* (2005) argue that there are three factors that are important to the consistency and robustness of the vision: (1) the degree of interpretative flexibility to a range of stakeholders, (2) the adaptive capacity to new developments, and (3) the coalition of stakeholders supporting the vision. In case the dominant system or regime is under stress, opportunities for alternative visions may emerge. These can, for instance, be advocated by outsiders, or by marginal parties in the present system that pursuit these new opportunities. Such alternative visions can be sustainable visions that are backed by alternative trend-breaking expectations about possibilities and may be based on different alternative worldviews. Such type of expectations is referred to as normative expectations by Berkhout (2006), which is different from the technological expectations proposed by Van Lente (1993).

Table 3.4 *Functions of visions for future system innovations (Smith et al. 2005: 1506)*

(1) *Mapping a 'possibility space'*	Visions identify a realm of plausible alternatives for socio-technical systems providing societal functions
(2) *Providing a heuristic*	Visions act as problem-defining tools by pointing to the technical, institutional and behavioural problems that need to be resolved
(3) *Providing a stable frame for target-setting and monitoring progress*	Visions stabilise technical and other innovative activity by serving as a common reference point for actors collaborating on its realisations
(4) *Providing a metaphor for building actor-networks*	Visions specify relevant actors (including and excluding) acting as symbols that bind together communities of interest and of practices
(5) *Providing a narrative for focusing capital and other resources*	Visions become an emblem that is employed in the marshalling or resources from outside an incipient regime's core membership

Elsewhere, Berkhout (2006) criticises the strong focus on guidance by future visions among transition management scholars. He rejects the guiding vision as *"...a device for specific desired end-states of particular socio-technical systems or regimes that are supported and brought about by an effective coalition of the willing working on processes of technological, institutional, cultural and behavioural changes on a trajectory towards the overall goal of the end-state"*. Instead, he defines visions as proposals ('bids') that are employed by actors in processes of coalition-formation and coordination. As a consequence, this requires interpretative flexibility of the vision enabling actors to align the vision with their own interests, their worldviews and their value systems. Articulation and diffusion of a vision can be successful if it is valid or attractive to a wide range of interests of a range of actors, or if it is backed by a powerful actor or group of actors that can enforce enrolment of others. In both cases involvement of new actors may also lead to modification of the vision.

3.1.5 Actor worlds and socio-technical scenarios

Callon (1986) has proposed the concept of the actor world. This can be interpreted as an actor-specific future vision, a desired end-state that an actor may want to achieve. It consists of expectations that support actor-specific actor worlds. Technical artefacts, knowledge and contributions of other actors are all part of the actor world, but they need to be realised. Thus, every actor could develop actor worlds, which may compete with one another. However, only a limited number of these visions can become shared among a larger group of actors. Such a shared image of a future world, in which an artefact functions and is used and produced, is basically the result of agreements among interacting and negotiating actors. As Callon writes elsewhere (Callon 1995: 313) on the agreements among negotiating actors on form, functions and design of entrance barriers for the underground in Paris: *"When compromise was reached, in fact they were agreeing on a particular (future) socio-technical network, which is often the result of tinkering with networks strongly defended by various actors.*" The agreement reached covered the artefacts, how these should be used, and how these are intended to influence (end)-users. The (future) socio-technical network of Callon is thus a shared actor world, which resembles the vision or Leitbild concept strongly.

The idea of the future socio-technical network has been used as a starting point by De Laat (De Laat 1996, De Laat and Larédo 1998) for connecting networks and technology development trajectories. As De Laat and Larédo (1998: 156-157) have put it: *"In technical research, implicit or explicit anticipation of new future worlds is the rule. This can be seen as an extension of Callon's socio-technical scenarios (Callon 1987) with a stronger emphasis on the trajectories defined by actors, and the conditions under which, according to these actors, these future worlds will be realised."* These authors argue that research projects presuppose or anticipate future networks including future production and use. Reaching agreements in the networks constituting these research projects also includes reaching agreements on the future worlds in which the artefact will function, as well as on the expectations supporting this.

3.1.6 Comparison of vision concepts

The previous sections have shown that visions, promises and expectations are important in the emergence of technologies and in system innovations and transitions towards sustainability. Several concepts of visions and expectations have been explored. Most concepts have a nested structure and can thus be related to different levels of aggregation, like niches or projects, networks, sectors and society at large (macro). An overview is given in Table 3.5.

Table 3.5 *Vision concepts and their meaning at different levels*

	Macro	Socio-technical systems	Meso (networks)	Micro
Leitbilder/visions (Dierkes et al. 1996, Mambrey and Tepper 2000)	*Not dealt with*	*Paradigms*	*Leitbild/Vision*	*Metaphors*
Promises and expectations (Van Lente 1993, Van de Poel 1998, 2003)	*The promise of technology at the level of society as a whole*	*Technological regime that allows promises and expectations that match with the rules of the regime*	*Promises for new fields or sets of new artefacts backed by a set of expectations*	*Promises for a new product or project backed by set of expectations*
Dominant and alternative visions (Grin 2000, Grin 2004, Grin et al. 2004)	*Competing macro visions each backed by large groups in society*	*Visions at the level of sectors can be adjusted and influenced*	*Emerging alternative visions*	*Emergence of really alternative visions to be explored in niche experiments*
Transition Management (Rotmans et al. 2001, Rotmans 2003, Kemp and Rotmans 2005),	*Part of the transition vision*	*Existing visions (business as usual)*	*There are different sets of transition visions to keep options open*	*Generating transition visions with small group of visionaries or a broader set of stakeholders*
Future actor world an shared socio-technical scenario (Callon 1987, Callon 1995, De Laat 1996)	*Part of the actor world*	*Part of the actor world*	*Shared socio-technical scenario is agreed on by networks of actors Different networks supporting different socio-technical scenarios compete*	*Actor world of particular actor, competing with actor worlds of other actors*

Several differences and similarities can be noted among the various vision concepts explored. Firstly, most vision concepts deal with emerging phenomena. The exceptions are promises and expectations within existing technological regimes and the existing visions in socio-technical systems and in society. Secondly, most vision concepts show several varieties, for instance at different yet nested levels, sometimes proposed by different authors, and sometimes developed by the same author. Thirdly, whereas some concepts focus on analysing and describing phenomena like promises and expectations and socio-technical scenarios, other vision concepts emphasise their guidance and the functions that can be provided. Examples of the latter are Leitbilder and visions in transition management.

What these concepts seem to have in common are bottom-up decentralised processes contributing to vision development, promise development and creating shared understandings or expectations, which cannot be controlled in a hierarchical top-down way. In addition, emergent normative visions or promises can guide, direct, coordinate, mobilise and stimulate actors and activities, also in a decentralised way. Visions and networks of actors involved evolve together in a co-evolutionary way and mutually influence each other. However, visions can be 'frozen' in negotiation processes by actors or coalitions of actors resulting in stronger and more hierarchical control, including a particular set of actors and excluding others, but this is a particular elaboration or design of a more generic vision or promise. Finally, although a bit neglected in most literature, emerging Leitbilder, visions and promises have in general to compete with existing dominant visions, and with other emergent visions.

3.1.7 Conclusion on visions

Visions are important in technology development, as well as in system innovations and transitions towards sustainability. Various concepts of visions have been identified in literature and have been compared. Most of them concern vision concepts that address emerging phenomenon like the development and diffusion of new technologies, the rise of new scientific fields and transitions towards sustainability. Many vision concepts have been elaborated at different levels, like the level of individual projects, the level of niches and emerging networks, the level of technological regimes, industries and entire socio-technical systems and the macro-level of a society. It has been found that emerging visions need to compete with other competing visions, as well as with existing dominant visions that are strongly entrenched in existing systems and supported by actors that defend existing interests. Finally, most vision concepts do not include explicit normative principles or so-called normative expectations.

In conclusion, the concept of Leitbilder seems particularly promising to be employed as a building block for conceptualising backcasting experiments and their follow-up and spin-off after five to ten years. This particular vision concept not only has a solid theoretical base and addresses various relevant aspects through its functions; it also covers both the early stages of emerging visions and the diffusion of the vision and growth of the network supporting the vision in later stages. However, the concept needs to be adjusted in such a way that it explicitly covers normative principles and that synchronisation can take place not only across different scientific disciplines, but also across different societal domains. These issues will be addressed more thorough when developing the conceptual framework in Chapter 4.

3.2 Learning

3.2.1 Introduction

Learning by stakeholders has been identified as another key element in participatory backcasting. The basic assumption is that learning leads to changes in the mental framework of stakeholders, enabling the space for behavioural alternatives, such as initiating follow-up activities and other effects. Conventionally, learning relates to gaining new, especially cognitive, knowledge by individuals or, as Van de Kerkhof and Wieczorek (2005: 734-735) have written (based on Hall 1993), "...*learning occurs when individuals assimilate new information, also based on past experiences and apply it to their subsequent actions*". The focus of this section is on concepts of learning at the level of groups of stakeholders or actors. It must be noted that learning in organisations and at the level of groups also takes place through individuals. However, it becomes relevant at the level of a group or an organisation when it is shared and internalised among groups of actors and individuals.

In the area of sustainability it is widely assumed that necessary changes and change processes require learning by stakeholders. Learning relates not only to new scientific knowledge, but also to any new insight resulting in changes in people's cognitive or mental framework. By contrast, the opposite is not automatically true; learning does not automatically result in observable changes. As Grin and Van de Graaf (1996) have argued, learning is an important condition but not a guarantee for change. Thus, learning about sustainability does not guarantee realisation of actions and activities supporting changes necessary for sustainable development or system innovations towards sustainability.

Learning at the level of groups has become an important subject in many fields and disciplines. Van de Kerkhof and Wieczorek (2005: 735) have found a range of conceptualisations for learning in the literature, which includes social learning, political learning, government learning, organisational learning and policy-oriented learning. Other conceptualisations can be found in other bodies of literature. For instance, Kamp (2002: 25-40) has reviewed learning in innovation studies and distinguishes, based on Garud (1997) and Lundvall (1992) four categories of learning (see Table 3.6): learning-by-searching (know-why), learning-by-doing (know-how), learning-by-using (know-what) and learning-by-interacting (know-who).

Table 3.6 *Learning in innovation processes (derived from Kamp 2002)*

Category of learning	Short term	Description
Learning-by-searching	*Know-why*	Systematic and organised search for new knowledge by R&D and study
Learning-by-doing	*Know-how*	Achieving knowledge to produce, includes tacit knowledge and counts for part of the learning curve in production
Learning-by-using	*Know-what*	Learning by users on how to use novelties
Learning-by-interacting	*Know-who*	Learning from users by producers through interaction

Several authors, including Van Mierlo (2002) and Brown *et al.* (2003), also refer to the learning framework of Rogers (1983, 1995), which explains the diffusion of innovations, especially the diffusion of new products among consumers in society. This relates to Kamp's category of learning-by-using. The framework of Rogers deals with individual choice or decision processes comprising several steps when deciding on purchasing a concrete specified option, which is mostly a tangible product or artefact. Although this framework manages to explain the diffusion processes of new artefacts among consumers very well, it is difficult to say whether it also applies well to actors and stakeholders.

3.2.2 First order and higher order learning

Learning in innovation studies strongly focuses on specific innovations, technical change, technical novelties and the capabilities to deal with them. It does not, or only to a very limited extent, make a distinction between small changes and huge changes like radical novelties and how this relates to different levels of learning. Such distinctions have been emphasised in other fields, for instance by distinguishing between first order learning and higher order learning (e.g. Brown *et al.* 2003). First order learning reflects new insights with regard to options in the case of a given problem and a given context. Higher order learning concerns new insights at a higher level with regard to problem definitions, norms, values, goals and convictions of actors, and approaches how to solve the problem. This is of great importance in the case of complex problems including actors with different mental frameworks or action theories (Grin *et al.* 1997, Grin 2000). The assumption is that when learning results in changes in mental frameworks or action theories of actors, this increases the space for behavioural alternatives available to actors or stakeholders (Grin and Van de Graaf 1996a). Higher order learning can thus be seen as a condition for implementing more sustainable solutions or getting these accepted and it accompanies (social) change processes. In case higher order learning involves larger groups in society or society at large, Wynne (1995) uses the term collective value learning.

There are differences in terminology, definitions and focus of higher order learning in different fields, which seems motivated by the wish to make it more applicable in specific fields and their research focus (see Table 3.7 for an overview). For instance, in policy-oriented learning[8] as described by Sabatier and Jenkins-Smith (1999), higher order learning involves redefining policy goals and adjusting problem definitions and strategies. In organisational learning it is called double-loop learning and it involves changes in norms, values, goals and operating procedures governing the decision-making process and actions of organisations. One difference is that organisations are relatively homogeneous in their mental frameworks compared to policy domains. In policy domains heterogeneous actors ally in so-called advocacy coalitions in policy sub-systems around particular solutions or problem definitions, while having greatly differing mental frameworks (Sabatier and Jenkins-Smith 1999). Grin and Van de Graaf (1996a) have proposed the term congruence for this. Congruence means that partial agreement is achieved in a way that stakeholder can cooperate or can take jointly a position, while value systems or mental framework differ considerably. It is emphasised that interaction among actors is important to achieve this (Grin *et al.* 1997). Congruence is different from consensus, as congruence does not assume full agreement on all aspects, as well as on many assumptions.

Hall (1993, see also Van de Kerkhof 2004: 77-78) distinguishes between three types of learning results in a given policy setting. The first type comprises the modification of the settings of policy instruments, while keeping the overall goals and the policy instruments the same. The second type leads to changes in policy programmes and policy instruments and likely their settings too, but overall policy goals are not altered. The third type comprises changes in the overall policy goals, which also leads to changes in both policy instruments and the settings of these instruments. Whereas the third type of learning leading to changes in overall policy goals is similar to higher order learning, double-loop or policy-oriented learning, the first and second types of learning resemble first order learning or single-loop learning.

Hoogma *et al.* (2002: 28-29) have put the distinction between higher and first order learning as follows. *"Learning can be limited to first order learning. That is when, in the niche[9], various actors learn about how to improve the design, which features of the design are acceptable for users, and about ways of creating a set of policy incentives which accommodate adoption. In [higher order] learning processes, conceptions about technology, user demands, and regulations are not tested, but questioned and explored. Opportunity emerges for co-evolutionary dynamics, that is, mutual articulation and interaction of technological choices, demand and possible regulatory options."* These authors also relate higher order learning to Wynne's (1995) collective value learning, because their concept of higher order learning is about clarifying and relating various values of producers, users and other third parties involved, such as governments.

Higher order learning is sometimes also referred to as interactive learning (Van Mierlo 2002: 30), and generative learning in teams or organisations (Senge 1990). Senge (1990) positions generative learning opposite to individual learning and emphasizes the role of shared visions, system thinking, an open dialogue and group commitment in teams for realising higher order learning when dealing with complex problems. Van Mierlo, elaborating on Wynne (1995), Hoogma *et al.* (2002) and Senge (1990), comes to a similar description of interactive learning with emphasis on shared vision development. Furthermore, it is often emphasised that higher order learning is difficult to achieve, lacking or very limited (e.g. Hoogma *et al.* 2002, Van Mierlo 2002).

Also elaborating on several theories of learning from different areas of research, Brown *et al.* (2003: 296) define higher order learning *"...as consisting of three interrelated shifts: (1) a shift in the framing of the problem and of the perceived solution (or a menu of solutions); (2) a shift in the principal approaches to solving the problem, and in the weighing of choices between desirable yet competing objectives; (3) a shift in the relationship among the participants in the experiment, including mutual convergence of goals and problem definitions. These shifts can occur among the participants of an experiment and their professional networks as well in the broader social sphere".* Brown *et al.* (2003) also distinguish between two types of higher order learning, namely between learning among the participants that are involved in an experiment and their immediate professional networks, and learning coinciding with further diffusion of outcomes of the experiment in society at large. In their evaluation of several sustainability oriented projects[10] in the field of mobility, they found considerable higher order learning among participants and their immediate surroundings, but not broader in society. However, achieving learning broader in society is very ambitious, both as a project goal and an evaluation criterion, because it assumes large efforts and different methods. This is because higher order learning starts among the individuals that participate in such projects, but subsequently needs to diffuse further in participants' organisations and associated networks. Further diffusion in society at large of learning results is definitely very important, but it is a complicated process.

Van de Kerkhof (2004) has applied learning in the context of a participatory backcasting experiment on options mitigating climate change. In addition to the distinctions between 'who learns' (the subject of learning), 'learns what' (the object of learning) and 'to what effect' (the results or effects of learning), she also distinguishes between first order learning and higher order learning and subsequently between first order and second order learning processes. Van de Kerkhof (2004: 78-80) relates first order learning to new insights among the actors involved into 'facts' and the expectations concerning a specific topic. Second order learning is related to gaining new insights into relationships between causal and normative reasoning, resulting in changes in norms and belief systems (or background theories) that guide stakeholders' their behaviour and perception of the topic or problem concerned.

3.2.3 What does learning affect in actors?

Some authors have related first and higher order learning to different levels in systems representing actors or actor groups. Sabatier and Jenkins (1999) have proposed a policy belief system of three levels for actors involved in policy domains. There are two levels at which learning can lead to change in the policy belief systems of advocacy coalitions, which are (1) secondary aspects and (2) policy core beliefs. In addition they distinguish a deep core level, which relates to fundamental worldviews and preferred social order, but changes at this level are extremely difficult to achieve and therefore not dealt with here.

Fischer (1980, 1995, as explained in Grin and Van de Graaf 1996a, 1996b), has proposed an appreciative system of actors consisting of four levels. There are two higher levels that may change due to higher order learning and reflection and two lower levels that change due to first order learning. The four levels are:

- The fourth and highest level concerns fundamental preferences about the social order in society (and relates to political ideologies).
- The third level concerns background theories and value systems.
- The second level concerns problem definitions, goal setting and strategy evaluation in a given background theory or value system.
- The first and lowest level concerns the evaluation of solutions, instruments or strategies in terms of costs and benefits using a given set of goals.

It is at the third level of background theories and value systems that professionals give meaning to situations and objects; it must be noted that within communities of professionals there can be different background theories or value systems inspired by different sets of fundamental preferences about the social order and society. The four levels together have also been referred to as frames of meaning. Grin and Van de Graaf (1996a) have elaborated such frames of meaning for technologists (technological paradigms) and policy makers (policy belief systems). Grin and Van de Graaf (1996a) have also coined the term congruence, which can be described as having agreement on issues, problem definitions and solutions, while differences exist with regard to background theories, world views or value systems. Learning can contribute to achieving congruence and shared opinions on certain issues, while considerable differences between actors remain.

A remaining issue is how to measure higher order and first order learning. Grin and Van de Graaf (1996) have shown for wind turbine development in Denmark that it is possible to reconstruct frame of meanings and evaluate changes with regard to this topic, for instance using in-depth interviews. Van Est (1999) has used similar schemes and has reconstructed policy belief systems of advocacy coalitions in wind energy politics in Denmark and California, which is at the level of actor alliances within policy domains or policy networks. Recently, Brown and Vergragt (2007) have used a similar scheme for evaluating learning processes by individuals in a team and on the team level in the case of a zero-energy residential building in Boston. They conclude that team learning consisted of participant turnover until congruence in worldviews and interpretive frames was achieved.

3.2.4 Comparison and conclusions on learning

The literature survey in this section has confirmed the relevance of higher order learning in change processes in different fields, as well as in settings of heterogeneous actors. It has been found that learning is a condition for change, but does not automatically result in achieving change. The findings confirm that learning by stakeholders is important in sustainable development, as well as in developing and realising system innovations towards sustainability. A range of conceptualisations of learning and the distinction between first order and higher order learning has been found, of which Table 3.7 presents a brief overview.

Table 3.7 *Different conceptualisations of learning*

	First order learning	Higher order learning	Locus of learning
Organisational learning	Single-loop	Double-loop	Within organisations
Policy-oriented learning (Sabatier and Jenkins-Smith 1999)	First order (changes in secondary aspects)	Higher order (changes in policy core beliefs)	Heterogeneous policy coalitions
Learning in teams and learning organisation (Senge 1990)	Individual leaning	Generative learning using shared visions, system thinking, open dialogue and group commitment	Within teams and organisations
Learning in innovation (Kamp 2002)	Learning-by-searching Learning-by-doing	Learning-by-using Learning-by-interacting (or interactive learning)	Within development projects, niche experiments, test markets
Learning in niches (Hoogma et al. 2002)	Learning in niches about improving design, attractive features and policy measures accommodating adoption	Technologies, user demands, and regulations are not tested, but questioned and explored, co-evolutionary dynamics	Various actors in niches and niche experiments
Collective value learning (Wynne 1995) Social learning (Robinson 2003)		Collective learning about values	Society at large, larger social groups
Learning in BSTEs (Brown et al. 2003)		Shifts in problem definitions, preferred solutions, principal approach, major priorities, relationships among groups of actors, as well as joint and congruent learning	Participants of experiments and immediate networks Broader diffusion & society at large

Several theories have been identified explaining the difference between first order learning and higher order learning at the level of particular actors. In conclusion, the conceptualisation proposed by Brown *et al.* (2003), which brings together elements from different concepts, seems especially interesting from the viewpoint of this research. This is not only because it integrates elements from different theories, but also because it connects learning to social experiments (which they call BSTEs or Bounded Social-Technical Experiments) that are bounded in time, resources and participants. This type of social experiments resembles backcasting experiments, although they strongly differ in the time horizon and the scale of solutions. Whereas BSTEs focus on rather short terms of development and introduction of concrete solutions or artefacts, backcasting experiments focus on time spans of 50 years and system innovations. In addition, the conceptualisation by Brown *et al.* (2003) distinguishes between learning in the experiment and the participating organisations on the one hand and diffusion of learning results broader in society and more distant networks on the other hand. This distinction seems useful in backcasting experiments too.

3.3 Stakeholder participation

3.3.1 Introduction

In this research backcasting is considered a participatory approach. Already the Brundtland report (WCED 1987) has called for sustainable development as a broad participatory process including bottom-up participation by citizens. Stakeholder involvement is a necessity, as sustainable development requires knowledge, support and actions from many actors in all societal groups like business, government, research, public interest groups and the general public, while the impact may affect them too. Various stakeholder concepts and stakeholder theories can be found in fields as diverse as strategic management, policy sciences and public decision-making, as well as in sustainability and corporate social responsibility studies. In technology assessment reference is made to social actors or a broad range of different actors (Grin *et al.* 1997) that need to be involved in decision making about technology development. Although not the term stakeholder is used, the meaning of involving social actors in addition to those that are already involved is very similar to various stakeholder concepts.

A well-known definition of a stakeholder has been provided by Freeman (1984) in the field of strategic management (for a recent review see Freeman and McVea 2005). Freeman (1984: 5) defined stakeholders as *"...any group or individual who is affected by or can be affect the achievement of an organisation's objectives"*. Although various other definitions can be found, the definition by Freeman can easily be reformulated in such a way that it can be used in other fields and contexts. A more general version of the definition could be the following: a stakeholder can be defined as an individual, organisation, group of individuals, or group of organisations that can affect or be affected by a certain topic (theme, decision, or achievement of an objective). For a discussion on the similarities and differences in definitions of stakeholder, see Van de Kerkhof 2004: 20).

Van de Kerkhof (2004: 26-27) has provided three major arguments for stakeholder involvement in (public) decision-making: (1) increased legitimacy of the decision, as more stakeholders have been involved; (2) Increased accountability, as the stakeholders involved have become co-responsible for the decision and related activities and action plans, (3) increased richness of the process, due to the input a wider range of viewpoints, interests, information and expertise about the topic under consideration. She continues (2004: 27-31) that stakeholder participation in science can contribute positively to dealing with uncertainties, contextualisation of knowledge and also for structuring and defining complex unstructured societal problems. Unlike structured problems, unstructured problems are more complex, have fuzzy boundaries and involve various perceptions on the problem, as well as various approaches to deal with it. In addition, possible solutions differ and may conflict, while stakes can be unclear and may change. Sustainability problems are clearly unstructured problems; system innovations towards sustainability contain a possible solution direction.

In the case of sustainability problems and system innovations towards sustainability, I would like to add that (1) stakeholders are experts in their own field and that this expertise is necessary for structuring the problem and finding possible solutions, (2) stakeholders may not only support to outcomes, thus provide legitimacy, but are also needed and indispensable for putting solutions into practice, as many system innovations towards sustainability require active contributions from government, companies, research bodies and public interest groups.

3.3.2 Degrees of participation

Van de Kerkhof has distinguished high, moderate and low degrees of stakeholder participation, building on the accounts by Arnstein (1969) for participation in policy-making and by Mayer (1997) on participation in science (also in Table 3.8). This involves both types and quality of participation. For instance, Arnstein's ladder of participation (Arnstein 1969, see also Van de Kerkhof 2004: 43-46) contains eight degrees of participation connected to degrees of decision-making authority, which is the degree of power by stakeholders to influence the decision-making process. Participation in science is slightly different from public or policy decision-making, as it has less to do with power and decision-making authority, and more with mutual learning and equality of the inputs by both scientists and non-scientists. Building on Mayer's (1997) framework Van de Kerkhof (2004) distinguishes between seven degrees of participation in science. All different degrees of participation distinguished are shown in Table 3.8. Clear explanations of all the terms related to different degrees of participation both in science and policy-making can be found in Van de Kerkhof (2004: 43-49).

Table 3.8 Degrees of participation (based on Van de Kerkhof 2004: 44)

Degree of participation	In policy-making	In science
High	Stakeholder control Delegated power Partnership	Mutual learning Co-production of knowledge Coordination
Moderate	Placation Consultation	Mediation Anticipation Consultation
Low	Information Therapy Manipulation	Information

Van de Kerkhof (2004: 49) adds *"When also considering the degrees of participation in the science domain, it turns out that it may be better to not narrow down participation to matters of political power and decision-making authority, since meaningful participation also relates to the meaningful flow of information, the transparency of the process, etcetera. In the classification of participation in scientific practice, participation refers to the level of interaction and debate between the stakeholders. According to this classification, participation is meaningful when the stakeholders have sufficient opportunity to articulate their knowledge, values, and preferences about the issue under consideration."* Van de Kerkhof (2004) also proposes three main degrees of participation, simplifying the more extensive schemes proposed by Arnstein (1969) and Mayer (1997). A distinction is drawn between a high, a moderate and a low degree of participation, while each category (see also Table 3.8) consists of several more detailed degrees of participation based on the original schemes.

A high degree of participation refers to control by stakeholders, power delegated to stakeholders, and partnerships in decision-making and to high levels of equality in knowledge generation reflected by mutual learning and co-production of knowledge. A moderate degree of participation is characterised by consultation of stakeholders, for instance by giving them some voice in a board at some distance (placation), or giving stakeholders an input to the research, which can include anticipations on the future, and stakeholder value or perception assessment by the researchers. A low degree of participation involves providing information or more active ways of convincing stakeholders through communication.

Finally, Teisman and Edelenbos (2004) have argued that in case of high degrees of stakeholder participation in change processes like system innovations towards sustainability, this requires the participatory processes to be sufficiently linked to formal (democratic) decision-making. High degrees of participation may also match better with strongly deliberative forms of democracy. This is an issue that deserves attention, whereas it is often to some extent neglected in sustainability issues.

3.3.3 Process management and settings

Van de Kerkhof (2004: 49-51) has also listed various possible problems related to stakeholder participation. These include irrationality of citizens, stakeholders defending their own interests, lack of stakeholder competence, aggravating the conflict, threatening legitimacy of representative democracy, reinforcement of the interests of the already powerful, and lack of objective selection criteria. In addition, participation can be further complicated by tactical and strategic behaviour by stakeholders in decision-making processes (De Bruijn et al. 2002). This includes stakeholders forming alliances, having hidden agendas, exerting influence on other stakeholders using existing contacts and dependencies, or trying to delay or de-legitimize outcomes.

De Bruijn et al. (2002) have proposed process management as an approach to dealing with stakeholder participation in decision-making processes, although it also has relevance to other types of interaction processes. Elsewhere, De Bruijn and Ten Heuvelhof (2000) briefly summarise process management as follows. The focus is on the interests of the most important stakeholders. Stakeholder support can be achieved by giving them influence in the design of an activity or an initiative in such a way that it becomes more attractive to them. The process design aims at solving the problem or issue at play. Process rules prevail and participating stakeholders need to agree on them. If the context changes, rules agreed on can become less important, and need to be replaced by new rules (that should also be agreed on). There are only a few rules with regard to content. The designer of the process and the process manager who facilitates the (decision-making) process and monitors a limited number of contextual conditions have the initiative.

According to De Bruijn and Ten Heuvelhof (2000) a well-designed process requires a sense of urgency and incentives for progress. It also requires openness and integrity, protection of core interests and core values of parties and conditions to come to sufficient substance. In addition, Grin et al. (2004) have pointed to the influence of settings to trend-breaking and rule-breaking activities, such as backcasting experiments. Settings can both stimulate and constrain trend-breaking and rule-breaking activities and are often connected to stable structures in organisations and networks. However, emerging networks have more possibilities to change the rules.

3.3.4 Conclusion on stakeholder participation

There have been strong calls for stakeholder participation in sustainable development, as well as in various other domains like science, policy-making and strategic management. Different concepts of stakeholder participation exist in different domains like policy-making, science and strategic management. In addition, the degree of stakeholder participation can vary considerable, from only being informed to being fully in control and being equal to scientists. A simplified scheme that distinguishes between high, moderate and low degrees of participation seems promising to characterise stakeholder participation in backcasting experiments. Stakeholder participation has various pitfalls, which calls active facilitation, adequate design and management of processes of stakeholder participation. In addition, the settings of a process of stakeholder involvement can both constrain and enable trend-breaking and rule-breaking activities.

3.4 System innovations and sustainability

3.4.1 Introduction

In innovation theory and innovation studies it is increasingly acknowledged that innovation consists of technical and non-technical change; the latter can comprise social, organisational, institutional changes, although not necessarily all of them. The mutual influence of technical and non-technical aspects has also been referred to as the co-evolutionary development of technology and society. Innovation can take place at different levels, not only at the level of artefacts or at the level of organisations, but also at the level of supply chains, industrial sectors and even society. In line with these insights a definition has been provided by Smits (Smits 2002b: 865): *"[innovation is] a successful combination of hardware, software and orgware, viewed from a societal and/or economic point of view... Innovation is a complex process that takes place at the level of specific products, businesses and sectors, as well as the level of national and international communities."* In this definition hardware relates to the material equipment, software refers to knowledge including tacit knowledge, while orgware refers to organisational and institutional conditions influencing the development and functioning of an innovation (Smits 2002b: 865). This definition explicitly refers to the level of socio-technical systems and the necessity of technical and non-technical innovations for realising successful innovation.

Various taxonomies of innovations have also been proposed, which in different ways distinguish between type, scope, and scale of innovations. For instance, Abernathy and Clark (1985) have distinguished two dimensions of innovations that can be both disruptive and conserving. These dimensions are: (1) customers and markets and (2) technological novelty, which results in a taxonomy of four types of innovations. In this taxonomy architectural innovations are disruptive on both dimensions, whereas any incremental innovations are conserving existing market linkages and technological competencies. Christensen (1997) distinguishes between disruptive innovations and sustaining innovations that are non-disruptive. Here, disruptive and sustaining are considered from the perspective of the value chain and the customer pool. Sustaining innovations also refer to existing technologies and existing technological trajectories. By contrast, disruptive innovations refer to new technologies having features attracting new and existing customers and may often involve outsider firms. This may result in firm entrants becoming new major players and in major changes in supply systems and in consumption systems. In fact, disruptive innovations as proposed by Christensen agree well with the category of architectural innovation by Abernathy and Clark (1985).

Freeman and Perez (1988) have proposed a taxonomy that covers an even wider range of levels[11]. These authors have distinguished between four types of innovation: (i) incremental innovations; (ii) radical innovations; (iii) changes in technology systems; and (iv) changes in techno-economic paradigms. As this research focuses at the level of socio-technical systems and their transformation, I focus on the latter two categories. Changes in 'techno-economic paradigms', also referred to as 'technological revolutions', concern changes that are extremely far-reaching in their effects and have a strong influence on the entire (global) economy, such as the rise of ICT in the last decades. This type of changes is basically at the same level as Schumpeter's long cycles and Kondriatiev's long waves. Freeman and Perez (1988: 47) write about the third category *"Changes of 'technology systems' are defined as far-reaching changes in technology, affecting several branches of the economy, as well as giving rise to entirely new sectors."* This type of innovation can include both radical and incremental innovations, in combination with managerial and organisational innovations, and generally speaking affect a large number of firms. Examples include the transition from the propeller engine to turbojets in aircrafts (Geels 2005), the arrival of oil platforms at sea, or the rise of supermarkets and hypermarkets in industrialised societies.

In conclusion, the type of system innovations as dealt with in this research, agrees quite well with the third type of innovation in the taxonomy proposed by Freeman and Perez (1988) and the category of architectural innovations proposed by Abernathy and Clarke (1985).

3.4.1 System innovations and sustainability

The previous section has dealt with several taxonomies of innovations, which included categories that can be considered system innovations. Other scholars have used the system innovation itself as a starting point, focusing on its specific nature, dynamics and change aspects. For instance, Geels (2005: 2) refers to technological transitions, as the change process from one socio-technical system to another. Elsewhere, transitions have been described as a long-term process of change towards purposely set goals during which a society or a subsystem of society changes fundamentally (Rotmans *et al.* 2001). System innovations and transitions fundamentally change both the structure of the system and the relationship among the actors in the system.

Quist and Vergragt (2000) have described system innovations as combinations of innovations at a system level combining technological and non-technical changes. They have characterised system innovations by: (i) the large number of variables; (ii) the large number of actors involved; (iii) the combination of different innovations (always including technological innovations); (iv) the so-called soft innovations required to meet cultural and institutional conditions (e.g. rules, legislation, paradigms, social structures, perceptions); and, (v) the system changes when implementing in society. This agrees well with De Bruijn *et al.* (2004: 3) who write *"System innovations ... are comprehensive innovations with a long time horizon, requiring (i) the efforts of many stakeholders, and (ii) a change of perspective and a cultural shift among these stakeholders."* Grin (2004: 14, my translation) sees system innovations as *"reflexive modernisation of knowledge intensive systems"*, thus relating system innovations to the concept of reflexive modernisation (Beck 1997, Beck 2006). Below in 3.4.3 I return to this issue.

Other terms have been used in the field of sustainability studies to describe similar phenomena as system innovations and transitions. These include transformations of socio-technical systems, industrial transformations (Olsthoorn and Wieckzorek 2005) and shifts to sustainable production and consumption systems. Different aggregation levels have been distinguished. For instance, several system innovations make up a more comprehensive transition (Rotmans *et al.* 2001, see also Chapter 1). Furthermore, not all system innovations need to result from a 'grand design', as has been argued by De Bruijn *et al.* (2004: 4): They may also develop as an emergent phenomenon resulting from decentralised and uncoordinated activities of many actors, as is illustrated by the rise of the Internet.

System innovations are not necessarily moving towards sustainability. In the past numerous transitions and system innovations took place in which a wide range of actors and social groups co-shaped changes and outcomes. However, there is certainly not yet a comprehensive understanding of system innovations, nor a full approach for guiding and managing system innovations and transitions towards sustainability. As a result, there is an emerging interest in the issue of transitions and system innovations towards sustainability (e.g. Rotmans *et al.* 2000, Elzen *et al.* 2004, Olsthoorn and Wieckzorek 2005, Voß *et al.* 2006).

3.4.3 System innovations and reflexivity

Above, it has been argued that regular system innovations differ from system innovations towards sustainability in their desired sustainable end-state, which is not static or in equilibrium, but dynamic and evolving further. Grin (2004) has linked system innovation to reflexive modernisation Beck (e.g. 1997) of knowledge-intensive systems, which resemble the socio-technical systems as dealt with in this thesis.

Reflexive modernisation is the response to regular or simple modernisation that has resulted in the present socio-economic system with its wealth in the North, its unequalled use of resources, sincere environmental problems and global inequity. Reflexive modernisation has been proposed as an alterative to regular ongoing modernisation, and assumes using reflexivity to prevent side-effects and unfair distribution of risks[12].

Sustainable development can be seen as a response to Beck's plea for reflexive modernisation, as it is a way of development that deals more adequately and timely with side effects, or, even better, prevents them beforehand. Constructive Technology Assessment has attempted this in technical change. System innovations towards sustainability can then be seen as doing this on a system level as part of sustainable development. As sustainable development is not automatically reflexive, reflexivity needs to be stimulated. Therefore, Voß et al. (2006) argue that 'reflexive governance' is needed for sustainable development. A distinction can be drawn between first and second order reflexivity. First order reflexivity relates to reconsidering and adapting actors' and individuals' own actions and behaviour. By contrast, second order reflexivity is more far-reaching and is about reconsidering and adjusting structures and institutions.

Changing socio-technical systems[13] or system innovations is even more complicated due to the fact that present institutions, structures and rule systems are guiding existing practices as part of existing socio-technical systems and are not equipped to solve so-called system crises or deal with disruptive and rapid changes. This has to do with what Giddens (1984) mentioned the duality of structure and action. This implies that structures and institutions emerge gradually from specific practices, are shaped gradually enabling specific practices and making these recursive, while constraining others. However, actions by actors exert influence on structures and institutions and can result in changing structures and institutions.

Niches and visions seem well-suited to deal with problems and issues for which no institutions and structures are available (e.g. Grin 2004, Grin et al. 2004). A niche can be seen as a space for experimentation in which the rules of the existing structures can be temporarily ignored and in which can be practiced with new rules or different sets of new rules. Then, the development of alternative sustainable visions in protected niches for experimentation can be a way of exploring new solutions and the possible (sets of) rules that are required for these. However, the major issue is how to gain support for the vision and rules outside the protected niches in such a way that it eventually results in an adjustment of structures and system innovations towards sustainability.

3.4.4 Conclusions on system innovations and sustainability

Studying system innovations can be seen as part of innovation studies. The phenomenon of system innovations towards sustainability has been connected to reflexive modernisation and the plea to enhance reflexivity in system innovations towards sustainability. In general, structures and institutions stimulate stability and existing practices, but changing these structures and institutions is possible through collective or group action by actors. Possible changes and system innovations can be explored in protected spaces or niches where the rules and structures can be ignored, while alternatives can be explored for which the development and evaluation of visions can be seen as the vehicle to test and evaluate. However, a major issue is how to escape from the niche and diffuse into society, resulting in system change.

Thus, when looking into the field of innovation studies, it is not only interesting to look into system innovation theories, their change aspects and change mechanism, but also to look into theories and concepts which explain how ideas and alternatives grow out of protected spaces and diffuse broader into society, resulting into changed socio-technical systems, or subsystems. Therefore, Section 3.5 looks into system theories in innovation studies.

3.5 System innovation theories

3.5.1 Large Technological Systems

The concept of Large Technological Systems (LTS) was proposed by Hughes (1983, 1987), and was based on a comparative historic evaluation of the emergence of electricity systems in the USA and Europe over several decades. Focusing on the emergence of a new infrastructural system, Hughes distinguished several phases in the development in which different type of leadership, competencies and strategies are required. An LTS is both socially constructed and society-shaping (Hughes 1987: 51), and contains heterogeneous components that can be both technical and non-technical. The latter includes legislative artefacts, financial organisations, manufacturing firms, organisational procedures, etc. See Van der Vleuten (2000) for a critical review of the LTS concept, its application and some developments.

The emergence of an LTS can be seen as a system innovation, which can grow by geographic expansion, as well as by dealing with each reverse salient that emerges. A reverse salient is an unstructured problem that prevents or constrains further development of a system and needs to be resolved before further development or growth can take place. Such a reverse salient can be either technical or organisational and either internal or external. Examples of external ones include like resource delivery security, market development, social pressure, and regulation. A reverse salient can be slumbering, as foreseeable problems that may or will arise in the future (Moors 2000).

Another interesting element in the LTS concept from the viewpoint of this research is the emphasis on system builder-entrepreneur who leads and manages development and further growth. This is a different type of person in different stages of development, like the inventor-entrepreneur in early stages and manager-entrepreneur or financier-entrepreneur in later stages. It may point to the possible crucial role of key individuals in system innovations.

3.5.2 Innovation systems

In the 1980s, the concept of innovation systems emerged, originally proposed by Freeman (1987) as National System of Innovation (NSI). The core of the NSI is the production of knowledge and the use of this knowledge by companies leading to competitive products and services to the market. This core needs to be facilitated by government regulation, transfer and dissemination of fundamental knowledge to knowledge that is applicable by companies and financial services including venture capital. Kern (2000: 6) mentions as the main function of a NSI *"...to enhance the competitiveness of firms and ultimately of a nation as a whole by means of innovation or in terms of knowledge advantages"*. End-users of the products generated by companies and social acceptance are largely neglected in this concept, although recently has been plead to pay more attention to producer-user interactions (Kamp 2002, Smits 2002b).

Actors, learning processes and institutions are emphasised as the key elements in an NSI. Institutions are in the NSI concepts not seen as explicit and implicit rules as I do in this research, but as organisations or socio-economic structures. Defining and bounding the NSI concept and its key elements is done differently by different authors. For instance, Lundvall (1992) distinguishes a narrow and a broad variety. The narrow variety focuses on the actors involved in producing and diffusing new knowledge and technologies, while the broad one includes all aspects and parts of the economy that affect learning and searching resulting in (new) knowledge that can be used in economic processes. The broad variety consists of the entire production system, as well as users of innovation, providers of funds, financial services, regulation, knowledge infrastructure, and even labour markets.

The NSI concept has evolved into several related varieties. This includes regional innovations systems (e.g. Silicon Valley), sectoral innovations systems (Malerba 2002) and Porter's (1990) related cluster concept, as well as technological innovation systems (Carlsson and Stankiewicz 1995, Hekkert *et al.* 2007). The latter category focuses on emerging innovation systems around specific technologies and has especially been applied to specific renewable energy technologies. Apart from technological innovation systems, most NSI theorists tend to focus on analysing and improving existing innovation systems and how they absorb new technological developments within existing institutions, while neglecting stronger dynamics and emergence of new industries and related innovation (sub-)systems.

What is interesting from the viewpoint of my research into system innovations towards sustainability? Firstly, innovation system approaches emphasise generating and using knowledge. Secondly, it requires involvement and contributions from very different actors to make an innovation system 'work', such as knowledge producers (research bodies), knowledge users (innovating companies), regulators (government), and many others. Thirdly, it points to the relevance of structures and institutions to make innovation systems work. However, it does not point to how structures and institutions may constrain major system changes. In addition, it focuses strongly on generating regular economic benefits, thereby neglecting to make innovation systems reflexive in the sense of reflexive modernisation, as was pointed to by Grin (2004, see also 3.4.3).

3.5.3 Technological regimes, regime shifts and niches

The concept of technological regimes (Nelson and Winter 1977, 1982) and the similar concept of technological paradigms (Dosi 1982) were originally developed with the aim of treating technology as an internal factor. It has been broadened by Rip and Kemp, who wrote (1998: 340) *"...a technological regime can be defined as a rule-set or grammar embedded in a complex of engineering practices, production process technologies, product characteristics, skills and procedures, ways of handling relevant artefacts and persons, ways of defining problems – all of them embedded in institutions and infrastructures."* Kemp et al. (1998: 182) explain elsewhere: *"Examples of such rule sets are the search heuristics of engineers, the rules of the market in which firms operate, the user requirements, and the rules laid down by governments, investors and insurance companies".*

Technological regimes emphasise stability and reproduction of existing structures, resulting in path dependencies and rather regular patterns of cumulative and incremental innovations. However, major shifts in technological regimes (Kemp 1994) or the transformation of technological regime (Van de Poel 1998, 2003) do occur and require a transformation process of several decades, like in case of the digital computer (Van den Ende and Kemp 1999). Van de Poel writes (Van de Poel 2003: 52): *"a transformation of a technological regime takes place when one or more of its core or constitutive rules change."* Core rules are at the level of guiding principles and at the level of promises and expectations.

Thus, changing a technological regime requires changing the rules. Kemp (1994) mentions three driving forces for radical regime shifts: new scientific knowledge, new needs (including both new market demand and wartime crises) and outsiders. Ashford (1994) mentions regulation as an important driving force. Van de Poel (1998, 2000) has strongly emphasised the importance of rule-breaking behaviour by outsiders, which he defines as those that do not share the core or constitutive rules of the existing technological regime. Van de Poel (2000) distinguishes three groups of outsiders: 1) industrial firms; 2) engineering and scientific professionals; and 3) public interest and societal pressure groups.

A widely supported idea behind regime shifts and the emergence of radical new technological regimes is that they develop in rather protected spaces or niches (Geels 2002: 1260). The proposed mechanism is that radical novelties develop in niches outside the technological regime and after sufficient maturing break out of the niche and become the new dominant regime (Kemp 1994, Rip and Kemp 1998, Kemp *et al.* 1998,

Van den Ende and Kemp 1999, Geels 2002, Geels 2005). These authors define niches as protected spaces where user requirements and demand are connected through multi-actor learning processes. In niches users can learn about using the new artefact in practice, but it may also lead to changing ideas about function and need fulfilment by users, like was reported by Hoogma and Schot (2001). These authors found in niche experiments with electric vehicles that users adapted their mobility patterns due to the possibilities and functionalities of the electric vehicles. Raven (2005) has studied this for the introduction of biomass based energy technologies in the existing electricity regime in the Netherlands.

It has been argued (Kemp *et al.* 1998, Schot *et al.* 1996, Rip and Kemp 1998, Hoogma *et al.* 2002) that governments should create and stimulate such niches for experiments with new technological artefacts that could contribute to government or societal goals (for which they introduce the term technological niche, by contrast to market niche). After becoming successful, protection should be ended. It is, however, also possible to identify niches emerging from bottom-up experiments by groups of civilians and public interest groups. This type of niche was characterised by Verheul and Vergragt (1995) as social experiments. Brown *et al.* (2003) have elaborated the idea of social experiments into the concept of Bounded Socio-Technical Experiments, in which alliances of companies and research bodies take the lead in experiments bounded in space, time and budgets.

The niche-based mechanism of change in technological regimes has been criticised by several authors. Van de Poel (2003) has evaluated eight transformations of technological regimes. He concluded that transformation can both originate from development and test niches inside and outside the technological regime. He also found evidence of a transformations not emerging from a niche, but initiated at the level of the regime itself (e.g. CFC alternatives) or through integration into next-generation characteristics. Elsewhere, Berkhout *et al.* (2004) have argued that the emergence of novelty and adjustment of regimes through niches is only one of possible mechanisms for regime transformation. Quist (2003) has argued that the concept of niches should be broadened beyond market niches and technological niches. If its core is a protected space for rule breaking behaviour, then a variety of niches is possible including pilot plants, R&D labs, R&D projects, bottom-up social experiments and also backcasting experiments or other visioning experiments.

3.5.4 Multi-Level Perspective and transition concept
The assumption that technical novelties develop and mature in niches and can result in regime shifts or replacement of technological regimes has been conceptualised in the Multi-Level Perspective (Rip and Kemp 1998, Kemp *et al.* 2001). The Multi-Level Perspective (MLP) consists of three inter-related levels. The lower micro-level consists of niches in which technological novelty develops and matures. The intermediate meso-level consists of technological regimes, which can under certain conditions be 'replaced' by expanding niches. The upper macro-level consists of society at large, which has been named socio-technical landscape. At this level broader socio-economic and socio-cultural developments evolve that influence the lower levels, while the meso-level of technological regimes also influences the macro-level. The three levels are nested; niches are embedded within technological regimes and technological regimes are embedded in socio-technical landscape. Niches are in this MLP considered to be the seeds of regime change (Kemp *et al.* 1998, Hoogma *et al.* 2002). From the landscape external pressures can arise that induce in instability (problems) at the level of technological regimes.

Geels (2005) has adjusted the MSL in several ways. Firstly, he has extended the concept of technological regimes into socio-technical regimes, which consists of parts from several different regimes (technological regimes, science regime, policy regime, user and market regime, and socio-cultural regime). This conceptualisation assumes that the parts extracted from this set of regimes share that their rules are 'aligned' and in this way can make up a socio-technical system that provides societal functions. Secondly, he proposes that regimes not only constrain radical novelties, but that ongoing developments at the level of socio-technological regimes and the macro-level also provide opportunities and may act as stepping stones for radical innovations. Thirdly, he has proposed four phases in technological transitions in which the interactions between the levels of the MLP differ.

These adjustments allow the inclusion of two more concepts in technological transitions. Firstly, the co-evolution of multiple technologies can become part of the transition. Secondly, the diffusion of a technical novelty through niche accumulation in which niches are succeeded by other (market) niches that can be conquered or developed, due to for instance improved performance or reduced costs. However, niches in the extended MLP as part of a technological transition comprise of international industrial sectors or substantial international markets, which differs strongly in scope and scale from test and experimentation niches like in Strategic Niche Management (for an explanation of SNM see Chapter 2).

The MLP is also a key starting point for transition management and the concept of transitions towards sustainability, as developed by Rotmans and others (e.g. Rotmans *et al.* 2000, Rotmans *et al.* 2001, Rotmans 2003, Kemp and Rotmans 2004, Kemp and Rotmans 2005). This concept is broader than Geel's concept of technological transitions. Transitions are transformation processes in which society or a complex subsystem of society changes in a fundamental way over a period of at least 25 years (Rotmans *et al.* 2000, Kemp and Rotmans 2004: 138). These subsystems are similar to the socio-technical systems as defined by Geels (see above). However, whereas Geel's technological transitions evolve around a central major technological breakthrough through a process of niche accumulation, Kemp and Rotmans (2004) see technological transitions as a special type that can be distinguished within a larger range of transitions. In transition management, the starting point is a complex societal problem, such as current sustainability problems that can be dealt with by a transition or a system innovation that addresses the problems and involves technology, but not necessarily as the central binding element. Furthermore, Kemp and Rotmans (2004: 138) distinguish between evolutionary transitions in which the outcome is not planned, which is similar to what De Bruijn *et al.* (2005) have called emergent system innovations, and goal-oriented (teleological) transitions, in which a (diffuse) goal or vision of the end state is guiding decision-makers in government and business. Kemp and Rotmans (2004: 144) write *"Transition management for sustainability tries to orient dynamics to sustainability goals. The long-term goals for functional [socio-technical] systems are chosen by society through the political process or in a more direct way in a consultative process.... The goals, and policies to further the goals, are not set in stone but should be constantly assessed and periodically adjusted in development rounds."*

Elsewhere, (Berkhout *et al.* 2004, Smith *et al.* 2005) it has been argued that regime change or transitions may originate from niches, but that this should not be seen as the only way of changing socio-technical regimes by inducing transitions, but as one of the possible mechanisms. These authors characterise transitions by (1) the locus of necessary resources, if these are internal or external to the socio-technical regime, and (2) the level of coordination. In the case of low coordination, the outcome is due to co-evolutionary behaviour of regime members and their interaction with other actors (Smith *et al.* 2005: 1498). This results in four different types of transitions:

(1) Purposeful transition (high-coordination and external resources), which is the type advocated by Rotmans and colleagues;
(2) Emergent transformations (low coordination and external resources), the type analysed by Geels;
(3) Endogenous renewal (high coordination and internal resources), and;
(4) Reorientation of existing trajectories in socio-technical regimes (low coordination and internal resources).

The distinction between internal and external resources seems in line with the findings by Van de Poel (2003, see also above) who concluded that transformation can start in niches inside, as well as outside the existing dominant technological regime. Smith *et al.* (2005) have also added that socio-technical regimes continuously face pressures and that the transition context determines the direction and form of how, in response to these pressures, the socio-technical regime changes.

3.5.5 Institutional theory

Institutional theory sees technical change in an organisation or a group of organisations, such as an entire industrial sector, as a change in institutions or as a process of institutionalisation. Institutions are rules prescribing behaviour, while three pillars (or aspects) of institutions can be distinguished that constitute the institutional environment. They are the cognitive, the regulative, and the normative pillar (Scott 2001). These three pillars also fuel three types of pressures exerted from the institutional environment, to which organisations tend to comply with in order to gain legitimacy. Mimetic pressures originate from competitors. Secondly, coercive pressures originate from powerful actors, especially regulation by the government. Thirdly, normative pressures originate form broader groups of stakeholders, for instance consumers, customers, citizen groups and public interest groups.

Institutionalisation is defined as *"...the process by which activities come to be socially accepted as 'right' or 'proper' or come to be viewed as the only conceivable reality"* (Oliver 1996: 166). Parallel to the institutionalization process certain practices may become outdated, resulting in the de-institutionalisation of both practices and related technologies (Oliver 1996, Scott 2001, Van den Hoed 2004).

Institutional theory is interesting from the viewpoint of system innovations, as system innovations may be conceptualised as processes of institutionalisation and de-institutionalisation driven by changes in the institutional environment that lead to the three types pressures mentioned above. Such processes may be identified in R&D or technologies strategies in industries before niches or regime shifts can be noticed, for instance by analysing what are seen as next generation technologies. For instance, Van den Hoed (2004) has used institutional theory to study the emergence of the fuel cell vehicles as a (more) sustainable future option in the global car industry. In addition, in institutional theory it is assumed that the three types of pressures do not affect the producers directly, yet that these are debated, backed, (dis)agreed on and prioritised in a multi-actor arena surrounding the industry that is called the organisational field. This is a part of the context of a particular industry in which actors are involved in a kind of diffuse and largely decentralised debating and negotiation process. In addition to the producers that may act both independently and collectively, a range of other actors can be involved, including government actors and public interest groups. The outcomes have been conceptualised as different types of pressures that affect the entire industry, although responses by specific producers are influenced by local contexts and firm specific characteristics.

The advantage of this framework over niche-based frameworks is that it includes R&D options that may become the starting point of system innovations before there is any niche in society, either a market niche or a technological niche. Van den Hoed (2004) has proposed five change factors: (1) shocks or crises, (2) new technologies or technological breakthroughs, (3) market changes, (4) relevant new entries or exits in the organisational field, (5) new practices and institutional entrepreneurship in the existing industry.

3.5.6 Conclusion on system innovation theories

This section has looked into several theories in the field of innovation studies that deal with innovation and change at the level of socio-technical systems. Such theories, concepts and conceptual perspectives are available and have been applied for explaining stability as well as for (system) change. However, these theories and perspectives offer limited relevance to evaluate changes on periods of five to ten years, which is the time span of this study on the spin-off and follow-up of backcasting experiments. After five to ten years changes may be visible at the level of niches and research activities or strategies, but they may be difficult to relate to system change or system innovation aspects and broader diffusion into society. In addition, the influence of external or contextual factors (Geels 2002, Geels 2005, Van den Hoed 2004) needs to be taken into account.

As a consequence, it seems necessary to look also into concepts and approaches focusing on smaller changes and shorter periods of time than is the case in the theories on system change and system innovation. This is done in the next section by looking into several network theories. At the same time system theories or specific elements need to be kept in mind when evaluating cases.

3.6 Networks

3.6.1 Introduction

In this research, networks are seen as consisting of actors and their relationships that together make up a social structure. In several fields of disciplines network theories have emerged aiming to understand the relationships among actors and how these influence actor behaviour and vice versa, how actor behaviour influences relationships. Networks can be used to couple developments at the micro level of actors to the intermediate level of the network, but can have a different focus in different disciplines. For instance, in anthropology the focus is on networks of relationships among individuals within a group of individuals. In psychology the focus is on the influence of networks on individuals and individual behaviour (Van Mierlo 2002: 36-37); Rogers' diffusion of innovation theory (1983, 1995) can be seen as an example of this.

In innovation studies, policy sciences and business studies the focus is on the relationships among heterogeneous actors. Actors can be organisations, groups of organisations, units of organisations, as well as groups of individuals and individuals. The share of individuals as actors is generally limited, but individuals can be powerful actors or key actors in a network. Networks provide a framework influencing behaviour by actors, thus influence the behaviour of actors in the network, but at the same time behaviour and actions by actors influence and shape both the network and the relationships. Often, networks develop around specific topics, such as specific policy topics, the development of technologies, or the supply of products. Progress and the direction of the central topic in the network are determined by composition, relationships (including dependencies) between actors, the interactions among actors and various other characteristics.

Networks of heterogeneous actors are considered here the most relevant elements to explore on building blocks and relevance for explaining and analysing spin-off and follow-up activities of backcasting experiments. Backcasting experiments themselves may be framed in terms of (temporary) networks too, but this is not attempted. As a next step, I look at network approaches from innovation studies and policy sciences looking for relevant notions and building blocks from the perspective of this study: these are (1) actor network theory in technology development, (2) industrial network theory, and (3) policy networks.

3.6.2 Actor network theory

A well-known network theory in innovation and technology development is actor network theory (Callon 1986, Callon 1987, Callon et al. 1992, Law and Callon 1992, Callon 1995, De Laat 1996, De Laat and Larédo 1998). Networks are considered as the whole of unspecified relationships (Callon 1995: 309). The key element of actor network theory is that every form of change is reflected in the network. If an actor enters or changes position, both the network and the technology are affected. On the other hand, if the technology is changed, this goes along with changes in the network. Furthermore, enrolment and translation of actors are important. Enrolment means becoming a member of a network. Translation here refers to the process that a particular actor influences another actor their values, interests, etc, aiming to move other actors in a more favourable position from the viewpoint of the influencing actor. The assumption is that negotiations on changes in the design of the innovation of the technology and related changes in the networks and the actors go on until a certain configuration stabilises resulting in agreements on actor's roles and the final configuration of an innovation. After actors have agreed (or have been translated), they start working together on achieving the shared goal.

Furthermore, this theory assumes that technology development and innovation is highly contingent and unpredictable. There is also little difference between human and non-human actors. An artefact that 'influences' the network, due to, for instance, path dependencies, is considered an actor, while a social actor without influence is not regarded as an actor. Although this is can partly be seen as a matter of definition, it has also been criticised, for instance it excludes socially relevant actors from the network if they cannot exert influence on the network.

Later, the actor network was broadened into what was called the techno-economic network (Callon *et al.* 1992). This has been defined as *"...a coordinated set of heterogeneous actors (....) which participate collectively in the development and diffusion of innovations and which via numerous interactions organise the relationships between scientific-technical research and the marketplace"* (Callon *et al.* 1992: 220). This elaboration extended earlier concepts in various ways:

(1) It brings development and diffusion together, which are both relevant and entail different sets of relevant actors;
(2) It introduces structural aspects and institutions in the network, acknowledging that these contribute to the direction and progress of negations;
(3) It proposes poles in the network that are all relevant for the development and diffusion. All poles are required in the network for becoming successful in both development and diffusion into society;
(4) It adds degree of irreversibility, length and degree of convergence as structural characteristics influencing strength and orientation of the network.

An implication of the definition above is that each innovation or technology is represented by an innovation-diffusion network, thus competitors need to build competing innovation-diffusion networks.

Although the focus in actor network theory is on the dynamics in developing and marketing a specific artefact or innovation, the poles have also been used as indicators to evaluate the embedding and diffusion of a new technology into society (Kruijsen 1999, Van Mierlo 2002). For instance, Van Mierlo (2002) added an additional pole representing government and society in line with De Laat (1996: 80) for evaluating embedding of photovoltaic in housing in the Netherlands. The poles applied by Van Mierlo were (1) Research and technology development; (2) product development; (3) use by both consumers and organisational customers; (4) government and society. This scheme seems very suitable for evaluating diffusion and embedding of a new technology, but it is doubtful whether this can be adjusted for evaluating the follow-up and spin-off of backcasting experiments; then networks are not likely to follow technology development poles and can be at a smaller level. Luiten *et al.* (2006) have dealt with the latter issue by distinguishing between micro-networks and technology networks; technology networks consist of several more narrowly bounded micro-networks. Another way of dealing with this issue is to include characteristics such as the heterogeneity of actors in the network and the number of actors involved, as was done by Mulder (1992) in a study on emerging technology networks in industrial fibres at various companies in different stages of the development of technology like R&D, product development, process development and up-scaling of production.

3.6.3 Industrial network theory

The starting point of the industrial network theory developed by Håkansson and further elaborated by others (Håkansson 1987, Håkansson 1989, Oerlemans 1996) is (1) that technical innovation is in general the outcome of mutual cooperation between firms and other actors in networks like suppliers (and their suppliers), customers (and their customers) and also non-market relationships, and (2) that firms are embedded in industrial networks and if these networks change that this affects the behaviour of firms (and vice versa). Håkansson (1987, 1989) views networks as the web of contacts that exists between suppliers, customers and producers in industry. Networks may include both companies producing complementary products and companies producing competing products. Due to the industrial setting Håkansson emphasises resources and activities related to exchanging and processing resources.

The industrial network theory therefore distinguishes three basic components (Håkansson 1987): (1) actors, (2) activities (3) resources. Actors perform activities and use or control relevant resources. Through activities resources are combined, developed, exchanged or created. An activity in full control of a particular actor is called a transformation activity; an activity that involves other actors is called a transaction activity. Transaction activities largely cover buying resources and selling products (or services), but include acquiring knowledge and other resources for innovation (Håkansson 1989, Oerlemans 1996). Resources include various types of assets, such as physical assets, financial assets, and human assets. The latter is broadly defined and includes labour, knowledge, access to networks and relationships.

Industrial network theory distinguishes three network functions: (1) contributing to the development of knowledge by actors, (2) coordination of the exchange of resources, (3) contribution to mobilising resources. Furthermore, the outcomes of this network and the dynamics in the network are influenced by the structure of the network, the actors involved, their relationships, power issues, and the particular combinations of resources and activities in the network.

3.6.4 Policy networks

A policy network can be defined as consisting of all the actors involved in a certain policy field (Grin and Van de Graaf 1994, cited in Van Mierlo 2002: 36). Kickert *et al.* (1997: 6) define policy networks as *"(more or less) stable patterns of social relations between interdependent actors, which take shape around policy problems and/or policy problems."* The key element is that actors in a network are interdependent, because they need the resources of other actors to attain their goals. Policy networks involve heterogeneous actors and can be either socially or cognitively closed (De Bruijn and Ten Heuvelhof 1995). Socially closed means that actors are (deliberately) excluded from interaction and thus from participation. Cognitively closed means that particular perspectives are excluded deliberately. Within the structure of a policy network, actors develop interaction and communication patterns that can become structural and are directed to particular policy problems or particular policy programs (Van Mierlo 2002). Within policy networks, competing alliances may operate such as the advocacy coalitions defined by Sabatier and Jenkins-Smith (1999).

Van Mierlo (2002: 36) considers innovation-diffusion networks for a particular technology as a special type of policy networks in case of desirable technologies demanded by society, or supported and stimulated by the government. Examples include renewable energy technologies like wind turbines or photovoltaic. However, innovation-diffusion networks for different renewable energy technologies have joint interests, but can also become adversaries in a larger policy network. The dynamics and embedding can be studied at the level of particular innovation-diffusion network framed as a policy network, as Van Mierlo has done, seeking for success and fail factors on this level. However, the dynamics at the level of a broader policy network and its context may be particularly relevant for explaining the impact of technology or innovation demonstration experiments.

3.6.5 Conclusions on networks

To conclude, network theories that focus on the relationships in networks between actors are most interesting from the viewpoint of evaluating spin-off of participatory backcasting experiments. With this assumption, three different types of network theories have been described, originating from different disciplines: actor network theory from innovation studies, industrial network theory rooted in economics and policy networks based in policy sciences. Each type has shared characteristics that match with insights and research problems in their fields of origin. On the other hand in all clusters adjusted versions can be found that either focus on a more micro level, or try to put the boundaries wider. Most varieties in the three clusters of network theories and concept tend to use a wide range of parameters for which detailed, sometimes longitudinal data are sought. The industrial network theory offers a promising basic structure consisting of actors, activities and resources without requiring an extensive set of parameters and considerable data for each parameter to characterise spin-off and follow-up of backcasting experiments. This would require investigating whether this theory can be adjusted in such a way that it can be applied in other domains as well, also when actors from these other domains like government, research, or public interest groups and society have the lead.

3.7 Conclusion of this chapter

This chapter has provided a critical exploration of theories and concepts related to three key elements of participatory backcasting: visions, learning and stakeholder participation. It has also explored system innovation theories and network theories for building blocks that can be used for a conceptual perspective that relates the backcasting experiment, its follow-up and spin-off as seeds for system innovations towards sustainability. Defining building blocks, constructing a conceptual perspective, and elaborating this in a research methodology for evaluating backcasting experiments and their impact, takes place in the next chapter.

Notes

1   I do not mean the Delta plan itself here, but the vision for making a secure Delta region in the Netherlands that existed throughout the first half of the 20th century and that guided the making and execution of the Delta plan in the 1950s up to the 1980s. It was after the disastrous flooding in 1953 that political support for this vision emerged and it was decided to develop and realise the Delta plan and allocate sufficient government funds.

2   The depletion of the ozone layer has also been referred to as a prediction or an expectation, because it was supported by scientific evidence and widely shared. However, I see it here as an undesirable vision of the future that has been responded to in order to avoid it.

3   The quote is as follows: "What distinguishes these studies ... is their fairly explicit vision on the future of technology and society. Given all the possible choices and the high level of uncertainty experienced nowadays both in politics and in the market place, we are in desperate need of such visions to guide our actions. It would seem a suitable goal for Technology Assessment to develop such visions, or at least contribute to the development thereof."

4   Following Fresco he distinguishes Shadow Thinkers (characterised by a strong pessimism and criticism regarding technological progress) and Light Seekers (characterised by an emphasis on the romantic appeal of fairness, serenity and simplicity of nature, body and mind), while he sees Fresco's personal stand representing the vision of well-known neopositivist faith in scientific and technological progress to solve societal problems for which Grin proposes the term True Believers or Rational Thinkers. These three visions can, for instance, be traced among actors involved in (the debates on) genetic food engineering and nuclear power and show at least three different 'shared' visions of the future at the level of major social groups, thus at the level of society.

5   Please note that a broader meaning of expectation is meant here than the meaning of technological expectations by Van Lente as dealt with in 3.2.3. Grin argues that any individual has expectations for any topic, while individual behaviour and choice processes in the present are also guided by expectations for the long term.

6   Rhetorics on emerging technologies and new scientific disciplines generally speaking are too positive and too optimistic (De Wilde 2000). This relates partly to strategic and tactical behaviour in setting technological and science agendas by actors having and defending their interests.

7   This may be a very relevant phenomenon. It implies that experimentation in a niche or experiment or project can be facilitated or stimulated by 'protection' from the top management levels. Elsewhere, this has been referred to as intuitional protection as part of the institutional settings (Grin et al. 2004), or in eco-design as commitment by top management (Van Hemel 1998).

8   These authors see learning as a cognitive activity that is not the only driver for policy change. Another driver is developments external to the policy like socio-economic developments or political developments. Yet another driver consists of changes in personnel (Sabatier and Jenkins-Smith 1999: 123).

9   Hoogma et al. (2002: 4) write "A niche can be defined as a discrete application domain (habitat) where actors are prepared to work with specific functionalities, accept such teething problems as higher costs, and are willing to invest in improvements of new technology and the development of new markets... If successful, a technology may move to follow-up niches, resulting a process of niche branching."

10   Brown et al. (2003) use the term bounded socio-technical experiments (BSTEs) for these projects. A BSTE has been defined as 'A BSTE attempts to introduce a new technology, service, or a social arrangement on a small scale... A successful BSTE creates a functioning, socially embedded new configuration of technology or service that then serves as a starting point for further innovation or diffusion, or that can inform the policy making process' (Brown et al. 2003: 291-292).

11   The typology of Freeman and Perez has been criticised for being technologically deterministic, but later Freeman and Louca (2001) have conceptualised their changes in techno-economic paradigms as co-evolutionary processes with periods of alignment and periods of misalignment between several social subsystems.

12   Preventing side-effects in current technology development practices has also always been a major aim of technology assessment, long before the concept of reflexive modernisation emerged, although both are rooted in similar social concerns. When technology assessment evolved into constructive technology assessment reflexivity has always been a major concern.

13   It is emphasised here that it is impossible to change socio-technical systems in a simple direct way, as it is the result of constructive interaction processes among actors involved, while it is sometimes also framed as the co-evolution between technology and society. The basic assumption here is that it is the result of social shaping processes and that it is possible to a certain extent to influence and guide such processes.

4

CONCEPTUAL FRAMEWORK AND RESEARCH METHODOLOGY

Based on the results of the previous chapters, the purpose of this chapter is to define the research focus (4.1), to develop a conceptual framework (4.2) as well as a set of propositions (4.3) for analysing backcasting experiments and their follow-up and spin-off. In addition, a multiple case study research methodology is developed for the empirical research (4.4), which includes the elaboration of an analytical framework and a discussion on the selection of the cases.

4.1 Defining the research focus

The main questions of this research as formulated in Chapter 1 are as follows:

A. *What factors determine the impact of backcasting experiments after five to ten years?*
B. *How should participatory backcasting be applied for exploring and shaping system innovations towards sustainability?*

The first question focuses on what factors determine and explain the follow-up, spin-off and broader effects of participatory backcasting experiments. It is a 'what' question that calls for a theoretical explanation. The second question is a practical 'how' question and it focuses on how backcasting practices can be improved, especially with respect to system innovations towards sustainability. These two main questions have been elaborated into a set of eight research question. This chapter deals with the third research question: *"How can backcasting experiments and their follow-up, spin-off and broader effects be conceptualised and evaluated five to ten years after completion?"*

A backcasting experiment has been defined as the activity or various related activities, in which backcasting is applied explicitly and in which stakeholders are involved. A backcasting experiment always involves stakeholder participation. If there is no stakeholder involvement, the term backcasting experiment is not used in this research, but the term backcasting study. A backcasting experiment can consist of a single project, but also of several successive projects.

The focus of this research is on analysing the follow-up and spin-off of participatory backcasting experiments after five to ten years, and to identify the factors that determine the degree of follow-up and spin-off. A distinction has been made between internal factors originating from the backcasting experiment and external factors originating from the context. A simplified model is presented in Figure 4.1. It shows the three main stages that can be distinguished when a system innovation grows out of a backcasting experiment. This is indeed a simplification, as system innovations are very unlikely to result from a single backcasting experiment, and many other actors and factors will be at play as well. However, this simplified model is used for the purpose of explanation. Figure 4.1 shows three phases:

-   The first phase consists of the backcasting experiment, reflected by the box on the left, which lasts one to three years.

- The second phase consists of the impact after five to ten years after completion of the backcasting experiment, which is shown by the box in the middle.
- The third phase consists of the long-term impact after 40 to 50 years, which is depicted by the box on the right, and which is also the time horizon of the system innovation towards sustainability envisioned in the backcasting experiment.

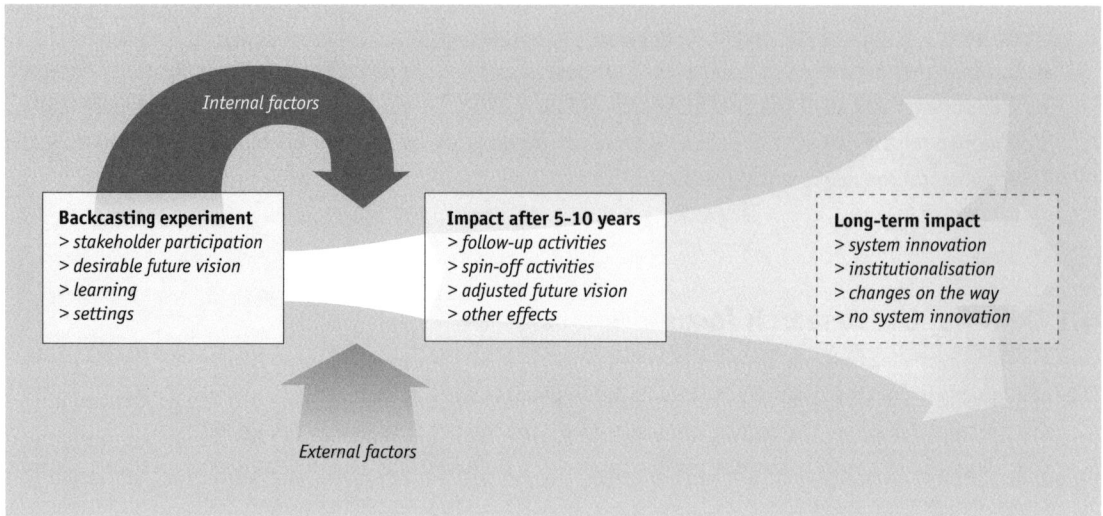

Figure 4.1 *The backcasting experiment, the impact after five to ten years and the long-term impact after 40 to 50 years*

Figure 4.1 also shows some of the key elements for each phase, which are elaborated below; in the next sections a coherent set of terms for describing these phenomena is introduced. My model of three phases, in which a system innovation towards sustainability grows out of a backcasting experiment, is simpler than the phases proposed by other authors like Geels (2005) and Rotmans *et al.* (2001). Generally speaking, these and other authors start with more extensive phenomena than a backcasting experiment in their first phase, while they also propose more phases on the way to a final phase of a system innovation or a transition. However, for the purpose of my research this simplified model of the three phases suffices.

Figure 4.1 shows a model how backcasting experiments may lead to system innovations towards sustainability and emphasises the time dimension. By contrast, it is also possible to view on the three phases as three separate, but nested levels. These levels vary considerably in terms of scale, the number of actors involved and the degree of institutionalisation. Then model can, for instance, be related to the Multi-Level Perspective (MLP) (see Chapter 3, or Geels 2005); in 9.2 I will return to the MLP and how it relates to the levels that can be associated with the three phases that I have defined.

Before focusing on the first and second phases, I briefly discuss the third phase, in which the envisaged system innovation will have come about after 40 to 50 years. First of all, this is not a static end-state but a dynamic one in which all kind of (change) processes will continue to take place. In many cases these dynamics will lead to adaptations or adjustments within the system, but it is also possible that these dynamics in turn will result in changes at the level of system innovation. Thus, it must not be ruled out that other system innovations will be 'on the way'. These may have started somewhere during the trajectory of the envisaged system innovation towards sustainability. One of these 'newly' emerging system innovations may even overrule the originally intended system innovation towards sustainability.

Secondly, visions, foresights, scenarios, or forecasts never (fully) come about, due to uncertainty and ambiguity, as has been shown in the study of past transitions and system innovations (e.g. Geels 2005). Thus, it is very unlikely that the long-term effects (fully) match with the desirable future envisioned in the backcasting experiment. It is more likely that it has changed course considerably. From the viewpoint of sustainability, this does not really matter, as long as the future vision has led or contributed to a system innovation towards sustainability. It can even be of crucial importance that adaptations and adjustments are made due to knowledge breakthroughs or emerging external developments. However, a possible undesirable outcome would be for no system innovation towards sustainability to come about, or that a system innovation would not result in any environmental improvement.

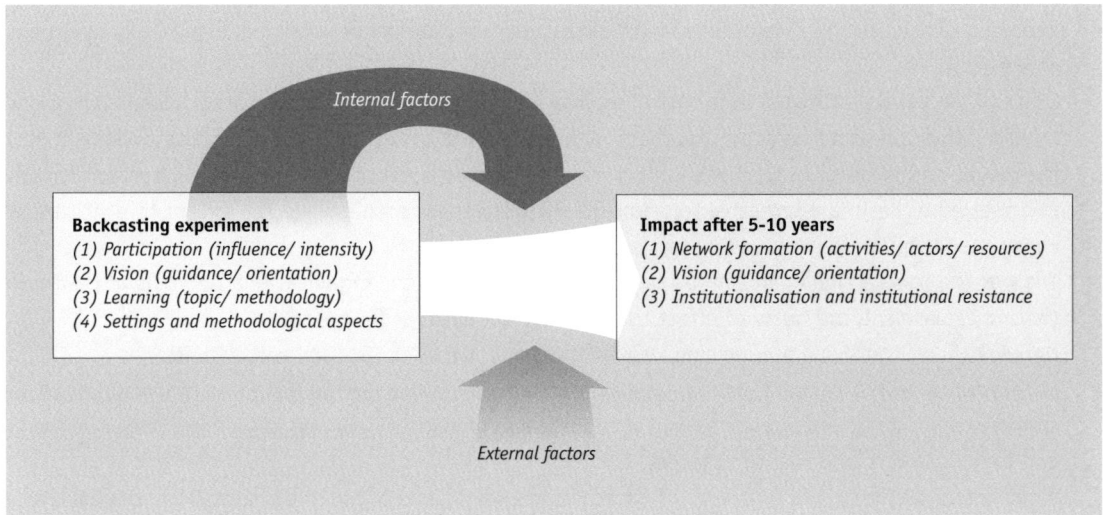

Figure 4.2 *The backcasting experiment and its impact after five to ten years*

My research interest is in the impact of backcasting experiments after five to ten years, as well as in relating the impact to the initial backcasting experiment. Therefore, I focus on the backcasting experiment itself (the first box in Figure 4.1), its follow-up and spin-off (shown in the second box in Figure 4.1) and the possible internal and external factors, as shown in Figure 4.2. Figure 4.2 also shows various elements for both phases, which appeared important in the previous chapters. In the next section these elements are conceptualised as building blocks of the conceptual framework. In the next section (4.2) I also introduce the exact terms and their definitions. As my interest is especially in the factors that influence the emergence of follow-up and spin-off, this is shown in Figure 4.2 by two groups of factors that can both be enabling and constraining to follow-up and spin-off. The first group of factors are internal factors that are characteristics or aspects of the backcasting experiment. The second group of factors are external factors that exert influence from the surrounding socio-technical system and its context.

4.2 Conceptual framework

4.2.1 Developing a conceptual framework

The following findings from Chapters 2 and 3 serve as a starting point of the conceptual framework:

- Broad stakeholder participation can help increase legitimacy and accountability, structure complex unstructured problems like sustainability problems, broaden the issue with a range of aspects and values relevant to groups of stakeholders, increase support and commitment, as well as support and involvement in realising follow-up of the backcasting experiment.
- New emerging future visions can become guiding images shared by groups of stakeholders that provide guidance and orientation to the supporting stakeholders in line with the future vision in a process of diffusion and further elaboration of the vision. Emerging visions face competition from other emerging visions and their supporters, as well as from the regular dominant vision supported by vested interests and established actors.
- Visions may have strong normative and ethical assumptions and be generated deliberatively by groups of stakeholders.
- Learning, in particular higher order learning, may encourage actors to reformulate problem definitions and shift their preferred ways and approaches to dealing with a certain problem. Increased insight into the values and views of other stakeholders may be another result. Learning may also lead to changes in preferred or desired alternative solutions to the (redefined) problem and the extent to which these changes and shifts are shared among groups of stakeholders.
- The way the backcasting approach has been applied, as well as the organisational settings of the backcasting experiment, are likely to affect the nature and degree of follow-up.
- Network theories provide a promising way for analysing follow-up and spin-off activities.
- Successful networks around follow-up and spin-off activities may lead to instances of institutionalisation in which institutions change, as well as to instances of institutional resistance from vested interests and backing actors who feel threatened.
- When actors decide to join follow-up of backcasting experiments or adopt elements from the backcasting experiment, they are expected to be influenced by external events and developments in the socio-technical system and its context.
- The emergence of follow-up and spin-off is influenced by both internal factors and external factors that can be both enabling and constraining.

The proposed conceptual framework comprises the two phases as shown in Figure 4.2. Both phases are conceptualised by building blocks that are derived from Chapters 2 and 3. The phase of the backcasting experiment is conceptualised through four building blocks: (1) stakeholder participation; (2) future visions; (3) stakeholder learning; and (4) settings and methodological aspects. I conceptualise the phase of follow-up and spin-off through three building blocks: (1) network formation, (2) future visions (3) institutionalisation. In addition, internal factors and external factors are distinguished. The conceptual framework is shown in Figure 4.3.

Both phases occur within a socio-technical system that has complex sustainability problems that cannot be solved through incremental changes (see Chapter 1). The backcasting experiment addresses the sustainability problems and the socio-technical system in which they occur. The socio-technical system can be defined differently, depending on the sustainability problem(s) in question; it can be a production and consumption system, an entire geographical region, or an industry. In this research the socio-technical system consists of four sub-systems or societal domains: (1) the research domain, (2) the business domain, (3) the governmental domain, and (4) the public domain of the general public, public interest groups and the general media. The context of the socio-technical system consists of other sectors and socio-technical systems in the Netherlands, as well as abroad.

The socio-technical system 'enters' the backcasting experiment through the participating stakeholders, but at the same time the backcasting experiment is to some extent an organised, albeit rather isolated space for experimentation within the socio-technical system. The socio-technical system also has a context consisting of other systems and activities. External factors as distinguished above are external to the backcasting experiment or to the follow-up and spin-off after five to ten years, but can originate either from inside the socio-technical system or from its context. In the phases 'follow-up and spin-off after five to ten years' I define the socio-technical system and its context in the same way. Note that the follow-up and impact may have diffused wider into the socio-technical system.

The broad arrow in Figure 4.3 connects the backcasting experiment and the follow-up and spin-off after five to ten years. This arrow depicts the process in which stakeholders are attracted to the future vision and the agenda generated in the backcasting experiment, and start turning them into action and activities. The process depicted by the arrow, which 'hides' the dynamics of how follow-up and spin-off grow out of the backcasting experiment is not conceptualised in the framework; I return to this issue in 9.2.

Figure 4.3 also shows that action and activities can take place in various societal domains that can be seen as subsystems of the socio-technical system; in this research I draw a distinction between (1) the research domain, (2) the business domain, (3) the government domain, and (4) the public domain (of public interest groups, the wider public and the general media). I put the general media in the latter domain, as it is an important source of dissemination to the wider public as well as an important locus of the public debate (like on the so-called opinion pages); professional and vocational media are part of the domain to which their target group belongs.

Each building block is also conceptualised by various aspects based on theoretical elements and concepts that may help explaining the extent and type of follow-up and spin-off (activities). The aspects are also used to develop the analytical framework and to define indicators, as part of the research methodology in 4.4. Finally, the building blocks are not independent from each other, but they influence each other and evolve together.

4.2.2 Building blocks of the backcasting experiment

Stakeholder participation

Stakeholder participation theory has provided evidence that stakeholder participation is useful because of their knowledge and expertise, their variety in basic values and worldviews. Also, they help achieve the desired legitimacy, accountability, credibility, support and commitment. It is also important for stimulating action or implementation as well as dealing with uncertainties. Stakeholder participation can contribute positively to various practices like:

(1) Public decision-making, especially in the case of complex unstructured problems (e.g. Van de Kerkhof 2004, Arnstein 1969).
(2) Scientific practice, especially in the case of research dealing with unstructured problems, such as environmental or sustainability problems (e.g. Van de Kerkhof 2004, Mayer 1997).
(3) Structuring and analysing complex problems, for instance regarding climate change (Van de Kerkhof 2004).

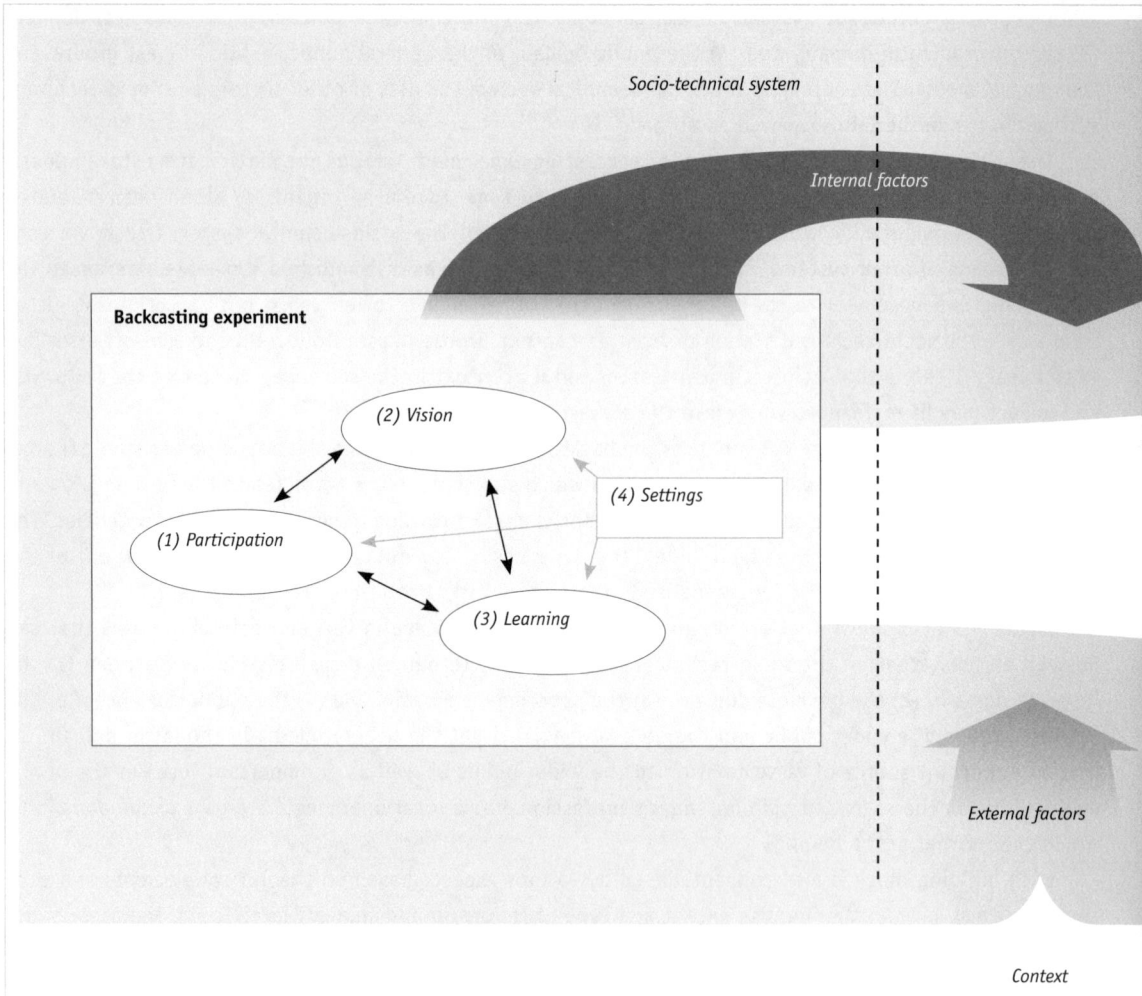

Figure 4.3 *A conceptual framework for the backcasting experiment and its impact*

These practices of stakeholder participation are relevant to system innovation towards sustainability and backcasting experiments as well; they relate to important topics like decision making on, and the science of, complex unstructured (sustainability) problems, as well as structuring such complex problems. Type, quality and variety of stakeholder participation cover a wide range of possibilities, which influence the extent of support and the extent of legitimacy that can be achieved. These relate thus also to important aspects of stakeholder participation in backcasting experiments and are used to conceptualise this building block.

The first important aspect that I include is stakeholder heterogeneity, which I define as the diversity of stakeholders from the four distinguished societal domains. The second important aspect is the degree of influence in decision-making and science (Arnstein 1969, Mayer 1997, Van de Kerkhof 2004), which expresses the influence that stakeholders can exert on content as well as process. The third aspect is the type of involvement, which I derive from economic network theory (e.g. Håkansson 1987, 1989). In economic network theory it is emphasised not only that involvement in the network is important, but also that resources within

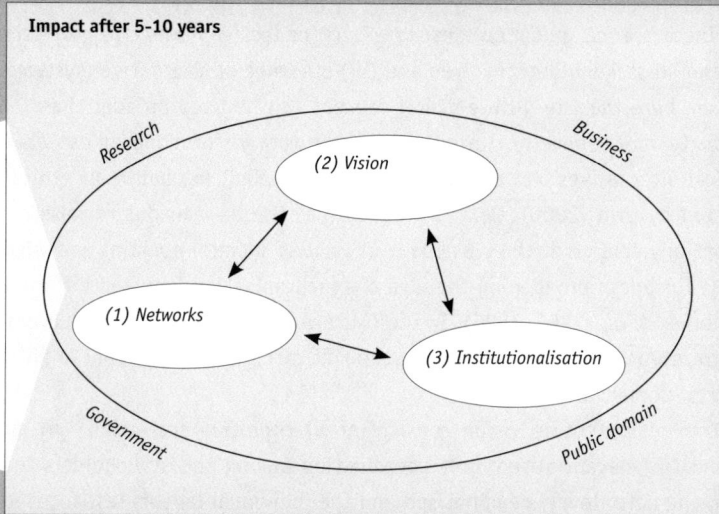

Socio-technical system
(5–10 years later)

Impact after 5-10 years

Research

Business

(2) Vision

(1) Networks

(3) Institutionalisation

Government

Public domain

activities
getting realised

Context

the network need to be mobilised. Sabatier and Jenkins-Smith (1999) also acknowledge the relevance of resources that actors can use or mobilise. This aspect reflects that involvement does not only relate to the capacity to participate, but can also take place through allocation of additional time, financial participation, moral commitment, as well as by providing knowledge. The fourth aspect is the degree of involvement, which reflects how much time participating stakeholders spend in the backcasting experiment under the assumption that a higher intensity of involvement positively contributes to process and learning results. It also allows for multiplicity in relationships, which means that participants develop relationships that are not related to the backcasting experiment. This includes private relationships and contacts in other projects and activities. Multiplicity contributes to developing trust (Mulder 1992: 24).

Future visions

As seen in Chapter 3, the Leitbild concept of Dierkes *et al.* (1992, 1996) shows several interesting characteristics with a firm theoretical foundation. I take the Leitbild concept as a starting point for conceptualising the future vision building block. Instead of the term Leitbild, I use the term future vision. The future vision concept has two main characteristics:

(1) The vision is shared and helps unite people and actors from different backgrounds, and;

(2) The vision guides the behaviour and actions of actors who share or support the vision.

A future vision is an emergent phenomenon that provides guidance and orientation. Guidance reflects 'where to go' and orientation reflects 'what to do', which I propose as two aspects in this building block. In line with Dierkes *et al.* guidance consists of a (i) collective normative projection or image, (ii) synchronisation among all stakeholders involved and (iii) presence of alternative (systems of) rules that do not yet exist. I assume here that the future vision concept can be used broader than Dierkes *et al.* (1992, 1996) have done for technical novelty. I argue that the future vision concept can also be used for sustainable socio-technical alternatives and that it may have an explicit normative or ethical component, as has also been suggested by Grin (2000). Next, I propose that the new visions for alternative future socio-technical system options emerge at the crossroads of various societal domains and various scientific disciplines. Consequently, I propose broadening the idea of synchronisation between different scientific disciplines, as argued by Dierkes *et al.* (1992, 1996). In the framework developed here synchronisation and alignment are not limited to representatives of different scientific disciplines, but apply to all the stakeholders from the various societal domains.

Orientation of the future vision consists of (i) cognitive activation, (ii) mobilisation of actors and resources, and (iii) decentralised daily coordination among the stakeholders involved. In comparison to Dierkes *et al.* the actor level is emphasised and the individual level is less focused on.

The third aspect I want to add in this building block are competing visions, which includes both the dominant vision and alternative visions that may emerge. This is in line with Grin (2000), who has pointed to the influence of existing dominant visions in existing sectors and socio-technical systems, which reflect the vested interests. For the emerging alternative and competing visions I use the idea that each vision can be seen as an emerging future socio-technical scenario (as proposed by Callon 1995, Callon 1987, De Laat 1996, De Laat and Larédo 1998). Particular socio-technical scenarios have been agreed on by involved actors, while competition takes place between various future socio-technical scenarios and their supporting actors. This is also in line with Berkhout (2006) who emphasises the competition between visions and their supporting actors and networks in processes of coordination and the formation of coalitions.

Stakeholder learning

Stakeholder learning is the third building block in the proposed conceptualisation of the backcasting experiment. The focus is on higher order learning (also called double-loop learning, second-order learning and collective value learning). Higher order learning leads to changes in the frames of actors and thus increases the space for actions and behavioural alternatives and allows for the formation of alliances or co-operation with other stakeholders. Higher order learning is also about actors who change problem definitions and perceived solutions, shift preferred ways and approaches how to deal with the problem and the extent to which these changes and shifts are shared among the participants (Brown *et al.* 2003).

In this building block I adopt the conceptualisation of higher-order learning as proposed by Brown *et al.* (2003). This concept combines various relevant characteristics from a range of theories on higher order learning and has proven to be applicable in temporary project-like settings dealing with innovations. Brown *et al.* (2003: 296) have defined learning as *"consisting of three interrelated shifts: (1) a shift in the framing of the problem and of the perceived solution (or a menu of solutions); (2) a shift in the principal approaches to solving the problem, and in the weighing of choices between desirable yet competing objectives; (3) a shift in the relationship among the participants in the experiment, including mutual convergence of goals and problem definitions.*" The three types of shifts are the aspects that make up this building block. The first and second types of shifts occur at the level of particular actors. The third type emphasises mutual convergence and congruence, and reflects learning at the level of a group of actors; I call the third type of shift joint and congruent learning. Please not that if divergence occurs with a group that is learning, this is considered to occur at the level of individual actors, thus to be part of the first or the second type of shift. It is also possible for two processes of convergence to take place among two groups of actors, while these two processes lead to divergence of goals and problem definitions. In that case two groups emerge that can be phrased in line with Sabatier and Jenkins-Smith (1999) as two different advocacy coalitions.

Finally, a distinction can be drawn between learning among the participants during the experiment and further diffusion of the insights after the experiment among a wider group of stakeholders; this distinction is relevant especially with respect to new stakeholders getting involved in follow-up or spin-off of backcasting experiments, but is not incorporated as an aspect in this building block.

Methodological aspects and settings

The fourth building block that I propose is made up of aspects related to the settings and methodological aspects of a backcasting experiment. Methodological aspects relate to how the backcasting approach has been operationalised. For this group of aspects I build on the methodological framework for participatory backcasting developed in Chapter 2 to select the following aspects (1) inter-disciplinarity, (2) five steps, (3) four different types of methods, (4) three different types of demands. All these aspects may influence the outcomes of backcasting experiments, as well as the follow-up and spin-off.

Settings include the aspect of institutional protection by top level management, which resembles the concept of niche protection (Van Lente 1993). A second aspect is the presence of a vision champion, which resembles concepts like system builder (Hughes 1987), network translator (Callon, 1987) and project champions or product champions in the product innovation literature. A third aspect concerns the main focus of the backcasting experiment. As backcasting experiments (see Chapter 2) always have several goals that can be different, it is possible that one (type of) goal is emphasised on or prioritised. A distinction can be drawn between a focus on: (1) generating content, (2) achieving follow-up, (3) the debate among participating stakeholders, or even (4) methodology development. A fourth aspect concerns the practical way in which the backcasting experiment is organised, which includes the type of management applied by the organisers of the backcasting experiment. For the latter the classification of De Bruijn *et al.* (2002: 21-32) is used as a starting point; these authors distinguish between (1) process management, (2) content-driven decision-making and management assuming the existence and use of objective information, (3) hierarchical command and control management, (4) project management, which emphasises milestones and deadlines.

4.2.3 Building blocks on follow-up and spin-off

Network formation

In Chapter 3 I concluded that network concepts and theories, which deal with relationships between (groups of) organisations as well as key individuals, provide a useful way for describing and analysing the follow-up and spin-off of backcasting experiments after five to ten years. I conceptualise this building block on network formation using the industrial network theory of Håkansson (1987, 1989). Following Håkansson's network theory I conceptualise networks as webs of relationships reflected by three aspects (i) activities,

(ii) actors, and (iii) resources. Relationships exist not only between actors. Håkansson also distinguishes relationships between activities, which he calls activity links, as well as between resources, which he calls resource ties. The three aspects are in line with Håkansson (1987, 1989), and they are related as follows in this research:

- Activities are conducted and established by actors;
- Actors mobilise different types of resources;
- Resources are heterogeneous and are needed to perform activities or have them performed;
- Mobilising resources and conducting activities requires relationships and co-operation with other actors, while different types of relationships are possible;
- Activities, actors and resources make up a network;
- Activities reflect relationships between actors.

As a next step I propose various adjustments to Håkansson's network theory as part of the conceptualisation of the network building block. Unlike Håkansson, I do not start with a focal company or group of companies, but instead introduce the focal activity as a starting point for determining the network, its actors and the resources mobilised for the activity. In my view Håkansson's network theory makes it possible to take the activity or the resources rather than the focal company or a group of companies as the starting point of a network analysis. As a consequence, I consider phenomena such as a research programme, an industrial R&D project, and the development or implementation of a policy programme all as activities; I thus ignore the fact that each of these activities can at the same time be thought of as consisting of a group of activities. Next, I suggest that such activities can take place in all four societal domains that I have distinguished: the research domain, the business domain, the government domain and the public domain. Activities are thus not limited to the business domain, on which Håkansson focuses, but can transcend domain boundaries, involve actors in different domains and can even be multi-actor multi-domain initiatives.

Finally, the relationships between the actors involved can take different forms, as is shown in abundance in the scientific literature on networks (see Chapter 3). Relationships can be both informal and formalised. Examples of the latter include contracts, written agreements, jointly submitted proposals, assignments, etc.

Future visions

This building block is similarly conceptualised as the vision building block in the backcasting experiment. It comprises the aspects guidance, orientation, and competing visions. Furthermore, while the vision emerges in the backcasting experiment, it diffuses and may be adjusted afterwards as visions and actor networks may influence each other. Therefore, flexibility as well as stability, or closure, of the vision is important. Flexibility is important (1) in terms of adaptive capacity (can it be reinterpreted and reformulated in such a way that it connects sufficiently to the goals and frameworks of particular stakeholders or aggregated stakeholders) and (2) interpretative flexibility (can it cover a range of meanings, while mobilising different types of actors at the same time). At the same time visions need to have some degree of closure or stability. Otherwise, guidance, orientation and the extent to which the vision is shared may fade away.

Institutionalisation

I conceptualise this building block in two aspects, which are based on institutional theory (see Chapter 3): institutionalisation and its opposite institutional resistance. I use the definition of institutionalisation from Oliver (1996: 166) as a starting point: "*the process by which a practice becomes socially accepted as 'right' or 'proper', or comes to be viewed as the only conceivable reality*". In this framework I use institutionalisation as an aspect for covering the wider and broader effects and entrenchment of follow-up (activities) and spin-off. In particular, the focus of this aspect is on changes in institutions in the sense of rules. In addition, as institutionalisation requires change processes, there will always be resistance due to existing interests, customs and institutions. I therefore define the second aspect of institutional resistance as the resistance to change on the part of existing institutions and the actors backing them.

4.2.4 Internal and external factors

In the conceptual framework a distinction is made between internal and external factors that exert influence on the emergence of follow-up and spin-off.

Internal factors

Internal factors are aspects or characteristics of the backcasting experiment. However, these have already been covered by the building blocks making up the phase of the backcasting experiment. As a consequence no further attempt of conceptualising them is done here. However, in Section 4.3 I develop a set of propositions that allow for evaluating possible relationships between characteristics or aspects of the backcasting experiment and the degree of follow-up and spin-off.

External factors

External factors can influence dynamics in defined systems and particular decision-making processes to a considerable degree. They can be either enabling or constraining. External factors can be broad and gradual developments in society, like aging and individualisation, or abrupt events with an immediate strong impact, such as elections, wars, and the 2004 tsunami. Many theories and theoretical perspectives take these factors in account. Geels (2005) conceptualises them as landscape developments and puts them at the highest level of his Multi-Level Perspective (MLP). In their Advocacy Coalition Framework (ACF), Sabatier and Jenkins-Smith (1999) distinguish three types of events external to the policy subsystem[1]. Van den Hoed (2004) identifies various external factors, which he sees as the seeds of institutional change.

As described in 4.1, backcasting experiments take place within socio-technical systems addressing complex sustainability problems of the socio-technical system. The context has been defined as the 'world' outside the defined socio-technical system. In this research I define external factors as broader developments or abrupt events outside the backcasting experiment or the follow-up and spin-off that exert influence on the emergence of follow-up and spin-off. External factors may emerge and evolve both in the socio-technical system and in its context; they may have both a positive or negative effect on the emergence and establishment of this follow-up and spin-off. When events or developments do not affect follow-up and spin-off I do not see them as external factors in this research. Finally, I adopt the five groups of external factors suggested by Van den Hoed (2004: 82-84, 93, 98, see also Chapter 3), I adapt slightly for their use in a socio-technical system and its context; also, I add a sixth one. The groups of external factors are:
(1) Major exits and entries by actors shifting the power balance in the socio-technical system or relevant subsystems;
(2) Shocks and crises in the socio-technical system;
(3) Technological or knowledge-related breakthroughs in the socio-technical system;
(4) Changes in markets and market demands;
(5) Changes in (institutional) practices by established organisations in the socio-technical system[2].
(6) International developments in similar socio-technical systems abroad or originating from supra-national levels.

Although developments and events outside the socio-technical system may also be of influence, I assume here that such events and developments trigger responses in the socio-technical system that fit the six types of factors. However, I add international developments as a sixth group of factors to take into account that similar related socio-technical systems may also be the source of factors, as well as events and developments at a trans-national level, such as the EU or the WTO.

4.2.5 Relationships between building blocks

Backcasting experiment

First of all, how do the future vision and stakeholder learning relate to each other? A key element is that the vision (and the process of generating, debating and investigating the vision) is a major source of learning by stakeholders. They learn both about the vision and from the vision. This includes learning on the changes that are necessary, the stakeholders that are necessary for these changes and their roles, and potential opportunities as well as possible threats for the own organisation. Stakeholder learning will also affect the future vision, as stakeholder learning and joint learning within the backcasting experiment may result in changes to the vision, or to further elaboration of the vision.

Secondly, how does the future vision relate to stakeholder participation and vice versa? To start with, participating stakeholders develop a future vision through a process of social interaction in the backcasting experiment. In this way they strongly influence the content of the vision. There is also an influence from the vision on stakeholder participation. The content of the vision appeals to stakeholders and mobilises or distracts them. Thus, the vision can become selective to certain (groups of) stakeholders and exert influence on those interested in entering, leaving or opposing the vision and the backcasting experiment.

Thirdly, how do stakeholder participation and stakeholder learning relate to each other? Clearly, participating stakeholders learn; this can take place both at the level of specific stakeholders and on the collective level of a group of stakeholders. At the same time, learning exerts influence on the participation of stakeholders, as learning results in an increasing awareness of the opportunities and possibilities for particular stakeholders or for co-operation among stakeholders. The learning process can also lead to an increased awareness of threats, disapproval, exit and even opposition.

The building block of settings and methodological aspects is portrayed separately, because the settings and the way the backcasting approach and methods are applied may affect learning, stakeholder participation, the way the vision is developed and its content. It also affects the conditions for learning, participation and vision development. It is possible that the other building blocks may 'affect' the settings, the use of specific methods and the way the backcasting approach is applied. However, it is not assumed here and even if it would occur, it would definitely depend on the settings.

Follow-up and spin-off

First of all, how do network formation and visions relate to one another? Networks may mirror the vision, which means that changes in the network and its actors may co-shape and adjust the guiding vision, while the actors in the network can use the vision to mobilise resources for the proposed activities. At the same time the vision also guides, shapes and structures the network, its members, their activities and the mobilisation of resources.

Secondly, how are network formation and institutionalisation related? Institutionalisation can be seen as an effect of follow-up and spin-off activities constituted by networks. Institutionalisation refers to changing institutions as the result of action through activities and network formation. Thus, network formation around activities may lead to entrenchment and to changing institutions. Institutionalisation may facilitate and stimulate extension of networks and activities, as well as increase the opportunities for new networks and activities, and the mobilisation of new resources and actors.

Thirdly, how are vision and institutionalisation related? Visions provide guidance and orientation to processes of institutionalisation, while institutionalisation stimulates the realisation of the vision.

4.2.6 Conclusion on the conceptual framework

A conceptual framework has been developed that comprises the backcasting experiment and its follow-up and spin-off after five to ten years. The conceptual framework is founded on theories that I have selected from a survey of relevant theories and bodies of literature in Chapter 3, as well as the developed methodological framework for participatory backcasting in Chapter 2. The theoretical framework consists of a description of building blocks, the aspects that make up the building blocks, and the mutual relationships between the building blocks in the same phase.

My next step is to make the conceptual framework operational in such a way that it is possible to analyse the cases and answer the empirical research questions formulated in Chapter 1. I do this in the next section by developing a set of propositions, after which I work out a research methodology. The propositions are based on theoretical notions, relate to different parts of the framework, and focus on the factors possibly enabling and constraining follow-up, spin-off and wider effects after five to ten years. Developing the research methodology includes the development of a set of indicators, which make up a framework for analysing the cases, as well as case selection.

4.3 Propositions

This section develops propositions focusing on the factors that may enable or constrain the emergence of follow-up and spin-off. For this I use the theoretical and methodological insights from Chapter 3 and Chapter 2, as well as the conceptual framework that has been developed. For various reasons I use propositions and not hypotheses. This is in line with Sabatier, who writes (1999: 5) *"..scientists should develop clear and logical interrelated sets of propositions, some of them empirically falsifiable to explain fairly general sets of phenomena."* He builds on Elinor Ostrom (Ostrom 1999) to distinguish between different stages of theory development and how this distinction relates to the distinction between propositions and hypotheses. Sabatier writes (1999: 5-6) *"..Ostrom has developed some very useful distinctions among three different sets of propositions. In her view a conceptual framework identifies a set of variables and the relationships among them that presumably account for a set of phenomena. The framework can provide anything from a modest set of variables to something as extensive as a paradigm. It needs not to identify directions among relationships, although more developed frameworks will certainly specify some hypotheses".*

The conceptual framework that I have developed in this chapter is not a fully developed theory, although for the purpose of this research it suffices. Therefore, I develop propositions rather than hypotheses. Developing hypotheses would suggest a higher level of maturity. The next issue is what the focus of the propositions should be, as propositions can be developed to test relationships within as well as between building blocks. Besides, the latter category can be divided into testing the relationships between building blocks within the same phase and those in two consecutive phases of the conceptual framework.

As pointed out earlier, my overall interest is in the validation why and how backcasting experiments lead to processes of learning, and subsequently result in behavioural changes, follow-up, spin-off and broader effects like institutionalisation. Within this scope I am particularly interested in the factors that enable or constrain the emergence of follow-up, spin-off and broader effects. When developing propositions, my principal focus is therefore on possible internal factors in the building blocks of the backcasting experiment. Each of the proposed building blocks of the backcasting experiment may contain factors that constrain or enable follow-up and spin-off.

Below, I present propositions for each of the four building blocks of the backcasting experiment. Each proposition suggests a possible relationship between one of the characteristics of the building block and the degree of follow-up and spin-off, using insights from the theoretical chapter (Chapter 3) and the methodological chapter (Chapter 2). In two propositions it has been suggested that relationships between the degree of follow-up and spin-off and a possible factor in the backcasting experiment can only be connected through learning. These propositions have been split in several related ones, which is further explained below. Nearly all aspects proposed in the four building blocks are dealt with in the propositions, though in some propositions aggregation of aspects has taken place.

Relevant factors are also expected in the building block of external factors and developments. However, no propositions are developed for this group of factors, as the variety among the types of factors is too large and theory development with respect to particular types of factors is too limited. The group of factors in this building block is only covered in an exploratory sense using a checklist presented in Section 4.4.

Finally, there are many possible interesting relationships between the building blocks of the backcasting experiment, or within the building blocks. However, my focus is on developing propositions for relationships between (parts of) building blocks and the degree of follow-up and spin-off, leaving other types of relationships to future research.

Propositions on participation

The first group of propositions consists of propositions that take the characteristics of stakeholder participation as their input. This group of propositions can be divided into two subgroups; whereas the first and the second relate stakeholder participation characteristics directly to the degree of follow-up and spin-off, the third and the fourth relate stakeholder participation characteristics to instances of higher order learning.

As described in Chapter 3, in stakeholder theory it is argued that high degrees of stakeholder influence on both content and process result in higher degrees of commitment and legitimacy, both in decision making and science. This includes influence on bounding the topic and object of analysis, problem definition, and constructing the future vision. Propositions P1A and P1B propose that a similar mechanism leads to a higher degree of follow-up and spin-off of backcasting experiments. A distinction is made between stakeholder influence on the topic (P1A) and on the process (P1B)

P1A: *High degrees of stakeholder influence on the content result in a higher degree of follow-up and spin-off.*
P1B: *High degrees of stakeholder influence on the process result in a higher degree of follow-up and spin-off.*

In Chapter 3 it was found that various types of resources are necessary for conducting activities (e.g. Håkansson 1987, 1989). One important resource is participation, in which stakeholders provide knowledge, expertise and opinions. However, participation that only consists of limited capacity can also be seen as a limited investment, and not necessarily indicative of serious commitment or support. Therefore, it can be seen as a positive signal if stakeholders provide additional resources, especially funding, but also other assets, substantial capacity, specialised knowledge, etc. This can also be a signal of (strategic) interest in or commitment to follow-up and spin-off, eventually resulting in a higher degree of follow-up and spin-off, which is reflected in Proposition P2.

P2: A greater variety in types of involvement by participating stakeholders results in a higher degree of follow-up and spin-off.

Learning theory as well as Constructive Technology Assessment (CTA) theory and public decision making theory (see Chapter 3) assume that increasing stakeholder heterogeneity enhances the number of aspects taken into account. This eventually may result in more instances of (higher order) learning. Proposition P3 proposes that broadening the backcasting experiment with various societal aspects through involving stakeholders from all domains, especially social actors, also contributes to more instances of higher order learning on the topic of the backcasting experiment.

P3: A high degree of stakeholder heterogeneity results in more instances of higher order learning on the topic among larger groups of stakeholders.

An important aspect of stakeholder participation is the degree of involvement. It is assumed that a high degree of involvement results in more extensive higher learning results. Proposition P4 proposes that this is also contributes to higher order learning in backcasting experiments.

P4: A high degree of stakeholder involvement results in more instances of higher order learning on the topic among larger groups of stakeholders.

Propositions on learning

Based on various learning theories, it can be argued that more substantial higher order learning extends the range of behavioural alternatives available to actors and stakeholders. If extending the range of behavioural alternatives leads to more actions aimed at putting the alternatives in practice, it may eventually result in a higher degree of follow-up and spin-off (Proposition P5). In this way the propositions P3 and P4, which relate stakeholder aspects to higher order learning, can be related to the degree of follow-up and spin-off through an additional step.

P5: More instances of higher order learning on the topic leads to a higher degree of follow-up and spin-off.

In various higher order learning theories (see Chapter 3) it is emphasised that higher order learning by one particular actor is not sufficient. When a learning result is shared by several actors, it enhances the probability that the related behavioural alternative is put into practice. Items of joint learning do not necessarily have to be similar or the same, but can also be congruent. Congruent means that different actors do not fully share the insight, but that they consider it as non-conflicting, like for instance win-win. This is proposed by Proposition P6, which is a more focused version of Proposition P5.

P6: More instances of joint or congruent learning on the subject leads to a higher degree of follow-up and spin-off.

Propositions on visions

A vision exerts influence; it guides and facilitates follow-up activities, spin-off and eventually moving towards realisation of the vision. In developing the conceptual framework I have used the German Leitbild concept for conceptualising the vision building block. The Leitbild concept claims and has shown historical evidence that if the guidance function and the orientation function are sufficiently provided by emergent visions, this results in the successful development, introduction and diffusion into society of novel innovative artefacts. I have assumed that in backcasting experiments the generated visions fulfil the same functions for system innovations towards sustainability. It is thus interesting to test whether higher degrees of guidance and orientation result in higher degrees of follow-up and spin-off, which is reflected in proposition P7.

P7: Visions providing high degrees of guidance and orientation result in a higher degree of follow-up and spin-off.

In Chapter 2 it was shown that backcasting experiments can focus on a single vision, but that there are also cases where two to four future visions have been developed and analysed. This may affect follow-up and spin-off. For instance, single vision backcasting experiments may on average have more resources (both funds and capacity) than multiple visions backcasting experiments. Also, the processes of domestication of the vision, of external communication and dissemination of the vision, and of attracting actors and resources may be simpler in single vision backcasting experiments than in multiple visions backcasting experiments. In multiple visions backcasting experiments the visions may even compete with one another, especially if they contain competing solutions. This may then result in competing groups of stakeholders within the backcasting experiment, each favouring one of the visions, and subsequently lead to a lower degree of follow-up. By contrast, single vision backcasting experiments may lead to a higher degree of follow-up and spin-off. This is reflected in Proposition P8, which links the number of visions to the degree of follow-up and spin-off.

P8: Multiple visions backcasting experiments lead to a less significant degree of follow-up and spin-off than single vision backcasting experiments.

In Chapter 3 it has been found that on the one hand various theories and models on future visions emphasise stability or closure of the vision in order to provide guidance and orientation. On the other hand various theories emphasise the need for flexibility in the future vision, which allows renegotiation and the entrance of new parties, as well as the incorporation of additional elements and insights. To test these conflicting claims, two propositions (Proposition 9A and Proposition 9B) are defined. It is unlikely that they will both be confirmed in this research.

P9A: Visions with high degrees of stability over longer periods of time result in a higher degree of follow-up and spin-off.

P9B: Visions with high degrees of flexibility over longer periods of time result in a higher degree of follow-up and spin-off

Propositions on settings and methodological aspects

As shown in Chapter 3, various theories claim that institutional or niche protection is important for R&D projects in industrial settings, as well as for strategic activities in various kinds of organisations. Such protection could thus also be important for backcasting experiments. This is reflected in proposition P10, in which this phenomenon is related to the degree of follow-up and spin-off.

P10: Institutional protection of the backcasting experiment leads to a higher degree of follow-up and spin-off.

Various bodies of literature emphasise the importance of project champions, product champions, system builders, network builders and change agents; the developers of the Leitbild concept also refer to persons or groups of persons fulfilling such a role in the emergence and diffusion of the Leitbild. Proposition P11 assumes that this phenomenon, which in this research is called a vision champion, can be related to a higher degree of follow-up and spin-off.

P11: The emergence of a vision champion in the backcasting experiment results in a higher degree of follow-up and spin-off.

Finally, proposition P12 assumes that there is a relationship between the applicability of the methodological framework to conducted backcasting experiments and the degree of follow-up and spin-off.

P12: If the conducted backcasting experiment has a better match with the methodological framework, there is a higher degree of follow-up and spin-off.

4.4 Research methodology

4.4.1 Multiple case study research

In this section a research methodology for the empirical part of this research is developed. The research methodology has to provide a framework for analysing the backcasting experiments, their follow-up and spin-off, as well as for evaluating the set of propositions that has been developed. The starting point is a multiple case study methodology. As Yin (1994: 14) has argued, a case study can be defined as an empirical inquiry that investigates a contemporary phenomenon, especially when the boundaries between the phenomenon under study and the context are not clearly evident. A case study approach also allows the in-depth identification of a large range of variables. Yin (1994) also recommends including external factors and using a multitude of sources. The latter helps realise a more complete picture of the cases and allows data to be validated more convincingly.

It is not likely that a single case study will show a sufficient amount of validated evidence of the possible factors, patterns and mechanisms. This calls for a multiple case study, which Yin (1994: 25) considers as a replication providing similar information on the phenomenon under study. A multiple case study also makes it possible to compare cases, which provides both a more complete picture of the phenomenon under study, and increasing validity to found similarities. The phenomenon under study in this research are the processes that start in backcasting experiments and bring about follow-up and spin-off, in particular the factors that constrain and enable these processes and their outcomes.

4.4.2 Analytical framework: the backcasting experiment

The next step is to elaborate an analytical framework for analysing cases that covers both (1) the backcasting experiment, and (2) the follow-up and spin-off, as well as the external factors. Below, indicators are proposed that are based on the aspects in the building blocks that make up the conceptual framework. Each building block consists of various aspects, which are translated into one or several qualitative indicators. If an aspect is turned into a single indicator, the result is a narrower description with regard to what exactly the aspect is being evaluated. The total set of indicators make up the analytical framework used to evaluate the cases on a broad range of aspects in the case chapters, and to evaluate the propositions in Chapter 8.

Thus, in the backcasting experiment four groups of indicators are distinguished. These four groups of indicators are shown in Tables 4.1 - 4.4. Table 4.1 shows four indicators for evaluating stakeholder participation that are derived from the four aspects making up the building block on stakeholder participation.

Table 4.1 *Indicators for stakeholder participation*

A1 Stakeholder participation	
A1.1 Heterogeneity	Extent to which participants from four domains are involved (business, research, government, public interest groups)
A1.2 Degree of involvement	Degree to which time is spent by the stakeholders involved
A1.3 Type of involvement	Presence of other types of involvement, besides the capacity to participate
A1.4 Degree of influence	Degree of stakeholder influence on content and process

Table 4.2 depicts the indicators for the three aspects that make up the building block future visions, which are guidance, orientation and competing visions. Each aspect has been elaborated into two or three indicators. The indicators for guidance and orientation are derived from the sub-functions proposed by Dierkes *et al.* (1996) and have been explained in Chapter 3 in combination with the adjustments proposed for the conceptual framework (see 4.2).

Table 4.2 *Indicators for future vision aspects*

B1 Guidance	
B1.1 Collective projection	Presence of a collectively shared normative projection
B1.2 Synchronisation	Presence of synchronisation between disciplines and among stakeholders from different domains towards the vision
B1.3 Alternative rules	Presence of alternative (system of) rules and institutions
B2 Orientation	
B2.1 Cognitive activation	Presence of cognitive activation of relevant scientific disciplines and stakeholders from various domains towards the vision
B2.2 Mobilisation	Mobilisation of actors and resources in line with the vision
B2.3 Coordination	Presence of decentralised daily coordination
B3 Competing visions	
B3.1 Alternative visions	Presence of emergent alternative visions
B3.2 Dominant vision	Presence of influence by the existing dominant vision

Table 4.3 proposes three indicators for the three aspects making up the building block of higher order learning. The indicators have been derived from Brown *et al.* (2003).

Table 4.3 *Indicators for higher order learning*

C1 Higher order learning	
C1.1 Shifts in problem definitions and solutions	Shifts in framing major problems and of perceived solutions by specific actors
C1.2 Shifts in principal approaches and priorities	Shifts in principal approaches toward solving the problems and in the way desirable yet competing objectives are weighed by specific actors
C1.3 Joint learning	Shifts towards congruent and joint opinions in any of the issues related to the previous shifts, or in relationships among stakeholders

Finally, Table 4.4 shows the indicators for the methodological aspects and aspects related to settings. In this group every aspect has been translated into a single indicator.

Table 4.4 *Indicators for methodological aspects and settings.*

D1 Backcasting framework	
D1.1 Inter-disciplinarity	Extent to which backcasting experiment and results receive input from various disciplines and are integrated
D1.1 Five steps	Presence of five defined steps
D1.2 Applied methods	Presence of four categories of methods (analysis, design, interactive, communication/coordination/management)
D1.3 Demands	Presence of normative, process and knowledge demands
D2 Settings	
D2.1 Institutional protection	Presence of support from top level management of stakeholders involved
D2.2 Vision champion	Presence of leading individual strongly committed to the vision and acting as a vision broker
D2.3 Major focus	Presence of a prioritised goal in the backcasting experiment
D2.4 Management	Type of management applied in the backcasting experiment

4.4.3 Analytical framework part II: follow-up and spin-off

For follow-up and spin-off, three building blocks have been distinguished, apart from the building block 'external factors'. The aspects of the three building blocks reflecting follow-up, spin-off and wider effects have been translated into indicators, resulting in three groups of indicators. Each aspect is elaborated into one or several indicators. The three groups of indicators are shown in Table 4.5. The three aspects of the building block network formation are each turned into a single indicator. The three aspects making up the future vision are each elaborated into two to three indicators. Institutionalisation and institutional resistance concerns two aspects that are both elaborated into a single indicator.

Table 4.5 *Indicators for network formation, vision aspects and institutionalisation*

E1 Networks	
E1.1 Activities	Presence of activities constituted by networks
E1.2 Actors	Presence of actors carrying out the activities in the network
E1.3 Resources	Presence of mobilised resources that enable the activities
F1 Vision: guidance	
F1.1 Collective projection	Presence of a collectively shared normative projection
F1.2 Synchronisation	Presence of synchronisation between disciplines and among stakeholders from various domains towards the vision
F1.3 Alternative rules	Presence of alternative (system of) rules and institutions
F2 Vision: orientation	
F2.1 Cognitive activation	Presence of cognitive activation of scientific disciplines and stakeholders from various domains towards of the vision
F2.2 Mobilisation	Mobilisation of actors and resources in line with the vision
F2.3 Daily coordination	Presence of decentralised daily coordination
F3 Competing visions	
F3.1 Alternative visions	Presence of emergent alternative visions
F3.2 Dominant vision	Presence of influence by the existing dominant vision
G1 Institutionalisation and institutional resistance	
G1.1 Institutionalisation	Instances of changes in existing rules and institutions
G2.2 Institutional resistance	Instances of resistance to change by existing institutions and the actors backing them

The diversity in external factors and the limited theory available on this topic make it difficult to develop indicators for them. Therefore, Table 4.6 provides a checklist for external factors in the socio-technical system, rather than a group of indicators.

Table 4.6 *Checklist for possible external factors*

H1 Possible external factors
(H1.1) Crisis/ shocks
(H1.2) Market and demand side changes
(H1.3) Technological breakthroughs/ radically new scientific knowledge
(H1.4) Exits/ entries in the socio-technical system
(H1.5) New practices in the socio-technical system
(H1.6) International developments

4.4.4 Case selection

Around twenty five backcasting experiments have been identified in the Netherlands where stakeholders were involved and participatory backcasting was applied to complex sustainability problems. An overview is given in Table 4.7. These backcasting experiments all have in common that stakeholder participation took place, future visions were developed, follow-up, agendas and necessary measures were discussed and learning took place among stakeholders involved. Within this group of backcasting experiments four clusters can be distinguished. Although on the one hand there are similarities among the backcasting experiments, there is also a substantial degree of variation. The differences include the organisational and institutional settings, the way the backcasting approach was exactly elaborated and applied, the intensity and degree of stakeholder involvement, and the degree of follow-up, spin-off and broader impact. Next, four brief descriptions of each cluster are given.

The first cluster comprises the backcasting experiments at the STD programme (e.g. Weaver *et al.* 2000). The STD programme, which introduced participatory backcasting to the Netherlands in the early 1990s, consisted of twelve backcasting experiments in the first term (1993-1997) and another four backcasting experiments in its second term (1998-2001, Andringa *et al.* 2001). The degree of follow-up and spin-off among the backcasting experiments varies (Coenen 2000).

The second cluster consists of two backcasting experiments on household nutrition and clothing care that were part of the international SusHouse project on sustainable households (Green and Vergragt 2002, Quist *et al.* 2001a, Vergragt 2000). This cluster also includes the backcasting experiments on industrial paint chains (1998-2001) in the Netherlands and Portugal, organised and evaluated by Partidario (2002). The SusHouse project (1998-2000), which was initiated as an international spin-off of the STD programme, included seven more backcasting experiments on household functions like clothing care, shelter, and household nutrition in four other European countries. All backcasting experiments in this cluster applied a similar backcasting methodology, but follow-up and spin-off seem limited, as well as mainly initiated by the researchers who conducted and organised the backcasting experiments.

Table 4.7 *List of backcasting experiments and their degree of follow-up*

Backcasting experiments per cluster		Degree of follow-up
STD programme (1993-1997)	Novel Protein Foods	+++
	Multiple Sustainable Land-use	+++
	Sustainable Closed-system Horticulture	-?
	Whole Crop Utilisation	?
	Urban Underground Freight System	+?
	IT-based demands-responsive public transportation	+?
	Mobile Hydrogen Fuel Cell	?
	Sustainable office	-
	Sustainable city renewal	+?
	Chemistry	++(+)
	Integrated urban & rural water chains	++
	Sustainable Washing	+
SusHouse & related projects (1998-2000)	Household Nutrition case	-
	Clothing Care case	+
	Industrial Paint Chain case in the Netherlands	?
COOL project (1999-2001)	Four national subprojects: Industry & Energy; Housing & Construction; Traffic & Transport; Agriculture & Nutrition	?
	One International subproject	?
DLO research organisation	Animal livestock programme (1998-2002)	++(+)
	Plant production programme (2001-2005)	++(+)

The third cluster of backcasting experiments consists of four subprojects in the Netherlands and the international subproject of the so-called COOL project (Climate Options On the Long term), which ran from 1999 to 2001 and focused on dialogue and debate on the options for mitigating climate change (Van de Kerkhof 2004). The four subprojects in the Netherlands dealt with (i) agriculture and nutrition, (ii) industry and energy; (iii) housing and construction; (iv) traffic and transport. These backcasting experiments have resulted in limited follow-up with regard to the topic.

The fourth cluster consists of two research programmes at DLO (currently part of WUR, Wageningen University and Research) in which the STD backcasting approach was adjusted and turned into a strategic tool to programme future(-oriented) research in agriculture and animal livestock breeding at the DLO organisation. A major difference with the other clusters is thus that these backcasting experiments were carried out within a particular organisation, albeit in close cooperation with external stakeholders who were given considerable influence. These backcasting experiments were carried out in two research programmes; one on the future research about sustainable livestock breeding in the Netherlands (e.g. Grin *et al.* 2004) and another one on the future of sustainable crop growing in the Netherlands (De Wolf *et al.* 2006). The former was completed in 2002, the latter late 2005. Follow-up and spin-off of both programmes is still continuing, or it is being initiated. An interesting difference with the other cases is that these backcasting experiments involved actors from various societal domains, but all operating within one large organisation (DLO, currently part of WUR) that has a high degree of influence on major portions of the financial flows to the follow-up projects.

Table 4.7 provides an overview of backcasting experiments from all clusters and some characteristics; the backcasting experiments of the second term of the STD programme have been left out. All backcasting experiments met the following criteria that were used for the initial selection:

-   Clear articulation in written sources of the term backcasting, the development and use of future visions and combining process and content. The backcasting approach had to be completed in terms of content and process (and not terminated before the scheduled end).
-   Broad stakeholder participation covering at least three of the four domains distinguished.
-   Availability of and access to multiple sources of information on the backcasting experiment

Evaluation of the backcasting experiments on two other criteria varied, which has been used for further selection. These criteria are:

-   The experiment must have been completed at least five years before analysing follow-up and spin-off, giving them time to emerge or be initiated.
-   Variation in the degree of follow-up and spin-off.

Table 4.7 shows that a considerable share of the backcasting experiments were completed around five years before the fieldwork in this study was conducted in 2004 and 2005. This would make cases from the STD programme an obvious choice. By contrast, it is considered more interesting to evaluate backcasting experiments from different clusters, as well as to include backcasting experiments showing different extents of follow-up and spin-off. These differences may be highly important in this research, which focuses on the validation of the way backcasting experiments can lead to processes of establishing follow-up, spin-off and broader effects, as well as on the factors that constrain and enable these processes. Backcasting experiments from different clusters have different settings; stakeholder participation is different, as well as how backcasting has been applied. It is interesting to see how this relates to the degree of follow-up and spin-off. In addition, including backcasting experiments that did not result in significant follow-up and spin-off may point to constraining factors, while backcasting experiments resulting in significant follow-up and spin-off are likely to reveal enabling factors. Due to the importance of substantiating how follow-up and spin-off may grow out of backcasting experiments, this study needs at least two cases with significant follow-up.

Another question is whether it is possible and would have added value to select cases from the same or related domains. Obvious choices are to take cases from different domains, or to take cases from the same or related domains. The former emphasises that the cases are fully independent. The latter may result in synergy, for instance because it involves the same or related socio-technical systems, similar actors, similar external factors, accumulation of knowledge from one case that is also useful in another case, and as a result possibilities for deeper understanding.

Table 4.8 *Selected backcasting experiments and some characteristics*

Case and origin	Period of backcasting experiment	Type of system
(1) Novel Protein Foods (NPF) case (STD programme)	1993 – 1996	Production and consumption system involving companies and consumers
(2) Sustainable Household Nutrition (SHN) case (SusHouse project)	1998 – 2000	Household consumption system and related supply system
(3) Multiple Sustainable Land-use (MSL) case (STD programme)	1994 – 1997	Spatial rural system involving agriculture and other functions like water, nature, leisure

Selecting cases from the same socio-technical domain seems interesting and may also provide additional focus and coherence to both this research and this thesis. In that case the obvious choice is nutrition and food production, as this topic was dealt with in all four clusters. Within the cases on nutrition and food production I prefer cases that (1) were completed least recently, and (2) showed diversity in the degree of follow-up and (3) involved different clusters of backcasting experiments as distinguished. The decision was made to select cases from the SusHouse and STD programme clusters, as the other cases had been completed too recently.

Selected cases and some of their characteristics are shown in Table 4.8. The selected cases varied in the degree of follow-up and are taken from two different clusters:

(1) The Novel Protein Foods (NPF) backcasting experiment carried out at the STD programme, which focused on sustainable alternative protein products for present meat products, its production and its consumption, which has led to considerable follow-up activities.

(2) The Sustainable Household Nutrition (SHN) backcasting experiment of the SusHouse project, which has not led to significant follow-up.

(3) The Multiple Sustainable Land-use (MSL) project from the STD programme, which dealt with making agriculture sustainable and integrating it with other functions in rural areas. This backcasting experiment has resulted in significant follow-up.

As Table 4.8 shows, the three cases focus on three very different but related subsystems of the socio-technical system of food production and food consumption. The Novel Protein Foods case emphasises industrial (food processing) activities and includes consumption. It points to industrial actors as leading stakeholders, to market and business innovation dynamics, while there is a more facilitating role for the government. The MSL case not only has to do with function integration, but also with spatial planning, which brings in the government as a key stakeholder, in addition to leading stakeholders for each of the functions at stake. The SHN case takes the household consumption domain as a starting point, while taking into account the related supply system. This points especially to consumers, citizens, their relevant representative organisations and public interest groups as key stakeholders. An interesting issue here is whether and how the type of system affects or conditions the emergence of follow-up from the backcasting experiment. Finally, I could partly build upon earlier work for the research reported on in this study, which is shown in Table 4.9.

Table 4.9 *Earlier activities on the three cases*

Case	Type and references of earlier collected data
NPF case	Activities as a junior member of the NPF project team (De Haan et al. 1996, Quist et al. 1996c)
SHN case	Backcasting action research as a Dutch case study of the SusHouse (Quist 2000a) and workshop evaluation (Quist et al. 2000)
MSL case	Case study on MSL (Quist and Vergragt 2001) as part of the Pathways to Sustainability project (Ashford et al. 2001)

4.4.5 Case study design and reporting

Each case has been studied in the following way:
- Through the use of (internal) documents and reports from the backcasting experiments, including papers in the scientific literature, as far as accessible during this investigation, and including internet sources.
- Through the use of (internal) documents and reports from follow-up activities and other spin-offs including internet sources.
- Through semi-structured in-depth interviews (mostly face-to-face, sometimes by telephone or e-mail) with a selection of key persons involved in either the backcasting experiment, or in follow-up (attempts) and spin-off, as well as with experts with broader views on relevant developments in the context of the case being studied. In addition, shorter telephonic inquiries were made to which is referred, if necessary, as personal communication. An overview of people who were interviewed and additional contacts is provided for each case in Appendix B. Appendix A contains the checklist used in the interviews.
- Use was made of data collection as part of earlier activities on the three selected cases (see Table 4.9). These data originate from externally funded research (SHN case and MSL case study) and involvement as a junior member in the project team of the NPF project.

Case descriptions in the case chapters are organised as follows:
1. An introduction of the topic and some background.
2. A process description of the backcasting experiment, as well as a brief account of its content results.
3. An analysis of the backcasting experiment in terms of (i) stakeholder participation, (ii) vision aspects, (iii) settings and methodological aspects, (iv) higher order learning.
4. Associated follow-up and spin-off of the backcasting experiment in four domains: research domain, business domain, government domain, and public domain.

5. An analysis of the follow-up, spin-off and broader impact in terms of (i) network formation, (ii) vision aspects, (iii) institutionalisation, followed by (iv) mapping possible external factors and selecting these that really exerted influence.
6. Conclusions and discussion on (i) the backcasting experiment, (ii) the follow-up and spin-off and (iii) broader effects and system innovation aspects.

Finally, this section has elaborated an analytical framework consisting of sets of indicators for the phase of the backcasting experiment and for the phase of follow-up and spin-off after five to ten years. In this section three cases were selected related to nutrition and food production. The cases are evaluated in the following three chapters.

Notes

1   *These are: (1) changes in socio-economic conditions, such as the rise of social movements or economic cycles; (2) changes in systemic governing coalitions, for instance due to elections, and; (3) policy decisions and effects of other policy subsystems. Changes in tax law, for instance, may have a major impact on many policy subsystems.*

2   *Van den Hoed 2004) refers to institutional entrepreneurs emphasising the role of key individuals, but in this respect I consider changes in institutional practices that can indeed involve so-called institutional entrepreneurs. An example of a changed institutional practice is the shift in the 1990s among major food multinationals to specialise either in end-products or in supplying ingredients.*

BACKCASTING FOR A SUSTAINABLE FUTURE: THE IMPACT AFTER 10 YEARS

5

BACKCASTING FOR SUSTAINABLE PROTEIN FOODS AND ITS IMPACT

This chapter deals with the Novel Protein Foods (NPF) case, which consists of the backcasting experiment on NPF at the STD programme and its follow-up and spin-off up to ten years after completion. Novel Protein Foods are meat alternatives or protein foods from non-animal sources that are highly attractive to consumers. This chapter starts with the emergence of the NPF topic at the STD programme (5.1), followed by a description of the process and some results of the backcasting experiment (5.2) and a further analysis (5.3). It also contains a description of related follow-up and spin-off activities (5.4), a further analysis of these activities (5.5) and conclusions (5.6).

5.1 Introduction

5.1.1 The Sustainable Technology Development programme

In the early 1990s, long-term concerns emerged among staff members at the Ministry of the Environment, which were strongly influenced by the so-called Brundtland report (WCED 1987). Using the IPAT formula[1] it was determined that human needs need to be met 20 times more efficiently in environmental terms in 50 years time (Weterings and Opschoor 1992, Weaver *et al.* 2000: 31-44). At that time, very little was expected from technology (e.g. CLTM 1990). Existing policies focused on the implementation of good housekeeping and environmental technologies. These policies would have an impact on a considerably shorter term (5-15 years), but would only enable an environmental improvement with a factor 2-5 (Jansen and Vergragt 1992: 27, Jansen and Vergragt 1993:180, Weaver *et al.* 2000: 66-69). Thus, there were no policies that would allow for an increase in the environmental efficiency with a factor 10-30 on the long term and that would deal with what were called sustainable technologies. This awareness eventually resulted in the initiative for the interdepartmental Sustainable Technology Development (STD) programme at the Ministry of the Environment (Jansen and Vergragt 1992). Supported by five ministries[2] the STD research programme could start in January 1993. It was placed outside the ministries with a small office and a limited number of staff. High-profile project managers would be recruited in relevant areas, while research to a very large extent would be commissioned to research organisations.

The mission of the STD programme was to explore sustainable technological alternatives together with major stakeholders from the research and business communities and from government and to illustrate that factor 20 technologies would be feasible. The purpose was also to increase awareness among decision-makers of leading stakeholders, as well as to motivate pioneers in science and technology to start working or extend their work on sustainable technologies (Jansen and Vergragt 1992, Vergragt and Jansen 1993, Weaver *et al.* 2000: 71-75).

In the STD programme societal needs and the future sustainable alternatives for meeting them were explored, using a participatory backcasting approach. This was done through fifteen backcasting projects in various fields, like nutrition, water, mobility, housing and chemistry (STD 1997, Weaver *et al.* 2000). Examples of factor 20 backcasting projects at the STD programme included the mobile hydrogen fuel cell (Vergragt and Van Noort 1996, Weaver *et al.* 2000: 247-276), urban underground freight transport (STD 1997), Novel Protein Foods (STD 1996a, Weaver *et al.* 2000: 119-150), multiple sustainable land-use (STD 1997, Quist and Vergragt 2001, see also Chapter 7), sustainable urban renewal in the city of Rotterdam (STD 1997), chemistry based on biomass (Weaver *et al.* 2000: 205-220), sustainable washing (Vergragt and Van der Wel 1998) and municipal water systems (STD 1997, Weaver *et al.* 2000: 151-170).

The backcasting projects dealt not only with radical technological innovations that met the factor 20 challenge, but also included identification of cultural and structural conditions for further successful development and implementation. In general, the STD programme was considered fairly successful in identifying alternative solutions with the potential to meet the factor 20 and in developing follow-up agendas and strategic research programmes (Weaver *et al.* 2000: 282-285). Many projects, though not all of them, led to follow-up activities and affected research agendas in the knowledge infrastructure.

The backcasting approach at the STD programme was inspired by the Swedish practice, but adjusted. A major adjustment was the involvement of a broad range of stakeholders and decision-makers. Other key assumptions of the STD programme included the factor 20 environmental improvement in the long term (40-50 years), the use of an integral and problem-oriented approach, a focus on meeting future societal needs in a sustainable way (in line with the widely accepted definition of the Brundtland report, WCED 1987: 43) and a focus on desirable sustainable future images. Achieving implementation and establishing follow-up activities was important as well. The factor 20 was partly meant as a metaphor symbolising the enormous increase in environmental efficiency that would be necessary[3] (Jansen interview 2004). Another starting point was that technology is part of society and that it is strongly intertwined with a society's culture and structure. These three elements influence each other and develop together, while technological change is also bounded by cultural and structural conditions (that are not static, but evolve over time). As the backcasting methodology had to be developed and there was uncertainty about the results, it was emphasised that learning-by-doing was another key element of the approach. The participatory backcasting methodology that was developed as a result of the STD programme has been explained in Chapter 2 (see also Weaver *et al.* 2000: 76-77).

5.1.2 The emergence of the Novel Protein Foods topic

Novel Protein Foods (NPFs) had already been mentioned as an interesting research topic in the proposal for the STD programme (Jansen and Vergragt 1992: 47). The basic idea was that if it would be possible to develop NPF type of foods that would combine consumer attractiveness with a low environmental burden, this could considerably reduce the environmental burden of protein intake. NPFs must thus be seen as part of the food product category of meat alternatives. This category contains several other types of products, such as tofu, textured vegetable protein (TVP) products, the myco-protein based product Qorn, and meat alternatives based on wheat protein, or on mixtures of soy and wheat protein (e.g. Engels interview 2004, Aiking *et al.* 2006: 53). Others, on the other hand, have argued that NPFs should be seen as a product category in itself in line with the distinction between margarine and butter. Margarine was first marketed in 1869 as an alternative to butter, fulfilling similar consumer needs. Nowadays, it is seen as a separate product category, because consumers and companies see them as different food products[4] (Linsen interview 2004).

In the early 1990s, meat production and consumption was increasingly considered a major sustainability problem in the Netherlands (e.g. Heidemij *et al.* 1997, Aiking *et al.* 2006). The country has a large intensive livestock production sector, which has huge environmental effects, due to emissions especially from manure, the inefficient conversion of vegetable protein to meat protein, considerable use of energy and spatial requirements for growing fodder crops both domestically and abroad. It contributes considerably to acidification, climate change, eco-toxicity, and nutrification or eutrophication of soils and surface waters, while it also prohibits using land and crops (biomass) for other applications such as bio-fuels or supplying raw materials to the chemical industry.

However, livestock was (and still is) an important economic sector in the Netherlands, as well as many other countries (STD 1996). Meat is also an important source of proteins in people's diet and fulfils a range of non-nutritional requirements among consumers, such as taste, habit, custom and status, which are strongly entrenched in national cultures and not easily changed. Many stakeholders have strong interests in meat and defend these by forming powerful alliances that influence government policies and other relevant arenas (e.g. Grin *et al.* 2004). In the Netherlands all this had made the environmental problems associated with meat extremely complex and persistent, as is the case in many developed countries. First regulatory steps were made to mitigate the environmental burden of livestock production, targeting cattle farmers.

In spite of the above-mentioned, meat alternatives were not at all on the food innovation and research agenda in the Netherlands in the early 1990s. On the agenda was, for instance, product diversification, adding value to foods and food components and chain (reversal) issues (e.g. Jongen and Meulenberg 1998). Attempts to put soy-based Texturised Vegetable Protein (TVP) foods on the market in the late 1960s had failed (Veldhoen and Van den Ende 2003: 61-65, Aiking *et al.* 2006: 8, 29). This failure involved Dutch food multinational Unilever, which as a result had developed a negative attitude with regard to this topic (Dutilh interview 2005).

Nevertheless, in the early 1990s the small market segment of meat alternatives was growing and entering the supermarkets, and the range of available products was increasing as well. To a certain extent this was driven by growing animal welfare concerns and livestock epidemics, rather than by environmental concerns. Environmentalists tended to persuade people to eat (considerably) less meat, rather than encouraging them to look for alternatives.

The NPF backcasting experiment took place between early 1993 and late 1996 (see Table 5.1 for the timeline). Interviews were conducted from November 2004 to mid January 2005 (see Appendix B).

5.2 NPF backcasting experiment

5.2.1 First activities: June 1992 - June 1994

The proposal for the STD programme that was issued in June 1992 contained a list of possible topics (Jansen and Vergragt 1992: 46-54). This list included the development of a meat-like product, which would meet consumer requirements as well as sustainability criteria, for instance meat-in-vitro[5] from tissue breeding (Jansen and Vergragt 1992: 47). The essence of this idea was that if a product or production method could be developed leading to protein foods with a low environmental burden but with similar characteristics and performance as high-quality meat products, consumers would be willing to buy and consume these products instead of meat. This would result in considerable environmental improvement.

Late 1992, this topic gained the support of the interdepartmental preparatory group of the STD programme, while the Ministry of Agriculture was willing to act as the leading ministry. This decision of the Ministry of Agriculture was strongly supported by the National Council for Agricultural Research[6] (NRLO), which at that time was shifting towards strategic foresighting and acting as a strategic think-tank for the future of the food industry and agricultural sector in the Netherlands. Because of the systemic environmental problems in agriculture and the shift towards strategic foresighting, the NRLO, consisting of prominent stakeholders from research and industry in the food and agricultural sector, was looking for more environmentally benign approaches and practices. The topic of meat-like products fitted very well in its recently initiated activities on sustainable nutrition, while the NRLO also had a strong methodological interest in the STD programme. The support from the NRLO and its staff was especially important in the early stages of elaborating the NPF topic, when the support of the Ministry of Agriculture was actively courted (Jansen interview 2004). However, there was also notable resistance to this topic at the Ministry of Agriculture and among some of the parties involved in the NRLO. One group of stakeholders within the NRLO was firmly opposed to exploring alternative protein sources. Instead, this group favoured focusing on how to improve the meat and livestock production sector in economic terms. The new topic was eventually accepted after substantive discussions and after the name of alternative protein sources was changed into Novel Protein Foods, which sounded more neutral (Linsen interview 2004).

Shortly after the STD programme started in January 1993, the topic of NPFs proceeded through five interviews carried out by Professor Leo Jansen, the initial director of the STD programme and a senior staff member of the NRLO (Jansen interview 2005, Linsen interview 2005). They spoke with scientific experts from public research institutes, R&D managers from Unilever Research and Gist-brocades and with Dr Hans Linsen, a council member of the NRLO, who was later to become the project manager of the Novel Protein Foods project. This group of stakeholders met in April 1993 and supported the idea of conducting a feasibility study as a next step (Jansen 1993). This study was commissioned to the consulting firm Arthur D Little (ADL).

The ADL consultants investigated technological, cultural, environmental and consumer-related aspects of NPFs through desk research, expert interviews and used the consultancy's network in the USA. The results included fifteen potential categories of (processing) technologies and proteins from non-animal sources, ranging from existing single cell proteins (SCP) and texturised vegetable proteins (TVP) to proteins from moulds, algae, tissue breeding and de-novo protein synthesis in the laboratory (ADL 1993: 9, Weaver et al. 2000: 131). Nearly all categories showed huge potential for environmental improvement, but they varied substantially in terms of technological maturity, estimated development time and costs, and the likelihood of technological and commercial success (ADL 1993, STD 1994b, Weaver et al. 2000: 121-134). Therefore, substantial R&D efforts would be required and as a next step a two year project of 3-3.5 million Dutch guilders (which equals € 1.4-1.6 million) was proposed. The project would not only deal with technology and consumer aspects, but also with the environment gain, business economics and economic implications, as well as with stakeholder communication and dissemination of results (ADL 1993, STD 1994b).

Next steps included a stakeholder analysis (STD 1994a), determining Terms of Reference (STD 1994b, STD 1994d, De Haan et al. 1995: 7) and securing financial commitment from ministries and sponsoring from industry. The latter was a requirement for funding by the Ministry of Economic Affairs. Gist-Brocades soon appeared willing to sponsor, while it took more efforts to elicit financial support from Unilever Research. Linsen, who was a former director of Unilever Research in the Netherlands and a member of the board of the NRLO council, became the project manager.

5.2.2 The NPF project: June 1994 – December 1996

In June 1994 the kick-off meeting of the NPF project took place, which was attended by the financing parties, the researchers of the nine research groups involved, various STD staff members and the project team[7]. The researchers included[8] consumer scientists, economists, food technologists and Life Cycle Assessment (LCA) researchers (STD 1994c). Together, this allowed for a multidimensional and multidisciplinary analysis. Most of the research started in the end of the year, due to delays in formal commissioning of the

Table 5.1 *Timeline of the NPF backcasting experiment*

Preparatory stages	
June 1992	Meat-like products as potential topic in proposal for STD programme (Jansen and Vergragt 1992: 47).
July – December 1992	Approval of STD proposal by participating ministries, start of interdepartmental preparatory committee, meat-like products supported by LNV (STD 1996b).
January 1993	Official start of the STD programme, office outside the government.
February – April 1993	Interviews with experts and companies who supported further exploration of NPF topic (Jansen 1993).
April – June 1993	Selection of Arthur D Little for feasibility study on Novel Protein Foods, after approval by steering group (STD 1996b).
July –November 1993	Feasibility study by Arthur D Little and final report (ADL 1993).
December 1993 – January 1994	Steering committee of the STD programme decides to start the first phase of a larger project on NPFs, stakeholder analysis (STD 1994a).
February – May 1994	Applying for funding from ministries and companies. Finding project champion (project manager). Selection of research proposals and project team.
Novel Protein Foods project	
June 1994	Kick-off meeting with researchers and sponsors, start of project team.
August 1994	Start of consumer research, involving participants for the 1st TvC meeting (for an explanation of the TvC method see 5.2.2).
August – December 1994	Commissioning research tasks to research groups (STD 1996b).
November – December 1994	Preparation 1st TvC meeting, which takes place in early December.
January – February 1995	First meeting Advisory Group, second TvC meeting.
January – May 1995	Majority of research activities, several plenary project meetings.
June 1995	Interim report by NPF project team (De Haan et al 1995). Support by Advisory Group. 'Go' decision by the STD steering committee (STD 1996b).
September 1995	Start next research phase of the NPF project.
October – November 1995	Defining and elaborating (clusters of) follow-up activities and 'embedding' activities by the project team. Last (3rd) TvC meeting (STD 1996b).
December 1995	Additional activity: development of a strategic action plan (RMC 1996).
January – June 1996	Communication activities by project team. Evaluation by project team.
April 1996	Completing research stage, shift to embedding and stimulating follow-up activities, fourth meeting advisory group, final report (STD 1996a).
August 1996	End of activities by project team.
October – December 1996	Evaluation study (Loeber 1997, Loeber 2004: 141-206).
January 1997	Symposium 'What will we eat in 2050', presenting results from the NPF project and two other projects in the sub-programme on Nutrition.

research. Meanwhile, the project team developed an incremental procedure to narrow down the number of NPF categories, using plenary project meetings and inputs from all research groups (STD 1994d, De Haan et al. 1995: 19-28).

The consumer scientists of SWOKA would apply a recently developed methodology called 'Future Images for Consumers (TvC)[9]' (Fonk 1994). This participatory methodology was based on insights from Constructive Technology Assessment (see Chapter 2) and consisted of (1) consumer research and technological research, and (2) several rounds of workshops involving a range of stakeholders from science, marketing and business, government and public interest groups. The TvC in the NPF project involved participants from business, government, research and public interest groups. They met during three workshops in which project progress, research results, opportunities and dilemmas were discussed. The focus was on consumer aspects in a broad sense, which includes a range of social concerns. The TvC was completed with a final statement by the participants containing their agreements as well as their disagreements with respect to opportunities and conditions for the future of Novel Protein Food consumption (Fonk and Hamstra 1996).

Technology researchers would not make real foods, but looked into different NPF options from various vegetable and microbial protein sources and different types of processing technologies required to extract proteins and turn them into protein foods. The economists and environmental researchers compared the NPF options using pork meat as a reference. Research included to narrow down the enormous number of NPFs, as all of the 15 categories that were originally identified (ADL 1993) consisted of numerous combinations of protein sources, technologies and types of food products. Tissue breeding and de-novo-synthesis were left out first, because of energy requirements and the embryonic nature of these technologies. In a next round fungi and seaweed were also ruled out, which left moulds, single cell micro-organisms and plants as protein sources to be investigated. The technologists subsequently defined 18 NPF options, using a range of selected protein sources, while also varying in possible processing technologies and types of food ingredients (such as fibrous, fermented, minced-meat-like, biscuit-like, sauce-like, dinks). These options were also studied in the other research lines, allowing for a so-called portfolio analysis leading to a further selection of seven NPF options based on three ingredients that had highest potential for meat substitution. These were minced meat-like, fibrous and fermented (resembling tempeh).

During the project a more detailed future vision was developed, based on a trend analysis[10] performed by (selected) researchers, project team members and a staff member of the NRLO (De Haan et al. 1995: 30-31, Baggerman and Hamstra 1995: 70-83). The early vision comprised the idea of protein foods from non-animal sources having a twenty times lower environmental burden than (pork) meat. The key of the more elaborated future vision was that Novel Protein Foods - meeting the factor 20 requirement - could replace 40% of Dutch meat consumption in 2035, while 5% substitution would be obtained in 2005. An important change in the vision was the focus is on NPFs as a separate food category. Another change was the shift from meat alternatives toward NPFs as food ingredients; it was assumed that NPFs would be particularly attractive when it is applied as an ingredient in assembled and processed foods. Processed foods are for instance burgers, sausages and minced meat. Assembled dishes and foods include pizzas, ready-made meals, soup, etc.

In June 1995, the mid-term evaluation prepared by the project team (De Haan et al. 1995) was approved by the financing ministries and the governmental steering group of the STD programme, which made the second part of the funding available.

As a result, in September 1995 research activities could be continued. Two different parts were distinguished. Firstly, there was the continued research on the seven selected NPF options[11]. This was referred to as the NPF analysis (STD 1995a), which was completed in April 1996 and was extended with a marketing study and the development of a 'strategic action plan' (RMC 1996). Secondly, the project team intensified its activities with regard to external communication and dissemination with the aim of achieving and stimulating follow-up and embedding (STD 1995b). The project team would also update the stakeholder analysis and develop a possible development and implantation trajectory and a follow-up agenda (Quist *et al.* 1996a).

During the last months of the NPF project and the second half of 1997 various meetings on the development of follow-up proposals and related cooperation took place, involving the project manager, research bodies and companies in various settings. The project team ended in August 1996, though the project manager stayed involved. An evaluation study was also conducted (Loeber 1997).

5.2.3 Results

Results included a set of seven NPFs analyses that were analysed with respect to consumer acceptance and benefits, environmental impact, production costs and socio-economic effects. It was concluded (STD 1996a, Quist *et al.* 1996b, Weaver *et al.* 2000), that the environmental burden of these NPFs would be 10 to 30 times lower than that of pork meat, that NPFs would be attractive to both consumers and producers. In addition, it was concluded that the development and large-scale introduction of NPFs in the future would be feasible, but that new knowledge and R&D were strongly required (STD 1996a, Quist *et al.* 1996b).

Table 5.2 *Action agenda for Novel Protein Foods*

---

(1) Communication with the general public and supply of adequate information
(2) Professional education and transfer of generated knowledge
(3) Consumer research and development of marketing instruments
(4) Fundamental research and chain organisation
(5) Novel Protein Foods product development (both as foods and as ingredients)
(6) Improvement of environmental impact of crop growing and LCA instruments for foods
(7) Supporting regulation and social measures (facilitating the growth of a novel protein food sector and the reduction of the meat sector)

Furthermore, results included the future vision, R&D-programmes to develop lacking fundamental and applied knowledge, and a development trajectory towards 2040 containing seven clusters of follow-up and implementation activities for the short term, as shown in Table 5.2 (Quist *et al.* 1996a). The activity clusters in Table 5.2 were seen as a policy and action agenda for development of Novel Protein Foods.

5.3 Analysis

5.3.1 Stakeholder participation

Stakeholder participation evolved in successive stages of the NPF backcasting experiment. Therefore, a distinction is made between (i) the preparatory stage of idea articulation, (ii) the NPF feasibility study and, (iii) the NPF project. Stakeholder participation is analysed in terms of stakeholder heterogeneity, the degree of involvement (reflecting how much time is spent on the backcasting experiment), the type of involvement (funding, capacity, or other types of resources,) and the degree of influence on both the content and the process of the backcasting experiment. The stakeholder heterogeneity is shown in Table 5.3. Table 5.4

summarises the other three indicators for the various groups in the three stages.

Table 5.3 *Stakeholder heterogeneity in the NPF backcasting experiment*

	Research organisations	Companies	Government	Public interest groups
NPF idea articulation	● (< 5)	●	●	-
NPF feasibility study	● (5-10)	●	● (3-7)	-
NPF project	● (>10)	●	●	●

During the preparatory stage, several staff members and the chair of the STD programme were highly involved, as were two senior staff members of the NRLO. Next, Unilever Research, Gist-brocades, a NRLO Council member, a professor in industrial microbiology and a leading researcher in mushroom growing became involved through a series of interviews and a joint meeting hosted by Gist-brocades. This meeting led to commitment by the Ministry of Agriculture for funding a feasibility study, thus leading to another type involvement.

The feasibility study included a number of expert and business stakeholder interviews in both the Netherlands and the USA. At this stage the involvement of the external stakeholders was rather low in terms of time spent, numbers and heterogeneity, although most of them operated at a decision-making level in their organisation. Stakeholders involved in the first round of interviews had considerable influence on the direction, scope and shape of the proposed project on NPFs. As a result, the follow-up proposal defined by the consultancy resembled a business R&D trajectory and emphasised the business-economic aspects. Another outcome was that two companies were willing to sponsor. Both results were in line with the philosophy of the STD programme to get companies highly involved.

During the next stage of the NPF project, the number of stakeholders involved increased considerably. Firstly, relevant research bodies carried out the research. The technological research groups did not charge for all their activities, as they considered the project of strategic importance, while the consumer researchers had funding from another (governmental) source. Thus, the degree of involvement of the research groups increased, while their type of involvement changed as well. Within the space bounded by the terms of reference, the aims of the project and the guidance by the project team, researchers to a certain extent could also influence the content of the research, for instance by developing criteria for the selection the NPF options. With regard to organisational matters, coordination, project management and external communication the project team members exerted most influence, most notably the project manager.

Besides, participation of three ministries and two major companies was guaranteed through funding. These business stakeholders (Unilever Research and Gist-brocades) were also members of the advisory board, which was extended with key persons from research and public interest groups. This included the NRLO, the Ministry of Economic Affairs, a leading nutrition scientist from WUR, a representative from the main Dutch consumer association and a representative from the livestock production section of the national farmers' union (LTO). The degree of involvement of these stakeholders was low and the influence of the advisory board was limited. The focus was more on informing relevant parties about the results than on incorporating

Table 5.4 *Characterising stakeholder participation in the NPF backcasting experiment*

Stakeholder group		Degree of involvement	Type of involvement	Degree of influence
NPF idea articulation and testing	Interviewees (from science and food companies)	low	Capacity	high (on content)
	Members STD/ NRLO	medium-high	Capacity	medium-high (on process)
	Ministries (VROM/ LNV)	low	Funding	low
NPF feasibility study	Interviewed stakeholders	low	Capacity	low-medium
	Contractor/ consultant	medium	Capacity	medium (on content)
	STD office/ steering group	low	Capacity	medium-high (on process)
	Ministries	low	Funding	low
NPF project	Funding companies	low	Some capacity, co-funding	medium
	Research contractors	medium	Capacity, co-funding	medium
	Project team/ STD office	high	Capacity	high (on process)
	Ministries	low	Funding, some capacity	low
	TvC participants	low	Capacity	low

Legend: low — low-medium — medium — medium-high — high

stakeholder feedback (Loeber 2004: 167). However, the companies could exert influence through additional bilateral contacts with the project manager.

In addition, the 'Future Images for Consumers' (TvC) meetings involved a range of stakeholders from various societal groups, among others the funding companies and Nutricia (nowadays Numico), Sara Lee, Campina and the Dutch supermarket chain Albert Heijn. It also included some of the technological research groups in the NPF project, governmental parties like the NRLO, the Ministry of the Environment, the Ministry of Public Health and various consumer groups and environmental groups, like the Dutch Association of Country Woman (NBPV), the Foundation for Nature and Environment (NM), the foundation for Consumers and Biotechnology (SCB), and a regional environmental organisation (MFZH). Both the degree of involvement and the degree of influence were relatively low. There were three meetings lasting a day and a half during which intermediate results of the NPF project and broader issues were discussed. This stimulated awareness and dissemination among stakeholders and generated feedback and comments. However, these were used by the project team to a limited extent only, as raising commitment among the participants prevailed (Loeber 2004: 166).

Finally, there were several additional ways of securing involvement through, for instance, consumer research, stakeholder meetings, articles in regular papers and vocational journals, and other types of dissemination, such a the project news letter. This resulted sometimes in establishing new ties. For instance, the company Boekos got acquainted with and interested in the project, while the company Avebe co-funded the last part of the NPF project.

5.3.2 Vision aspects

Vision: inception

The future vision on NPFs started as an idea mentioned in the proposal for the STD programme in 1992, which was called in-vitro meat (Jansen and Vergragt 1992). After the programme was approved, the idea was tested and elaborated through interviews. Afterwards, the idea was described as a future meat-like product derived from either in-vitro tissue breeding or plants. These products would have a factor 20 lower environmental impact than meat at that time and should be attractive to consumers and producers (Jansen 1993). In the feasibility study (ADL 1993) this early vision was extended with a portfolio of 15 categories each combining protein sources and technologies analysed on their technological maturity, development costs, and environmental improvement potential. During the NPF project a more detailed future vision was developed. The key of the elaborated future vision was that Novel Protein Foods could replace nearly 40% of meat consumption by 2035 and should be seen as food ingredients in processed and assembled foods.

Looking from a backcasting perspective, the vision implied that food technology had to be improved considerably. This would enable the production of protein foods similarly attractive to consumers as meat, both in nutritional value and non-nutritional aspects. This implied not only technological change, but cultural changes as well. Cultural changes would not only relate to the role and image of meat and meat consumption, but also to the role and image of novel protein foods as a distinct food category separate from both crop sources and microbial sources. This future vision also implied structural changes, as the meat sector would decrease, while new crop-based and microbial protein food chains would emerge. Consequently, based on this future vision NPF options were further elaborated and analysed.

The vision had multiple roles. It was used to align the research and integrate the results. The future vision and the assessments were also used for internal, as well as external communication.

Vision: guidance

Guidance has been analysed in terms of (1) collective normative projection, (2) alignment and synchronisation of the interactions among different disciplines and among stakeholders from different backgrounds, (3) presence of alternative (system of) rules and institutions.

Collective normative projection

The future vision of Novel Protein Foods replacing a substantial share of current meat consumption offered a clear collective normative projection toward the future. However, a substitution of 40% in 2035 was not widely shared (the assumption being that no exceptional government support would be provided). Nevertheless, the idea that NPFs would play a much more significant role in future diet and that 5% replacement on the short term would be easily achieved, was strongly supported.

Synchronisation

The future vision, including its more detailed elaboration during the project, delivered substantial synchronisation and alignment among the researchers from different disciplinary backgrounds. For instance, technologists used the vision to translate perceived quality demands into technology and product characteristics. Economists investigated what the consequences for the supply system would be, assuming a certain share between vegetable sources and microbial/fungal sources and performed cost-price calculations using inputs from the technologists regarding processing. The vision made it possible to look at the same phenomenon from different perspectives and backgrounds, as well as carrying out assessments using the same set of (normative) assumptions. It thus shaped and structured the activities in the backcasting experiment, but the vision was also shaped by these activities, the researchers, the project team and the interactions with other stakeholders. Furthermore, the future vision not only provided synchronisation and alignment among the researchers from different disciplinary backgrounds. It also provided synchronisation and alignment in the discussions with stakeholders involved, for instance during the TvC meetings. The TvC methodology to a certain extent provided a platform for exchange of normative aspects of NPF development and meat consumption, such as related genetic modification or world protein security, thus enabling synchronisation among stakeholders from very diverse societal groups.

Alternative rules

The future vision contained numerous alternative rules, such as the factor 20 environmental improvement, protein foods meeting similar requirements as current meat does, changes in production and consumption patterns, changes in consumer attitudes, and new roles to a range of stakeholders. The future vision also contained an alternative production and consumption system, which would require a separate set of rules.

Vision: orientation

Orientation has been analysed in terms of (1) cognitive activation, (2) mobilisation of actors and resources, and (3) decentralised daily coordination.

Cognitive activation

The future vision provided substantial cognitive challenges and resulted in activities dealing with these challenges. For instance, cognitive challenges were translated into research questions and research activities. In addition, cognitive activation also took place by defining a trajectory or pathway that would make it possible to put the future vision into reality, which could also be used to determine follow-up activities and a broader follow-up agenda containing several clusters of activities. The future vision also allowed for dialogue and feedback, for instance at the TvC meetings, which also contained cognitive elements.

Mobilisation

The future vision was used to mobilise actors and resources. Three ministries and three companies funded the NPF project. This made it possible to engage research groups with relevant expertise and to commission research to them, while some groups provided additional capacity or had already raised funds from other sources. The vision and resources made it possible to establish a project team, among whom especially the project manager was willing and capable to mobilise his personal network in food research and business. Several dozens of stakeholders provided capacity by allowing individual people to participate in activities like the advisory board or the TvC meetings. In the course of the project, the number of 'mobilised' actors grew, while dissemination activities contributed also to a wider acquaintance of the project, its topic and the future vision. As a result, stakeholders emerged that wanted to be involved in the project, like the companies Boekos and Avebe. There was interest in the project from national and specialist written media, which resulted in a substantial number of articles.

Decentralised coordination

There was a certain level of decentralised day-to-day coordination among the parties involved, provided by the future vision. Most activities took place in a decentralised way at different locations in the Netherlands. However, this coordination was also partly due to the project structure, which included goals, research tasks, plenary research meetings and management by the project team. Coordination was also partly constituted by the available funding and the project structure, which included commissioning contracts and hiring people.

Competing visions

Other future visions were present in the context of the project. For instance, there was the regular dominant vision on meat production and meat consumption supported by the vested interests. The actors backing this vision preferred the status quo and were in favour of improving the 'economic health' of the cattle and meat processing sector, as well as to reduce the emissions of intensive livestock production by improving stables and fodder conversion. The actors backing this vision paid hardly any attention to the NPF future vision. However, they opposed it when the vision became 'within reach', as happened in early meetings at the NRLO. Resistance was also expected from within the Ministry of Agriculture. However, on average these actors did not consider the NPF vision a seriously competing and threatening vision. For instance, the chair of the Product Boards for Livestock, Meat and Eggs (PVV) referred to meat alternatives in his New Year's address (Loeber 2004: 193, citing Tazelaar 1997) as a niche market that could also be served. This was, for instance, exactly what Boekos did, as it produced meat and meat products, as well as meat alternatives.

In networks of environmentalists and vegetarians other visions were present as well. The vision among environmentalists included reducing meat consumption by targeting and educating consumers as well as shifting consumption towards eating organic meat by incorporating all externalised costs of meat. This vision included to decrease the meat sector by regulation, to enact further environmental regulation of intensive livestock production and to stimulate organic livestock production. There was also a vision of a meat-free society among vegetarians and animal welfare activists, but despite the growth in vegetarianism, this was still a marginal vision in society as a whole.

These alternative visions were occasionally voiced and used to comment on the business orientation and the focus on technology in the NPF backcasting experiment, or to argue that the options to eat less meat and to target the meat sector were neglected. This happened, for instance, in the TvC meetings (Loeber 2004: 166).

5.3.3 Learning

In line with Brown *et al.* (2003), higher order learning[12] is defined in terms of (1) a shift in framing major problems and of perceived solutions by specific actors; (2) a shift in the principal approaches to solving these problems and in shifting priorities by specific actors; and (3) joint learning, a shift in congruence and joint opinions among groups of stakeholders in any of the issues related to the previous shifts. In this research a distinction is made between learning about the topic of NPFs, for instance as a shift in dealing with the existing sustainability problem of meat consumption and production, and learning about the participatory backcasting approach. It is assumed that higher order learning about the topic of NPFs mainly involves the 'content' stakeholders, while higher order learning about the approach to a large extent involves those organising and managing the backcasting experiment. The purpose is not to be comprehensive in this respect, but to provide some empirical evidence. In addition, there has also been considerable first order learning in the NPF backcasting experiment, which can, for instance, be found in research reports (e.g. STD 1996a including the references).

Higher order learning about the topic

The NPF backcasting experiment helped increase awareness among food technology researchers in the Netherlands about the environmental impact of foods and food processing. For instance, developing certain food supply systems also determines the type and magnitude of their environmental impact (Van Boekel interview 2005). Before, a majority of Dutch food technologists in the Netherlands related the environmental impact of foods to primary agriculture, thus leaving food processing outside the problem boundaries. This is an example of a shift in the framing of a major problem. It also became clear that working on sustainability problems and sustainability solutions requires a multidisciplinary approach (Van Boekel interview 2005), which is a shift in principal approach. During the NPF project the insight emerged that they not be seen as end-products, like fresh meats, vegetarian burgers, or sausages, but that NPFs should be seen as food ingredients that are components of assembled or processed end-foods. This is both a shifted priority and an example of joint learning resulting in adjusted joint opinions within (part of) the backcasting experiment.

Joint learning also took place with regard to the opportunities for alternative protein foods, which until then were seen as non-existent. This concerns a shift in a perceived solution to the problem of meat production and consumption. For instance, the knowledge and technology needed to make making NPFs hardly existed in the Netherlands, while it could be a challenging research theme and thus embodied research opportunities (Van Boekel interview 2005). Joint learning also took place as a result of the TvC stakeholder meetings. The participants issued a final statement that included a list of issues on which agreement was reached, like the relationship with global food security (Fonk and Hamstra 1996: 67-71), which can be seen as joint learning, or at least as congruence. In addition, there were topics on which no agreement was reached among the participants, in other words where no joint learning had taken place[13]. The NPF backcasting experiment has contributed to the awareness that food companies can contribute to sustainability, that some of their activities could already be characterised as sustainable and that business and sustainability can be reconciled (Linsen interview 2004, see also Loeber 2004: 188). In one of the interviews the following was also mentioned (Van Boekel interview 2005), *"Very important about the project on Novel Protein Foods was that it brought together a number of people who saw opportunities and were willing to continue and work together in developing follow-up research"*. This statement reflects a shift in priorities and in relationships, and reflects joint learning.

Higher order learning about the approach

Both in the interviews as part of this study and in the evaluation by Loeber (2004: 187) it became clear that the participants of the NPF backcasting experiment considered the participatory backcasting approach an innovative and attractive novel way of dealing with sustainability problems (e.g. Van Boekel interview 2005). This relates to a shift in principal approaches to solving problems.

It was emphasised that the backcasting experiment on NPFs could have been improved, for instance by a broader dialogue, more time for debate and a more extensive strategic orientation in an early stage of the backcasting experiment (Jansen interview 2004, Linsen interview 2004). This would have enhanced and broadened the stakeholder debate on the topic of NPFs as a potential contribution to dealing with the sustainability problems of meat production and consumption. Loeber (2004: 177, 202) also pointed to this issue. She mentioned that the fixed objectives and limited flexibility in the NPF project constrained joint learning among the TvC participants. They wanted to broaden the debate with alternative options and visions, while there was no room for this. This example of a shift in priorities by the organisers of the backcasting experiment was executed in other backcasting experiments at the STD programme (Jansen interview 2004), like the one on sustainable chemistry (Aarts 1997: 119-124) and in the participatory domain analysis on nutrition, which ran parallel to the NPF backcasting experiment (Aarts 1997: 33-38, see also Chapter 7).

5.3.4 Methodological aspects and settings

A methodological framework for participatory backcasting has been developed in Chapter 2. This framework includes (i) inter-disciplinarity, (ii) five steps, (iii) four categories of methods, (iv) three types of demands, and (v) stakeholders heterogeneity (for the latter, see 5.3.1).

Backcasting framework: inter-disciplinarity

The NPF backcasting experiment was clearly a inter-disciplinary endeavour. It comprised of a range of disciplines, including economics, environmental studies, technology, consumer studies and communication. Furthermore, researchers used data from the other disciplines, while results from all disciplines were all integrated in the analysis of NPF options, as well as in the elaboration of the future vision and the possible development trajectory.

Backcasting framework: steps

The five steps as proposed in the methodological framework can more or less be identified in the course of the backcasting experiment. However, in general the steps were not well articulated, and nor was the order followed exactly.

For instance, the preparatory stage consisting of idea development and the feasibility study included elements of problem orientation. During the NPF project a fair number of activities were also carried out that can be marked as strategic problem orientation, such as a stakeholder analysis and testing the idea among relevant stakeholders.

The NPF backcasting experiment actually started as an idea for a sustainable solution which was then gradually developed and explored further. A basic future vision was also developed in the early preparatory stage, while a more detailed elaboration was developed much later in the backcasting experiment after considerable analysis had been conducted. The more elaborated future vision was also important for the integration of different analyses.

Backcasting took place relatively late in the project, after the more detailed elaboration of the future vision had taken place. However, it can be argued that defining the project proposal as part of the feasibility study and proposing research activities in five research lines, involved some kind of backcasting. The proposal outlined the activities that were considered necessary for developing NPFs. However, it is difficult to determine with hindsight what exactly was the result of backward looking and what of forward planning. After the midterm evaluation in the NPF project, researchers were asked to deliver data for the trajectory back from the future vision, while two different trajectories leading from the same future vision were distinguished. As a consequence, this development trajectory, as it was called, was partly the result of backcasting activities and partly the result of looking forward and projections.

Further elaboration, consisting of both design and analysis, and working on embedding of results and stimulating follow-up can easily be identified in the entire NPF project. Elaboration and embedding have been defined in the methodological frameworks as step 4 and step 5, respectively. However, in the NPF project these two steps overlap. In fact, embedding results and topic among stakeholders was an ongoing activity throughout the backcasting experiment, both in the early stages and in the NPF project.

Backcasting framework: methods

Methods from all four categories as distinguished in the methodological framework were applied in the NPF backcasting experiment. The results, although they are not comprehensive, are shown in Table 5.5. Explanations of particular methods can be found elsewhere (ADL 1993, STD 1995a, STD 1995c, STD 1996a).

Table 5.5 *Applied methods in the NPF backcasting experiment*

Analysis
> Life-cycle analysis
> Stakeholder analysis
> Consumer research (in-depth interviews, survey)
> Economic input-output analysis and modelling
> Cost-price calculations
> Technology analysis
> Portfolio (multi-criteria) analysis of NPF options
> Process evaluation (in-between and afterwards)
> Trend extrapolation analysis
> Market analysis

Design
> Design of novel protein foods, processing routes and supply chains
> Project/ process/ research design
> Development of future vision
> Design of potential development trajectory
> Design of action/ research/ policy agenda
> Development communication plan
> Marketing action plan
> Development of ideal-typical process (STD 1996c)

Participation/ Interaction
> Early expert and stakeholder interviews
> Future images for consumers (TvC)
> Project advisory board meetings
> Stakeholder interviews and meetings by project team
> Government steering group meetings

Management, communication, coordination
> Regular project management
> Internal project communication and coordination
> Plenary research meetings
> Dissemination activities (media interviews, papers specialist journals, lectures, project news letter)
> Setting Terms of Reference (ToR)

It appeared that is was not always easy to link every method to a specific category, as particular methods can be related to several categories, or contain activities in more than one category. For instance, developing a marketing plan can be seen as a design activity, while putting it into practice requires activities that would match the category 'management and coordination' better, while market research may precede the marketing plan. In addition, if market research involves focus groups or in-dept interviews, it has participatory elements as well. Similarly, process or project evaluation should be seen as analysis, while project and process are designed in advance, and managed and coordinated during its operational period.

Design has been taken in a broad sense and refers to all activities that involve any kind of design, also when that was not acknowledged as such, or when no formal design methodology was followed[14]. For instance, in this evaluation further development of future vision and development trajectory are considered design activities, while no clear design methodologies applied and the people involved did not view these activities as design activities.

Backcasting framework: demands

In the methodological framework a distinction has been made between normative demands, process demands and knowledge demands. These demands can, for instance, be found in project aims or so-called Terms of Reference.

The basic assumptions of the STD programme, as mentioned in Section 5.1 provide several normative demands for the NPF backcasting experiment. These included the factor 20 environmental improvement (normative demand), involvement of stakeholders (also a process demand), a focus on follow-up and implementation (also a process demand), and learning-by-doing with respect to the backcasting approach together with the stakeholders involved (both a process and normative demand). Normative demands specific for the NPF backcasting experiment were for instance a substantial future market share for NPFs, and approaching NPFs as ingredients in processed and assembled foods. Additional demands can be extracted from the so-called Terms of Reference. These included attractiveness to producers and consumers (both a process and a normative demand) and feasibility of technological options (knowledge demand). Knowledge demands with respect to research and research activities were not clearly articulated. This seems to be left to the responsibility of the research institutes. However, there was a focus on quantification and in this way making a convincing case for business and the environment, which is clearly a knowledge demand.

Settings: institutional protection

The STD programme had institutional protection from top level management at the ministries. In case of the NPF project, this included support from the Ministry of Agriculture and the NRLO, as well as from high-ranking R&D officials in major food companies just below board level. However, this did not coincide with strong influence and active control with daily matters and daily organisation. As a consequence this protection provided room for experimentation in the backcasting experiment, but bounded by the project aims, success criteria and Terms of Reference.

Settings: vision champion

The project manager Dr Hans Linsen was considered and acted as vision champion in the field of food business and food research. In addition, Professor Leo Jansen, the initiator of the STD programme, was also a strong supporter of the NPF topic and had been highly involved in the early stages of the backcasting experiment. Jansen promoted NPFs as an interesting topic for follow-up activities within his network (see also 5.4). Thus, he acted also as a vision champion, but in different networks than the project manager.

Settings: focus and management

A major focus was on realising follow-up. In addition, the project was managed by a project team responsible for integration of results, embedding of outcomes and stimulating follow-up. The project team applied a rather strict project management type of approach. The project team also used budget control to adjust research and perform additional activities like carrying out a market analysis and developing a marketing action plan. As a consequence of the strict project management approach, the focus on quantification and the tight time schedule, there was limited room for initiatives by the researchers, or to incorporate stakeholder feedback from, for instance, the TvC meetings, in the project activities. The approach of the project team was reinforced by the objectives (and ToR) that emphasised feasibility, content and endorsement, and not, for instance, quality or richness of the dialogue process.

5.4 Follow-up and spin-off

5.4.1 Research domain: Profetas programme

The most substantial research effort established after the completion of the NPF backcasting experiment was the multidisciplinary research programme Profetas (www.profetas.nl, Profetas 1998, Aiking *et al.* 2006), which stands for Protein Foods, Environment, Technology And Society. This programme dealt with technological and crop breeding issues, socio-economic aspects and cultural and consumer-related aspects of the production and consumption of NPFs. The central issue was whether a partial transition from animal to plant protein foods in the diet of consumers in the North would be more sustainable, technologically feasible and socially desirable (www.profetas.nl, Profetas 1998, Aiking *et al.* 2006). Profetas started late 1999 and a final symposium took place in October 2004 (Profetas 2004).

Profetas was a joint endeavour by two research schools in the Netherlands, the research school for Socio-Economic and Natural Sciences of the Environment (Sense) and the research school for Nutrition, Food Technology, Agro-biotechnology and Health Sciences (VLAG[15]). The programme consisted of 15 PhD and postdoctoral research projects and involved research groups[16] from three Dutch universities and five research institutes in food sciences and agriculture[17]. It involved nearly 50 people (Aiking *et al.* 2006: 218).

The Profetas programme was funded by regular academic research councils, as well as by various ministries and companies, in all around 3 million Dutch guilders. The central board of NWO, the Netherlands Organisation for Scientific Research, contributed 1.1 million guilders (Aiking 2002). The Social Science Research Council (MaG), the division of NWO for social and environmental sciences, and technology foundation STW, which is the research council for technological research, together funded a similar amount of money (Aiking 2002). Furthermore, the Ministry of Agriculture contributed 400,000 guilders. Wageningen Centre for Food Sciences (WCFS), which is one of the five Technological Top Institutes (TTIs) in the Netherlands, and five Dutch companies co-funded the programme for 300,000 guilders (Kemper interview 2005). The five companies were Boekos, a producer of meat (products) and at that time the main supplier of meat alternatives in the Netherlands, Cebeco, a cooperative agricultural firm also active in crop processing, Quest International, an ICI subsidiary producing mainly flavours, DSM Food Specialties and the Unilever Food Research Institute.

Because of the impossibility to create a joint financial funding framework and because of non-matching organisational guidelines between STW and MaG, Profetas consisted of a sub-programme dealing with the socio-economic part funded by MaG and a sub-programme dealing with technical and consumer aspects funded by STW. A mandatory user group for the STW sub-programme was established for the co-funding parties (Kemper interview 2004), in addition to a programme committee for the entire programme. The programme committee was chaired by Linsen, the former project manager of the NPF project, and also involved the former STD chair (Waaijers interview 2004, Linsen interview 2004).

Unlike the NPF backcasting experiment, the Profetas programme concentrated on a single crop, the green pea (Aiking interview 2004, Vereijken interview 2004), thus leaving out other crop options and the microbial options that had been studied in the backcasting experiment. Consumer-oriented product development was applied, which is an emerging novel approach to product development and innovation in food research and food industry and a major research focus at WUR (Van Boekel interview 2005).

Profetas: conception

In 1995, the board of NWO decided to initiate a programme for innovative multidisciplinary research on an environmental issue. Possible topics were discussed in a small group of prominent Dutch scientists that included the chair of the STD programme and the director of the Institute for Environmental Studies (IVM). This group suggested to elaborate upon the first results of the multidisciplinary STD programme, which included Novel Protein Foods (Jansen interview 2004). Next, the board of NWO invited the research schools SENSE and VLAG to elaborate a proposal on a nutrition related topic (Aiking 2002, Aiking interview 2004). This involved leading researchers from WUR that were also working on NPFs. As a result, the topic of NPFs appeared high on the agenda of the programme (Van Boekel interview 2005). However, it took further substantial efforts, such as a preparatory study, the development of research project proposals, several rounds of reviews and adjustments, workshops with researchers and with stakeholders, dealing with non-matching guidelines of MaG and STW, etc. This lasted until early 1999, when MaG and STW both approved the programme proposal (Aiking 2002, Van Dam-Mieras 2002: 7).

Profetas: outcomes

The Profetas programme delivered results and outcomes at various levels (for details, see Aiking *et al.* 2006). Academic results included scientific knowledge, product-concept development, scientific models, technological and environmental tools, and consumer knowledge.

Opinions on stakeholder involvement in Profetas vary. For instance, stakeholder involvement was too limited, apart from the separate STW user group meetings (Aiking interview 2004, Linsen interview 2004, Jansen interview 2004, Van Dam-Mieras 2002). However, it was also emphasised that funding by the Ministry for Agriculture and five companies must be perceived as strong support from crucial stakeholders (Van Boekel interview 2005). Three companies left during the programme, but this was due to developments outside the Profetas programme. For instance, Boekos, a major producer of vegetarian foods in the Netherlands and a firm supporter of the Profetas programme, was taken over by the Zwanenburg Food Group, after which all non-meat activities were terminated. Cebeco made a strategy change and decided to quit all food processing and internal research activities, and sold its subsidiary 'Pelmolen', which was involved in pea processing and pea valorisation. Furthermore, the protein activities of Quest were sold to the Irish company Kerry Ingredients. Although Kerry Ingredients is strong in protein ingredients, especially in wheat proteins, the company left the Profetas programme.

5.4.2 Research domain: other activities

At WUR a novel research programme on system innovations towards sustainability has been initiated (www.scheppenvanruimte.nl), which includes a project on new protein foods (e.g. Van der Knijf *et al.* 2004). The focus of this project has shifted from vegetable and microbial protein foods for human consumption towards exploring the potential of these crop-derived protein sources for substituting fishmeal in fish breeding and substituting milk in veal breeding[18]. The rationale is that consumer acceptance of meat alternatives is difficult to achieve, while substituting dairy proteins in the diet of veal, or substituting fishmeal in fish-breeding lowers the environmental strain as well. In 2005 the Innovation Network Green Space and Agro-cluster (INGRA) commissioned a study to determine the feasibility of a floating fish farm where natural gas-based single cell protein would be fed to fish, which is currently done in Norway (EBS 2005).

In 1997 TNO Nutrition formed an R&D alliance with DSM Food Specialties and Boekos (Loeber 2004: 185, Coenen 2000: 74, see also 5.4.3). The Agrotechnology and Food Innovations institute has attempted the same, but so far without success (Vereijken interview 2004).

It has been suggested that NPF research was a theme at the technological top institute WCFS (Coenen 2000: 74). However, WCFS focuses on bulk food categories and not on niche specialties such as NPFs. Research themes at WCFS like protein-structure interactions have relevance to the development of NPFs, but are not due to the NPF backcasting experiment (Vereijken interview 2004).

Finally, no substantial multidisciplinary academic research on NPFs was found abroad. This was confirmed by the lack of international academic interest, when attempts were made to organise an international multidisciplinary scientific conference on NPFs (Linsen interview 2004).

5.4.3 Business domain

Initially, five companies were involved in the Profetas programme, including two producers of end-products to consumers or the food service industry (Unilever and Boekos), and three suppliers of ingredients (Cebeco, Quest, DSM Food Specialties).

DSM has been working on a NPF ingredient-based on mould protein since 1997, when Gist-brocades, which at that time had just been taken over by DSM, started a joint research project with TNO Nutrition and Boekos. Funding was largely provided by a government innovation funding programme. This alliance was attractive to both companies. DSM Food Specialties was specialised in supplying ingredients like enzymes produced by micro-organisms like yeasts and moulds. Boekos was a major producer of vegetable protein foods and expected meat consumption to go down in the future. DSM considered NPF ingredients based on mould proteins as an interesting business opportunity in an emerging market with successful brands like Qorn and Tivall. NPFs fitted well in their reformulated strategy that emphasised high added-value specialties (Loeber 2004: 182). The research project included compliance with the strict Novel Food regulation of the EU, which had been enacted in 1990 and requires extensive food safety data. The project has advanced towards product development (Van Egmond & Sein interview 2004), which is presently being carried out in cooperation with a new company downstream that produces foods for end-users. (Van Egmond & Sein interview 2004). However, market introduction was postponed in the end of 2004, due to increased competition among supermarkets in the Netherlands. The so-called supermarket war started late 2003 and resulted in rock-bottom meat prices, which made it hard to compete with meat. At the same time, short-term profit contribution has become more important in the company, which limits the possibilities for radical innovations like the mould protein-based NPF ingredient (Van Egmond & Sein interview 2004). Finally, NPF activities represent a rather small share of total R&D activities at DSM Food Specialties.

Unilever was interested in Profetas in a general sense, but was not working on protein ingredients for vegetarian protein foods or meat alternatives (Dutilh interview 2004, Zwijgers interview 2004). Unilever does not see itself as a producer of vegetable protein foods, but as a producer of major categories of foods for end-users (both consumers and food-service providers). Besides, Unilever finished its upstream food ingredient activities in 1997 and it would require a global supplier producing the ingredient if it were to become interested in meat alternatives. However, no such a global supplier was represented in Profetas (Zwijgers interview 2004). Unilever is already active in alternative protein sources, such as soy milk in Argentina, where cow milk is too expensive for a majority of the people. Furthermore, meat alternatives as an alternative protein source would be put on time horizon of 4-7 years, as it concerns radical new products (Zwijgers interview 2004).

Until it was taken over in 2001, Boekos had been very active in NPF developments ever since the company got acquainted with the NPF backcasting experiment in early 1996. At that time Boekos was the main producer of vegetarian protein foods in the Netherlands. NPFs related to its core business, though it was still a small share compared to the meat activities. Boekos initiated a range of activities dealing with different aspects of NPF and vegetarian protein foods. In addition to their participation in Profetas and their cooperation with TNO and DSM in R&D projects, Boekos worked on new food product development and initiated joint activities with TNO Kathalys that worked on the sustainable product development of appliances. The latter activity focused on novel kitchen appliances and the domestic use of NPF in cooking (Goekoop 2000, see also Chapter 6). Boekos also developed (together with other parties) a programme proposal for the NIDO Contest focusing on optimising the protein food production chains and kitchen chains (Boekos Food Group *et al.* 2000, see also Chapter 6).

There were also companies outside the Profetas programme that deployed NPF-related activities. For instance, the Dutch leading cooperative dairy firm Campina[19] launched the meat alternative Valess[20] in February 2005. Due to the dairy proteins, this product has superior features regarding taste and texture compared to meat alternatives based on vegetable proteins (Van Calmthout 2004, Scholtens 2005, Van Boekel interview 2005). The company has started a € 3 million marketing campaign, its most expensive one to date (Azough 2005). The campaign focuses on non-vegetarian consumers and addresses aspects like variation and healthiness. Van Boekel (Van Boekel interview 2005) sees it as a positive signal that such a leading dairy firm recognizes (business) opportunities in the market of NPFs and vegetarian protein foods and wants to become a player and an early entrant in this emerging market.

Avebe, the cooperative potato-derived starch producer in the North of the Netherlands, for several years has been working on extracting proteins from Lucerne and grass that could be used as a source to NPFs. The aim was to establish production activities in the off-season of potatoes and to offer new crop opportunities to the farmers (Loeber 2004: 188-189). However, these R&D activities were abandoned in 2002 (Linsen interview 2004).

There are also substantial activities dealing with NPFs and present meat alternatives in catering chains, which involve the companies Sodexho, DeliXL, and Schouten Europe, as well as the research institute LEI and KDO Consultancy (Kramer[21] interview 2005). Sodexho is a major international catering company. Deli XL is a business unit of Ahold that provides wholesale and distribution services to catering companies and restaurants. Schouten Europe is a medium-sized producer of meat alternatives that has developed new products for testing at a range of catering facilities. The activities are partly funded by AKK, the Dutch organisation on agro-food chain knowledge (www.akk.nl, Zimmerman *et al.* 2006) and by a Senter-Novem programme on improving energy efficiency (www.epz.novem.nl, Londo and De Kuijer 2002). It was concluded (Zimmerman *et al.* 2006) that the testing of meat alternatives at several catering facilities suffices follow-up activities by the three companies involved.

In addition, there were the activities by producers of 'regular' vegetable protein foods in the Netherlands. In general these are SMEs, who do 'their thing independent of NPF developments' (Engels interview 2004). An exception is the entrepreneur who established the trade company Planet Green and Aurelia, a market research consultancy in organic and vegetarian foods. This entrepreneur saw business opportunities for vegetarian protein foods in the mid-1990s (Engels interview 2004) and established Planet Green in 1997, which offers a range of vegetarian protein food products for diner and bread meals. The brand achieved a substantial market share in most Dutch supermarkets. However, the brand could not survive independently during the Dutch supermarket war that started in 2003 (Engels interview 2004) and was sold to another producer. Activities of Aurelia include market research on vegetarian protein foods, preparing the start-up of a national product office for vegetarian protein foods, similar to the existing Commodity Boards, monitoring of alternatives for meat and fish products, and organising events, such as symposiums (Aurelia 2002) and V-day, the public event on vegetarianism and vegetarian food (together with the Dutch Union of Vegetarians (NVB).

Finally, the Product Board for Livestock, Meat and Eggs (PVE) includes meat alternatives in its regular studies on the consumption and image of meat and meat products (Van der Lans 2001, Anonymous 2005).

5.4.4 Government domain

The Ministry of Agriculture contributed around 400,000 guilders to the Profetas programme in 1999. However, the opinion at the ministry recently changed. The current opinion is that NPF has been supported sufficiently and can be left to the market (Huizing personal communication 2004). This is (partly) due to a growing awareness at the ministry that NPFs may compete with meat and meat products. In October 2006 this was confirmed by an official viewpoint of the Minister of Agriculture (2006) on the results of the Profetas programme that was sent to the Parliament.

At the Ministry of the Environment the unit for (sustainable) products and consumption was triggered by the NPF backcasting experiment. There were many interactions between the STD programme and the ministry, while NPFs had for instance also been a case (see Quist et al. 1996c) in a background report to the advice for sustainable consumption by the Council for Environmental Management (RMB 1996). This advice emphasised the importance of sustainable products and services as part of stimulating sustainable consumption, which was also part of the strategy outlined in the NMP3, the third National Environmental Policy Plan (VROM 1998). Stimulating NPFs and meat alternatives (Brand 2005, personal communication) was seen as an opportunity to develop policies for sustainable consumption and sustainable products. However, the results from the NPF backcasting experiment were considered both too academic and too elaborate, thus as not yet appropriate for policy-making on sustainable food consumption. The relevant policy question was whether and how production and consumption of (more) sustainable meat alternatives could be stimulated (De Kuijer and Wielinga 1999: 2). A study was commissioned that recommended stimulating the demand side and dealing with bottlenecks, such as the image of meat alternatives and the resilience among the environmental movement to GMO[22] (De Kuijer and Wielinga 1999: 59-65).

Several other activities were initiated at the Ministry of the Environment in order to stimulate meat alternatives and Novel Protein Foods. This included supporting development of the demand side, facilitating producers like Boekos and a dialogue with the environmental movement and other public interest groups on the issue of GMO and how to guarantee the supply of GMO-free vegetarian foods. The topic of meat alternatives was also integrated in activities dealing with sustainable nutrition in a broader sense. For instance, NPFs and meat alternatives were put on the agenda of the ministry's stakeholder platform on nutrition, which involved a range of business and non-business stakeholders. Events promoting vegetarian foods like symposiums (e.g. Aurelia 2002) and V-day in September 2004 were supported financially. Meat alternatives were also included in the exploration of the nutrition domain (Van der Pijl and Krutwagen 2000), in which meat alternatives were positioned as a more sustainable meal alternative. This study proposed a policy agenda, which included stimulating meat alternatives and vegetarian meals not only at home, but also in the food-service sector.

However, the interest in sustainable products and consumption has decreased significantly at the Ministry of the Environment since the last National Environmental Policy Plan (VROM 2002). Priorities have shifted to transition management, climate change and sustainable use of space, Furthermore, the topic of food products and food consumption has been transferred to the Ministry of Agriculture in 2001, after which meat alternatives for human consumption have become neglected. By contrast, the Ministry of Agriculture is strongly interested in alternative protein sources, but as an ingredient in fodder. After the BSE incidents in the early 1990s animal protein sources were banned, this has resulted in an increased dependence on soy proteins in fodder. As a result, alternative protein sources have become a priority.

Finally, the MNP (formerly known as RIVM), the Dutch governmental planning office for the environment and nature, has used NPF as a case for testing an approach to monitoring transitions (Van Wijk and Rood 2002, Rood et al. 2002).

5.4.5 Domain of public interest groups and the public

Animal welfare groups and the Dutch Union of Vegetarians (NVB) are very positive about any meat or fish alternative, as this contributes to the consumption of vegetarian foods and vegetarianism. However, they do not consider promoting vegetarian food products as their core business, but rather a vegetarian lifestyle. Nevertheless, animal welfare groups and the vegetarian movement are increasingly interested in vegetarian products (Timmerman interview 2004). For instance, the NVB and several animal welfare groups have developed an action plan for monitoring fish and meat alternatives in supermarkets. The NVB is also involved in the plan for a product and promotion agency for vegetarian foods and it wants to enhance its activities for the international vegetarian label for which they are responsible in the Netherlands. This label is currently granted to well known food industries for foods not containing any meat (components). The NVB and the former alternative consumer association[23] have launched the vegetarian 'Schijf van Vijf / Good Food Diet Practice' in 2004. The NVB was also the main organiser of V-day in 2004, the public event aimed at promoting vegetarianism and vegetarian foods (Timmerman interview 2004).

Environmental organisations are less involved in the topic of meat alternatives and vegetarian foods, as environmentalists consider organic meat a good alternative to regular meat (Remmers personal communication 2004). Among environmentalists there is also a dislike of both industrial foods and the possibility of genetic modification of Novel Protein Foods (Engels interview 2004). However, vegetarian foods are occasionally taken into account by the organisation for Nature and Environment (NM), which has included meat alternatives in its activities towards supermarkets. This includes the 'dialogue with supermarkets on corporate social responsibility' (Remmers 2004: 6, 42-43). In this activity several public interest groups have been involved, including animal welfare groups, fair trade organisations and Oxfam Netherlands (also known as Novib).

Finally, involvement in the TvC meetings has resulted in increased attention to meat alternatives in an environmental organisation dealing with environmental campaigns targeting the general public (Loeber 2004: 192). A few years later this organisation became part of Milieu Centraal, an independent organisation providing environmental information to the public, which still provides information on meat alternatives and their environmental impact.

5.5 Further analysis

5.5.1 Network formation

In the previous section a considerable number of activities have been described dealing with the topic of NPFs or meat alternatives or that could be related to the NPF backcasting experiment. In this section these activities are analysed in terms of network formation. This analysis uses the indicators (1) activities, (2) actors and (3) resources and takes the activities as starting points to map the actors involved and the resources mobilised (especially funds). Results are summarised in Table 5.6 and grouped in the four domains distinguished: the research domain, the business domain, the government domain and the public domain, which involves both the general public and public interest groups.

Table 5.6 shows that in all four domains NPF-related activities were found. The most extensive activities in terms of resources were found in the research and business domains. They include R&D at or commissioned by companies, as well as public research. A landmark was the Profetas programme, which involved more than 50 people, including fifteen fulltime researchers. The financial resources of the research activities were provided mainly by ministries, by government innovation programmes like BTS and AKK, or by research councils such as NWO, MaG and STW, whose budgets are also provided by the government, However, there was also substantial co-funding by companies.

The table shows that the number of (product) development and production investment activities that were fully financed by companies is limited. This type of activities includes 'regular' product development and product introductions by Boekos, regular production and product development activities of Planet Green, Campina's investments in product development and marketing of Valess, and the product development trajectory of the mould protein ingredient by DSM food specialties and an unknown partner downstream. Furthermore, follow-up activities by the parties involved in testing meat alternatives in catering are likely to be fully covered by their own investments as well. However, when companies decide about production and market introduction, large budgets are reserved.

A major contribution of the government thus consists of funding research through, for instance, innovation programmes, 'special' funding like the contribution of the Ministry of Agriculture to the Profetas programme, or academic research councils. Further activities in the government and public domains, with the aim of stimulating present meat alternatives are considerably more modest in terms of resources. However, the Ministry of the Environment organised quite a number of activities and financed NPF-related activities by public interest groups. Although budgets are considerably more modest compared to the research and innovation budgets, these activities may have (had) a considerable direct and indirect impact among public interest groups and the public. For instance, they may have contributed to supplying adequate information to public interest groups, tackling the concerns regarding GMO at these organisations, raising awareness among the general public, or stimulating the demand side. Besides, meat alternatives have become important at organisations providing environmental information to consumers, like Milieu Centraal.

Each activity comprises a temporary or more structural set of actors who are involved in the activity, constitute that activity and in many cases have been able to mobilise the resources for the activity. When there are successful joint applications for (external) funding, then actors also have contracts that tie them to the activity and each other. The temporary network of the NPF backcasting experiment has thus evolved into several networks surrounding particular activities for which resources have been mobilised successfully. It has clearly not evolved into a single large NPF network in the Netherlands, but into a number of decentralised networks that are sometimes connected, and sometimes not, or just a little. Most activities started

Table 5.6 *Networks around NPF-related activities*

	Activities	Actors	Resources
Research domain	*Profetas Programme (1999 – 2004)*	*Three universities (VU, Twente, WUR), research schools VLAG and SENSE, WCFS, several research institutes, five companies (Boekos, Cebeco, Quest, DSM, Unilever Research)*	*Funding by NWO board, STW, MaG, LNV, WCFS, participating companies, totally around 3 Million Dutch guilders*
	TNO research programme on the future of protein foods	*TNO Nutrition and various companies*	*Funding by TNO, companies, innovation programme funding*
	R&D project on mould based protein food ingredients (1997–2001)	*TNO Nutrition, DSM Food Specialties (formerly Gist-brocades), Boekos*	*Funding by BTS governmental innovation programme, company investments*
	WUR strategic research programme (2002-2006)	*WUR (former DLO part), WUR research institutes*	*Internal budgets from the central organisation of WUR*
	Development of programme proposal on sustainable kitchens and food chains (2000)	*Boekos, TNO-Kathalys, TU Delft, WUR, Kleine Aarde and various others*	*Proposal for € 450,000 was submitted to NIDO contest but not approved for funding*
Business domain	*Product development of mould-based protein food (ongoing follow-up of the BTS funded R&D project)*	*DSM Food Specialties (formerly Gist-brocades), unknown successor of Boekos*	*Company investments*
	NPFs and meat alternatives in catering chains (including product development and catering tests)	*Sodexho, Deli XL, Schouten Europe, LEI, KDO Consultancy, test facilities at seven organisations*	*Funding by AKK, Novem and companies, totally several million euros for several related projects*
	Product development at Boekos	*Boekos, probably customers*	*At least partly internal investments*
	Workshop and proposal on development domestic appliances for NPFs (2000)	*Boekos, TNO, Philips DAP*	*Proposal submitted to EET programme, but not approved*
	R&D projects on extracting proteins from lucerne and grass (1997-2002)	*Avebe*	*Internal funding and probably other sources*
	Development and market introduction of Valess (2005)	*Campina*	*Internal funding, € 3 million alone for the marketing campaign*
	Establishment of Planet Green and Aurelia Marketing & Research	*Both companies involved the same entrepreneur, Planet Green used various specialised suppliers*	*Payments by customers*
Government domain	*Various activities at the Ministry of the Environment stimulating vegetarian foods on the short term*	*Ministry of the Environment, environmental organisations, research contractors*	*Internal capacity, existing policy programmes, allocation of budgets at the ministry*
	Funding Profetas (1999)	*Ministry of Agriculture*	*Internal budgets at the ministry*
	Transition monitoring research using NPF as case study (2002)	*MNP, Ministry of the Environment*	*Regular and project based governmental funding*
Pubic domain	*V-day & other public oriented activities, symposiums*	*NVB, Aurelia, producers, environmental organisations, animal welfare organisations*	*Partly funded by the Ministry of the Environment*
	Dialogue on CSR in supermarket using meat alternatives as a vehicle	*NM, supermarkets, other NGOs*	*Internal/ external sponsoring*
	Incorporating meat alternatives in public environmental campaigns	*Milieu Centraal*	*Internal budget, provided mainly by government*
	Monitoring meat and fish alternatives in supermarkets	*NVB, Aurelia, animal welfare groups*	*Unknown*

as alliances preparing proposals and mobilising actors and funds. In several cases this did not succeed and the attempted activities were abandoned. In other cases activities were abandoned by stakeholders because expectations were not met, or due to strategy changes.

It is possible to identify network clusters that consist of several activities that overlap in actors or show similarity in type and scope of activities, although not all activities can easily be positioned in such clusters. The clusters that have been identified to a large extent adhere to the four domains, albeit not completely. A first cluster is the research network around Profetas, which also involves a number of public and private financing parties. Although this activity was also supported (and co-funded) by companies, there has been less interaction and overlap than in the NPF backcasting experiment and less than initially strived for. In fact, the Profetas programme was essentially an academic research endeavour. Nevertheless, the companies involved considered the Profetas programme a serious endeavour that directly and indirectly supported and legitimised their activities in the field of NPFs and meat alternatives.

Secondly, there is a business cluster involving R&D in large food companies in the Netherlands. Although this cluster includes R&D activities at Avebe, Boekos, Campina and DSM Food Specialties, their activities are not closely related, but have similar scopes. This cluster also includes testing NPFs and meat alternatives in the catering and related product development in which the catering company Sodexho has the lead. This activity consists of various related projects and involves a medium-sized producer, a research institute and a consultancy.

Thirdly, there is a cluster of SMEs involved in the production of meat alternatives, which includes the activities of Planet Green. This cluster does not seem very much influenced by the NPF backcasting experiment, but they do 'their own regular thing' in a gradually growing market.

A fourth cluster concerns the activities by the Ministry of the Environment, which consists of various policy-making activities, stimulating dialogue on the topic of meat alternatives and related funding of activities by public interest groups. This cluster is strongly related to the various activities by public interest groups. These are, to a certain extent facilitated and funded by the Ministry of the Environment, while their constituting networks overlap and are rather strongly connected.

Compared to original backcasting experiment, the total numbers of actors and individuals, involved in follow-up and spin-off activities has increased (see Table 5.6), as has the total amount resources that was mobilised for NPF-related activities. In addition, the number of activities has grown considerably, from the NPF backcasting experiment to a range of activities in the four domains distinguished. Most activities also fit in the follow-up agenda proposed at the end of the NPF project (see Table 5.2), although remarkably enough this is not due to central(ised) coordination. The activities emerged in a decentralised way. Regarding financial resources, the vast majority of funds originated from public sources like institutional academic funding organisations like NWO, MaG and STW, and several ministries.

5.5.2 Vision aspects

Vision: guidance

Guidance of the vision is analysed in terms of (1) collective projection, (2) synchronisation and alignment of the interaction among stakeholders from different backgrounds and different scientific disciplines, (3) alternative (system of) rules and institutions.

Collective normative projection

The core of the future vision on Novel Protein Foods could be found among many of those involved in identified follow-up and spin-off activities. The core included the business opportunities of NPFs and meeting consumer demands regarding taste, texture and other characteristics using new technology, and that NPFs and improved meat alternatives will play a significant role in future diets. This was also articulated by stakeholders and individuals who were new entrants and had not been actively involved in the backcasting experiment. In the case of the Profetas programme these included research groups, some of the companies and academic research councils.

By contrast, in most follow-up and spin-off activities the collective projection has evolved, resulting in adjustments, due to ongoing insights, inputs by new actors, or by adding domain or organisation specific elements. For instance, in the Profetas programme, the future of NPFs was considered to be on a global scale, including global protein security in the long term, while the envisaged system innovation was also connected to transitions to future sustainable fresh water supply and future sustainable energy supply on a global scale. The future vision was thus broadened in scope and in topics, making it more complicated, but also adding world protein security as an additional future goal. The ambition of 40% replacement of meat in the Netherlands, which was already challenged during the NPF backcasting experiment, was abandoned in the Profetas programme. The emphasis was put on crop-based NPFs using green peas as a model, thus leaving NPFs from microbial sources outside the scope of the programme.

Adjustments of the future vision on NPF also took place in other activities and their constituting networks. At companies, it became part of a business orientation, thus focusing on perceived product opportunities in the market. At the Ministry of the Environment, the long-term and abstract future vision was considered not to be adequate for short-term policy-making that should target current consumption, products and production and the existing stakeholder target group meetings on food. As a result, the vision was reformulated into facilitating vegetarian foods and new vegetarian food products on the short term, raising consumer awareness, stimulating the demand side and dealing with potential barriers, such as the GMO concerns among environmentalists. Vegetarian organisations and animal welfare organisations were attracted to the vision of NPFs, because it matched their future visions of a meat-free, animal friendly society.

Synchronisation

Did the vision still lead to synchronisation and alignment among the actors from the various domains and disciplines involved in the different networks, and was synchronisation affected the adjustments in the vision in different clusters? The answer to the first part of the question is an affirmative one; in each of the activities or clusters of activities synchronisation and alignment between actors from different disciplines and domains has been achieved, which can, for instance, be seen in the multi- and inter-disciplinarity of the Profetas programme.

Regarding the influence of adjustments, the changes in the future vision in the Profetas programme were stimulated by the research groups involved. These changes made the vision also more attractive for these groups to become involved and extended the number of views and disciplines involved in the programme. Another example is that NPF ingredients were connected by the technologists to the novel approach of consumer oriented food product development. This was an emerging research topic at WUR and fitted nicely to the radically new food products containing NPFs. It can thus be said that within this particular activity synchronisation over all these actors was established and resulted in adjusting of the future vision. Could these changes that enhanced synchronisation across various scientific disciplines have affected the synchronisation with other actors, for instance companies? Involvement of companies in the Profetas programme was rather low in terms of the number of interactions, while some changes in the vision may have been unfavourable for synchronisation with this group of actors. For instance, companies considered the knowledge fundamental and not yet suitable for product development (Van Egmond & Sein interview 2004), or they considered peas a less interesting source (Zwijgers interview 2004).

Also, the adjustment in the collective normative projection at the Ministry of VROM enabled to synchronise across the target stakeholders groups of the appropriate unit at the ministry, which were food companies, environmental organisations, and organised vegetarians and animal welfare activists. This matched the short-term policy at the ministry with regard to sustainable food consumption, as well as the short-term orientation of their stakeholder target group. If the long-term projection had not been adjusted, it would have been more difficult to achieve synchronisation and alignment with those stakeholders. Regarding the pilot projects in the catering sector, these may have been inspired by the long term collective normative NPF vision, but the short-term goals set are probably equally important to achieve synchronisation and alignment for successfully conducting these activities.

Alternative rules

Changes in the future vision also affected the alternative rules (system) embodied in the future vision. For instance, narrowing the range of sources to green peas has huge consequences for the supply system and the players in such a system. When moving the issue to a global level to include oil-containing crops, many new actors get involved and other existing rules, like WTO trade agreements, are challenged and may need to be altered. Another example concerns the meat alternative chains that were tested in the catering company Sodexho, as this required new agreements, contracts, etc.

Vision: orientation

Orientation of the future vision can be defined in terms of (1) cognitive activation, (2) mobilisation of actors and resources, and (3) presence of daily decentralised coordination.

Cognitive activation

Both the core of future vision and the adjustments in different activities and supporting networks provided cognitive challenges and facilitated cognitive action towards the vision in research, policy-making, business and public interest groups. This is not only about generating new knowledge through research as took place in the Profetas programme and in other R&D projects. It is also about realising opportunities and developing them cognitively in such a way that they can support further action by actors in all domains. For instance, after adjustment of the NPF vision at the Ministry of the Environment several studies were commissioned leading to policy recommendations, which were (partly) further elaborated and put into action. The R&D projects in companies also required making business plans, cost price calculations, etc. All these activities require cognitive activities, in addition to other types of activities.

Mobilisation

Both the generic future vision and its network-specific varieties were used successfully to mobilise actors and resources, as is shown Table 5.6. For instance, the number of companies participating in the Profetas programme had initially increased to five (while there were three in the NPF project). The number of research groups increased as well, while there was also a broadening in disciplines. Furthermore, the Profetas research alliance successfully mobilised resources from various research councils and other sources, though it took considerable efforts and time.

Initiators of other activities were also capable of using the future vision or adjusting the future visions for mobilising actors and financial as well as other types of resources. For instance, the adjustments at the Ministry of the Environment mobilised a range of public interest groups, while the available resources provided by the ministry enabled that these groups extended their activities on this topic. Not all attempts were successful. Sometimes, actors were mobilised and formed alliances that submitted proposals to get access to funding but did not succeed. This was, for instance, the case with the proposal to the NIDO programme contest by an alliance led by Boekos, or the proposal championed by TNO Kathalys on domestic appliances for NPFs. On average, however, the numbers of actors, as well as the amount of resources have considerably increased.

Mobilisation took place, for instance, through the involvement of leading researchers in the early stages of the Profetas programme who were also involved in the NPF backcasting experiment. The chair of the STD programme also proposed NPF as an interesting topic within his network, for instance when he was involved in the discussion on multidisciplinary research at NWO. With regard to most activities, it is possible to identify individuals or stakeholders (or both) that were both involved in the NPF backcasting experiment and were important for initiating follow-up activities and to attract new participants and resources. All of them used the future vision to mobilise resources and actors.

Decentralised coordination

Did the future vision provide daily coordination in the different networks? A conclusion could be that this was indeed the case, as it provided sufficient stability to establish the activities summarised in Table 5.6 and to mobilise sufficient resources and actors to realise them. However, there was also a dynamic at work in the content and framing of the future vision resulting in several adjusted visions that had the same core but were also a bit different, while there were exits and entries of actors in the various networks. It may be that the precise content of the future vision mirrors in a way also the constellation of the network. In that case the future vision must have enough flexibility to survive these adjustments and to enable entries, which must be followed to provide sufficient stability for the ongoing activity. Furthermore, dynamics can also be created by exits and entries when activities are going on, like Cebeco, Boekos and Quest leaving the Profetas programme. However, this did not affect the ongoing activities, but may be of influence when such a major activity as Profetas is completed and actors and resources again need to be mobilised. Another example is the policy topic of nutrition that was transferred from the Ministry of the Environment to the Ministry of Agriculture, and the changing viewpoint at the Ministry of Agriculture. Such changes might affect the stability of ongoing activities and networks.

Competing visions

The dominant vision of regular meat production and consumption is still strongly present, although the environmental burden and animal welfare issues are ongoing concerns.

Several alternative and possibly competing visions could be traced. Firstly, there is the competing vision of organic meat and eating less meat that is supported by a large part of the environmental movement. This vision is often connected to a dislike of industrialised foods, which may include NPFs. On the other hand, activities by environmental organisations sometimes promote existing meat alternatives as sustainable alternatives, thus showing at least congruence with the vision on NPFs. Furthermore, a vision of a meat-free society was supported among vegetarian and animal welfare organisations. As the vision on NPFs can be seen as a step towards this further-reaching vision, it was judged positively by the vegetarian NVB.

Stakeholders who supported particular visions could not only resist a competing vision, they could also learn from it and adjust their vision, or even combine visions or look for joint elements and joint goals, which has been named congruence. This allows cooperation and coalitions among stakeholders with different value and belief systems from different societal groups. Such congruence has not actively been looked for in this study, but several examples have been found, such as the cooperation among vegetarians, animal welfare groups and environmental organisations.

An interesting synthesis of various visions was given by Aiking *et al.* (2006: 211) reporting on the results of the Profetas programme: *"What consumers should ideally do in the context of a protein transition may boil down to (a) eating one third less protein (the average Dutch over-consumption), (b) replacing one third by plant-derived proteins, and (c) replacing the remaining third by extensively produced meat (such as most beef and lamb)"*. This can be seen as a cognitive synthesis and has not been the outcome of social process of deliberation and learning by stakeholders. However, it integrates several existing visions identified in this research, and may be able to compete seriously with the regular dominant vision on meat consumption and production, which favours system optimisation instead of dealing with its global sustainability impacts in a more far-reaching way.

5.5.3 Institutionalisation and institutional resistance

A distinction is made between institutionalisation and institutional resistance. Institutionalisation has been defined in Chapter 4 as the process by which a practice becomes socially accepted as 'right' or 'proper', or comes to be viewed as the only conceivable reality (cf Oliver 1996: 166). Institutional resistance has been defined as the resistance to changes by existing institutions and the actors backing them (see also Chapter 4).

Institutionalisation

A clear example of institutionalisation is the fact that the Profetas programme had a broader effect on the food industry and food knowledge infrastructure in the Netherlands (Van Boekel interview 2005), while generating such a knowledge base is accompanied by processes of institutionalisation among the actors involved as well in the acting networks. At WUR, NPF-related topics have become part of various other research programmes, which points to further institutionalisation. Institutionalisation did not only concern NPF-related knowledge and technological results. As a result of this, companies were interested in the scientific results and the potential business opportunities (Vereijken interview 2004). It was probably also the reason why Campina wanted to become a player in the field of meat alternatives (Van Boekel interview 2005). Institutionalisation also refers to research practices; within WUR the Profetas programme was regarded as an interesting and positive multidisciplinary research effort involving technical, social, economic and environmental sciences within a single programme (Vereijken interview 2004, Van Boekel interview 2005). A related example of institutionalisation is the fact that the food industry considered Profetas a positive example of the emerging approach of consumer-oriented food product development (Van Boekel interview 2005). The research and product development activities at DSM Food Specialties and an unknown customer goes along with institutionalisation among the actors involved, which is also the case with the development and pilot cooperation by Sodexho and the producer Schouten Europe.

Meat alternatives in the sense of products have become more important at various public interest groups, such as Milieu Centraal, the vegetarian NVB and animal welfare groups, which can also be seen as a form of institutionalisation, as well as the gradual growth in the consumption of meat alternatives. Another example is that meat alternatives are increasingly seen as protein foods by the Product Boards for Livestock, Meat and Eggs (PVE), and included in their consumer surveys on meat and meat products.

Institutional resistance

Examples of institutional resistance can also be found in the case results. For instance, the Ministry of Agriculture has shifted its position with regard to NPFs. It wants to leave the topic to market parties, arguing that there is a niche market and sufficient fundamental knowledge is available for product development trajectories funded by companies. This changed view also seems fuelled by a wider concern at the ministry that meat and meat products are economically important and that competing meat alternatives should not be nurtured too much, which points to some kind of institutional resistance. A different example is the uneasiness of Unilever about focusing on the pea in the Profetas programme, which may partly be explained by institutional resistance at Unilever against sources other than soy, because the company has such a large knowledge base about and experience with processing soy.

5.5.4 External factors

So far, this chapter has focused on the emergence, diffusion and institutionalisation of NPF-related activities, using the backcasting experiment as a starting point. However, the decision by individuals or stakeholders to start or to join a particular activity generally speaking is also influenced by other factors as well. As an additional check this section deals with external events and developments in the context of the activities identified that may have influenced the emergence of the NPF-related activities, which can have been both constraining and enabling. A distinction is drawn between external factors that have been of immediate influence and broader ongoing developments in the socio-technical system of (protein) food production and consumption that can have a more diffused influence. Ongoing developments and events have thus been looked for, mainly using the interviews as sources and focusing on the immediate context of the NPF-related activities in the socio-technical system on (protein) foods.

Results are presented in Table 5.7 structured by the checklist proposed in Chapter 4. The question is whether and how these developments and events have served as external factors to the NPF-related follow-up and spin-off activities, and have influenced their emergence and shape. The question is not if these developments have the potential to influence, because they all have, but whether they have been relevant and taken into account by those involved.

For instance, the EU regulation on Novel Foods has influenced the R&D projects on the mould-derived protein ingredient at DSM, as it enhances costs and development times. Although this is a constraining factor increasing costs and development times, it has not blocked the R&D project at DSM Food specialties. The supermarket war has been another constraining external factor in the short term, as it was the major reason for postponing the market introduction of products based on this novel ingredient. The supermarket war has led to the exit of Planet Green as an independent company[24].

The specialisation in the food industry in either end-foods or food ingredients both enables and constrains cooperation. This would explain why Unilever was not very positive about the choice that was made in Profetas for the green pea, as there was not a major international supplier of pea-derived proteins involved. It also explains why the company would favour soy-based products, as is it has huge experience and established contacts with suppliers. Ongoing concerns about genetic modification in soy crops have led to a decrease in the use of soy protein in regular meat alternatives (Engels interview 2004). This has affected existing producers and the ingredients in existing meat alternatives, although it was also targeted with by the Ministry of the Environment when it organised a dialogue project on these issues, targeting the environmental movement.

The steady market growth to some extent seems to be a relevant factor in the NPF-related activities. For instance, it supported the start-up and expansion of Planet Green and it likely influenced decision-making

Table 5.7 *Possible external factors of the NPF case*

Shocks and crises
> Several outbreaks of livestock and chicken diseases temporarily affecting level of meat consumption and enhancing concerns about animal welfare and health aspects of foods
> Increasing novel food regulation in the EU constraining development and introduction of protein food ingredients from novel sources

Market and demand side changes
> Gradual growth in consumption of meat alternatives in the Netherlands, annually with 10-15%, although its share is still below 5%. Gradual growth of organic foods including organic meat
> Supermarket 'war' starting late 2003 leading to lower prices and revenues and reinforcing the trend towards own brands by supermarkets
> Steady growth in catering services, and to a lesser extent in eating out

Knowledge and technological breakthroughs
> Meat alternative from dairy proteins developed by Campina resulting in a meat alternative of superior quality.

Entries and exits
> SME producers of meat alternatives disappear, or are taken over
> Food multinationals, such as Heinz, Nestle, and Kraft include meat alternatives in their portfolio, often by take-overs of international medium-sized companies or brands, such as Tivall, Quorn and Linda McCartney
> Soy multinationals like ADM, Cargill, and DuPont joint venture Solé/ PTI develop soy food product lines based on health claims requiring considerable daily intakes of soy protein[ii]

New practices
> Among food multinationals specialisation in either end-products or crop processing and food ingredients supply (since mid-1990s)
> Shift towards consumer-oriented food technology and product in research bodies and large food firms
> Ongoing concerns about GMO in foods; increase of GMO free chains
> Health-based arguments in food innovation and food marketing, such as obesity
> Adding value to foods and food ingredients, due to market saturation in European countries (e.g. by using health claims or adding healthy additives)
> Corporate Social Responsibility (CSR) and licence to operate are of increasing importance in food industry

International
> Growing concerns about increasing global meat consumption (e.g. in China and Brazil) and possible increase in meat prices (as is already the case in parts of China)
> Growing concerns about decreasing global protein security for instance at the Food and Agricultural Organisation (FAO) of the United Nations

i   *Although Unilever seems to be a major exception, the company positions meat alternatives on a longer time horizon, as it assumes it requires more radical and more time taking innovations.*

ii *In the USA soy-based health claims have been approved by the authorities. This has resulted in many product development activities and rather aggressive marketing campaigns targeting meat consumption by soy multinationals like Cargill, Dupont (Sole/PTI) and ADM. These health claims are presently also under consideration in the EU.*

at Campina regarding the Valess product line. However, here is also contingence at play, as an independent inventor sold his patent to Campina.

On average, the influence of the external factors on the emergence and performance of the identified activities seems rather modest, although it is very likely that they may exert a greater influence in the near future when decisions have to be made with respect to R&D projects on NPFs and new NPF product introductions in the market. The soy-based health claims that have been approved by the authorities in the USA have resulted in many product development activities and rather aggressive marketing campaigns targeting meat consumption by soy multinationals like Cargill, Dupont (Sole/PTI) and ADM. These health claims are presently also under consideration in the EU, but have not affected marketing strategies in the EU yet.

The question is then what other factors may have influenced the emergence of the identified follow-up and were considered significant by insiders. For instance, the Profetas programme would not have become possible in its final shape and extension without the ambition of the General Board of NWO to initiate a multidisciplinary research programme on sustainability, as well as the willingness of MaG and STW to cooperate on this issue. Furthermore, the research schools and research groups involved saw research opportunities that matched their mission and ambition. For instance, WUR aimed at incorporating environmental issues and sustainability aspects into its research, while IVM was interested in extending its activities in multi-disciplinary approaches on agriculture and food. Food technologists at WUR also saw an opportunity to apply and further develop their novel approach on consumer-oriented (food) technology development (Van Boekel interview 2005), while IVM was interested in extending its research into food and agriculture (Aiking interview 2004).

In addition, several other technology policy and innovation funding programmes have been relevant factors in establishing NPF-related follow-up. For instance, the R&D cooperation between DSM Food Specialties, Boekos and TNO Nutrition was financed from the BTS innovation programme. Various activities in which catering company Sodexho, wholesaler Deli XL, and producer of meat alternatives Schouten were involved were (partly) funded by SenterNovem and AKK. Also, the Ministry of the Environment used a funding programme for public interest groups in cases where these groups could submit for funding activities.

Finally, strategic changes, often due to mergers or take-overs within business, appeared to be decisive factors to leave or finish activities, such as in case of Boekos, Cebeco, Quest and Avebe.

5.6 Conclusions

5.6.1 Backcasting experiment

The NPF backcasting experiment has resulted in a normative future vision and in involvement and commitment among a wide range of relevant stakeholders, which also led to mobilising sufficient resources from different sources. In addition, it has led to higher order learning among participating stakeholders with regard to the problem definitions, alternative solutions, and shifting principal approaches and priorities. Among the organisers of the backcasting experiment learning took place on how to apply and improve the backcasting approach.

The future vision has gradually developed during the course of the backcasting experiment. It also had different functions at the same time, such as a tool for integration and analysis, or as a shared vision providing guidance and orientation to the backcasting experiment and its activities. The vision offered guidance in the sense that it served as a collective normative projection to the future, allowed for synchronisation and alignment among researchers from different disciplines and among participating stakeholders from different domains. Alternative rules were also derived from the future vision and guided activities and discussions. The future vision also provided orientation that facilitated and stimulated cognitive activities in research and non-research topics. The vision was also used to mobilise (financial) resources and to attract stakeholders that could enter the backcasting experiment at several moments, or could remain engaged at a lower level of intensity. The vision in combination with the project structure and the available resources also provided daily coordination among the temporary network.

With regard to the methodological aspects, the NPF backcasting experiment did not follow the exact steps as defined in the methodological backcasting framework. However, all steps could be identified in the course of the backcasting experiment and its activities. Methods from all four categories in the backcasting framework could be identified, as could the three different types of demands. It appeared that two persons acted as a vision champion, though in different domains and partly in different stages of the backcasting experiment. Institutional protection was provided by top level management in the government and relevant companies.

5.6.2 Follow-up and spin-off

Significant follow-up and spin-off associated with the topic of NPFs have been identified in all four domains: the research, business, government and public domains. Follow-up and spin-off include activities that were initiated by stakeholders who had been actively involved in the backcasting experiment, mostly together with new stakeholders. Examples include the Profetas research programme, R&D activities at several companies, and the policy related activities at the Ministry of the Environment. Follow-up and spin-off also included activities initiated by stakeholders who were inspired in other ways by the outcomes and the future vision and who had not actively participated in the backcasting experiment and activities that were already taking place or were being initiated, which took on board minor or major elements from the NPF backcasting experiment. Examples of this category include the establishment of Aurelia market research and the SME Planet Green and activities by vegetarian and environmental organisations, as well as the development and market introduction of Valess.

The activities that have been identified can be framed as networks, consisting of activities, constituted by actors and enabled by mobilised resources. Several separate and decentralised networks have been found around the identified activities, while several clusters of networks could be determined as well. Furthermore, the volume and number of resources have increased since the backcasting experiment. The numbers of actors involved and the degree of their involvement in the activities have increased as well, although several stakeholders left or finished NPF-related activities. The vast majority of resources has been provided by academic research councils and ministries. Activities funded by other sources include the development and marketing investments by Campina on the Valess product line and the product development on a mould protein ingredient by DSM Food Specialties and an unknown customer downstream. The latter project was preceded by an R&D project that was funded by a government innovation programme.

The vision has been adjusted and to some extent redefined during most activities. For instance, in the Profetas programme it was redefined from a national setting to a global setting. This was partly because of the entrance of new actors, while the adjusted vision was widely supported by major stakeholders. However, a stable core of the future vision has remained that assumes that sustainable non-meat protein foods will become important in people's future diets. The decentralised nature of the identified activities makes it possible for varieties of the future vision to be at play in different clusters of activities. This points to decentralised (daily) coordination across clusters of activities and across domain by the vision in its various adjusted versions. It is also likely that the adjustments result in a better fit with the belief systems of the stakeholders involved and thus can provide guidance and orientation in a better way. As a consequence, adjustment may be a condition for further diffusion of the vision into series of new activities in line with the vision. In other words, new entrants learn from and are influenced by the vision, but also add their interpretations, perceptions and priorities and influence the vision.

The follow-up and spin-off that were found have come along with wider effects and instances of institutionalisation, while also examples of institutional resistance have been found.

External factors are often important and can both enable and constrain the emergence and extension of follow-up and spin-off of the backcasting experiment. For instance, the general board of NWO had the ambition to establish multidisciplinary research programmes, which got connected to the initiators of the Profetas programme and eventually led to funding by NWO. Constraining and enabling developments in the socio-technical system under study have been identified too, such as the enhanced competition among supermarkets in the Netherlands, the EU regulation on Novel Foods, the growth in the segment of vegetarian foods, and concerns about GMO and industrialised foods.

A key mechanism in establishing follow-up activities starts when participating stakeholders see opportunities in the future vision enabling behavioural alternatives. These stakeholders are represented by individuals who either need support from higher levels in their organisation or personally have the capabilities to initiate activities in their own organisation. In many cases cooperation evolved around specific activities or proposals for activities (and raising funds for these through other external channels). This happened, for instance, in case of the Profetas programme, the policy-related activities at the Ministry of the Environment and the public interest groups. Interestingly, not all these people were strongly involved in the NPF backcasting experiment. Sometimes, stakeholders got acquainted with the backcasting experiment and its outcomes by 'regular' dissemination activities or by persons who had been involved in the backcasting experiment.

A possible threat can emerge when government-funded activities are completed and the funding actors want, for instance, to leave it to market parties. It can therefore not be determined if established activities and their networks, for instance around the Profetas programme, will continue or will disappear. This was a major issue during the interviews late 2004. By mid-2006, there is still no certainty about this, though the Ministry of Agriculture has taken the stand that a protein transition cannot be dictated by the government (Minister of Agriculture 2006).

5.6.3 Wider effects and system innovation aspects

A final question is to what extent the follow-up and other effects are the signs of an emerging system innovation. The present activities are still at a niche level, although networks and alliances are growing. Also, a knowledge base has been established in the innovation system that may eventually result help turn the generated knowledge into products. The consumption of meat alternatives is slowly, but steadily growing in the Netherlands. One signal is the recently launched meat alternative by Campina that is based on dairy proteins. This points to both a growing awareness of the opportunities of NPFs and changes in the rules of the existing socio-technical system. All these signals can be seen as seeds for a system innovation. They relate to the initial phases of the conceptual perspectives proposed by Rotmans et al. (2001) and Geels (2005).

Thus, despite significant follow-up and spin-off of the NPF backcasting experiment in different domains and involving various actors from those domains, there are no signs of an imminent system innovation towards sustainability yet. There are potential future drivers, like the dynamics among food multinationals, a possible trend of increasing meat prices over 10 to 15 years and an increasing protein deficiency in other parts of the world. Although these factors may become decisive, they have not become so yet, while the vested interests of the meat sector still play an important role as well. Therefore, more time is needed before it can be evaluated if the participatory backcasting experiment on NPFs in the Netherlands will have contributed to a system innovation towards sustainability in which meat alternatives and NPFs have substituted a considerable share of the consumption and production of meat and meat products. Finally, it may also be necessary to initiate additional facilitation in order to make a next step. But if and when that will happen, who will have to take the lead?

Notes

1   The IPAT formula has been explained in Chapter 1. The assumptions made at the Ministry were that over 50 years the population will have grown by 50 to 100% (factor 1.5 to 2), that global wealth would have increased four to eight times (which relates to an annual economic growth of the global GDP between 2.5% and 4.5%) and that the total global environmental impact would be kept at the same level or be reduced by 50% (Vergragt and Jansen 1992: 23). From these estimates it can be determined that the increase in eco-efficiency must be between 10 and 30 times.

2   These were, using their official names at that time, the Ministry of Agriculture, Nature Management and Fisheries (LNV), the Ministry of Economic Affairs (EZ), the Ministry of Education, Culture and Science (OCW), the Ministry of Transport and Public Works (V&W) and the Ministry of Housing, Spatial Planning and the Environment (VROM).

3   For instance, elsewhere reference was made to a factor 10 to 50, depending on potential and possibi-lities (Jansen and Vergragt 1992: 28, Jansen and Vergragt 1993: 179).

4   However, it took more than half of a century and the Second World War before the transition from butter to margarine really took place in the Netherlands (Aiking et al 2006: 7, Stuijvenberg 1969). For a comprehensive account of the development and rise of margarine and the margarine industry, see Stuijvenberg (1969).

5   This idea has been abandoned after the NPF feasibility study, because of the embryonic nature of this technology and its limited potential for meeting the factor 20 (ADL 1993).

6   In late 1990s after a term of five years, the NRLO was turned into the Innovation Network Green Space and Agrocluster. Next to the task for strategic foresighting for agriculture and food industry, it had to function as a breedery for system innovations too.

7   In addition to the project manager, the project team included an expert in the field of food processing and micro-organisms, a communication consultant and a project secretary.

8   More specifically, consumer research was carried out by SWOKA Institute for Consumer Studies. Environmental assessments were carried out by TNO-MEP and the Centre for Environmental Studies (CML) of Leiden University. TNO Nutrition, ATO-DLO and the Food Technology department of Wageningen University did the technology research; ATO is presently part of the Agrotechnology and Food Innovation Institute. LEI Institute for Agricultural Economics, which was also a DLO research institute, conducted research on business economics and structural aspects.

9   In Dutch, TvC stands for 'Toekomstbeelden voor Consumenten'. Its purpose is to enhance interaction in the early stages of technology development, on the one hand between social actors and consumers, and on the other hand producers, technology developers and researchers. This should ensure that consumer aspects, social concerns and the perspectives of future users are much better integrated in the early stages of the R&D process (Fonk 1994).

10   The expert trend analysis and extrapolation resulted in an estimated growth over the next 40 years of the meat products segment between 45% and 75%, while the share of meat would decrease from 55% to 25%. Based on the consumer study results and the results from the TvC meetings, it was estimated that NPFs

would, if quality and consumer demands would equal those of meat, would have the potential to substitute half of the meat products segment (sausages, minced meat, meat in assembled meals, etc), but nothing from the segment of steaks and associated meats.

11 *The seven NPF options covered a range of protein sources, which included pea, Lucerne, Spirulina (cyano bacterium), Fusarium (mould), pea fermented with Rhizopus (mould) and lupine fermented with Rhizopus.*

12 *Learning was investigated by asking the interviewees about major (positive and negative) learning experiences, both personally and with respect to their own organisation and other stakeholders.*

13 *Full consensus is often not necessary, provided sufficient congruence can be achieved.*

14 *By contrast, a more 'narrow' approach to design can be distinguished which includes applying formal(ised) design methodologies.*

15 *In Dutch VLAG stands for 'Voeding, Levensmiddelentechnologie, Agrobiotechnologie en Gezondheid'. It is part of Wageningen University and Research Centre (WUR) and strongly embedded in the Dutch food innovation system.*

16 *The official list is as follows (www.profetas.nl). Free University Amsterdam: the IVM Institute for Environmental Studies and the Centre for World Food Studies. University of Twente: the Department of Policy Science and Policy Analysis. Wageningen University: Integrated Food Technology, Food Chemistry, Human Nutrition and Epidemiology, the Laboratory of Plant Breeding, Environmental Systems Analysis, Environmental Economics and Theoretical Production Ecology. TNO Nutrition as well as the WUR institutes Agrotechnology and Food Innovations (AFI), LEI Agricultural Economics Research Institute and Plant Research International (PRI) were also involved.*

17 *WUR, Wageningen University and Research Centre is a holding, which consists of Wageningen Univer-sity and the former DLO research institutes. DLO was originally the research division of the Ministry of Agriculture in the Netherlands, but was placed outside the ministry and merged with Wageningen Agricul-tural University.*

18 *There is an interesting analogy with the TVP affair of Unilever, where TVP also ended as a substitute for milk in the diet of veal, as cow milk is for human consumption and is applied as an additive in, for instance, sausages, which improves juiciness (Dutilh interview 2004).*

19 *Campina had also been involved in the first TvC meeting of the NPF project, but no confirmation was found if there was a relationship with the development of Valess.*

20 *Valess consists of a line of novel meat substitutes in different varieties based on milk-derived curd. This consists of coagulated milk fibres of protein, which are turned into a matrix of sodium bounded by phosphate, to which methyl cellulose and protein from chicken egg is added. It was an invention of a Dutch independent cook who sold his patent to Campina. Researchers from Campina worked for four years, before it could be patented and introduced to the market (Scholtens 2005).*

21 *This interview was conducted as part of the household nutrition backcasting case (presented in Chapter 6).*

22 *During one of the interviews (Engels interview 2004) it was mentioned that halfway the 1990s producers of meat alternatives were looking for alternatives to soy proteins, because of the increase in GMO soy crops,*

23 *The current name is 'Goede Waar & Co', previously the name was 'Alternatieve Konsumtenbond'.*

24 *Since April 2005 Planet Green has also disappeared as an independent label and has been integrated in the existing vegetarian food label Vivera (www.planetgreen.nl, visited August 2, 2006). Vivera is a brand of the Enkco Food Group that produces both meat alternatives and meat products.*

BACKCASTING FOR A SUSTAINABLE FUTURE: THE IMPACT AFTER 10 YEARS

6

THE IMPACT OF BACKCASTING FOR SUSTAINABLE HOUSEHOLD NUTRITION

This chapter describes and analyses the Sustainable Household Nutrition (SHN) case, consisting of a participatory backcasting experiment in the Netherlands and its follow-up and spin-off after five years. This chapter introduces the SusHouse project and its international subproject on nutrition (6.1) and provides a description of the backcasting experiment (6.2), as well as an analysis (6.3). It also contains a description of related follow-up and spin-off activities (6.4), a further analysis of these activities (6.5) and conclusions (6.6).

6.1 Introduction

6.1.1 The Sustainable Household project

The EU-funded SusHouse (Strategies towards the Sustainable Household) project was concerned with developing and evaluating strategies for system innovations to sustainable households. It used stakeholder involvement, normative scenarios and backcasting. A key starting point of the SusHouse project was that a combination of technological, cultural and structural changes is necessary to achieve a factor 20 environmental gain in the next 50 years through system innovations, while taking into account both consumption and its interconnection with production through products and product usage. Another important starting point was to involve stakeholders in the process of redesigning the fulfilment of a household's needs in such a way as to make it compatible with the concept of sustainable development (Green and Vergragt 2002, Vergragt 2000, Quist *et al*. 2001a).

The SusHouse project was very much inspired by the STD programme (Weaver *et al*. 2000, see also Section 5.1) and it can be seen as a spin-off applying a similar backcasting approach on a different topic. Instead of focusing on supply chains and sustainable technologies like the STD programme did, the focus in the SusHouse project was on households and household consumption. This was motivated by the fact that the environmental impact of household consumption is huge (e.g. Noorman and Schoot Uiterkamp 1998), while households and consumers are also important actors in sustainable development through their decisions on how, what and when they consume (Quist *et al*. 2001a, Green and Vergragt 2002).

The methodology was explored earlier in the Sustainable Washing project at the STD programme (Vergragt and Van der Wel 1998), but it was further developed and tested during the SusHouse project. Another difference with the STD programme was that it would be applied at the same time in five European countries, while the emphasis in the STD programme was on the national context of the Netherlands (for a more detailed comparison, see Vergragt 2005). The SusHouse project was thus an international academic endeavour for which an international alliance of research groups received funding of around € 2 million from the EU DG 12 Environment and Climate RTD programme. Other major changes compared to the STD backcasting approach included that (Green and Vergragt 2002, Vergragt 2005):
- Less emphasis was put on technology as the main agent for sustainable development; instead, a combination of technological, social, and cultural changes was envisaged.

- Greater emphasis was put on the participation of non-governmental stakeholders.
- A design orientation was chosen, rather than a policy-making orientation.

Three household functions were studied (1) Clothing Care, (2) Shelter and (3) Nutrition[1], while six research groups[2] in five European countries were involved (Vergragt 2000). Each research group had specific expertise regarding a specific part of the overall methodology, such as stakeholder workshops, scenarios construction, or one of the three scenario assessments (see Table 6.1). Each household function was studied in a subproject by an international research group organising backcasting experiments in three countries. In a fourth subproject research methodologies were developed, coordinated and evaluated (see also Table 6.1). Conducting the research, like analysing scenarios, was part of the backcasting experiments. Each subproject included: (1) normative scenario construction for the sustainable fulfilment of functions endorsed by a range of social partners, (2) scenario assessments and (3) development of follow-up agendas including implementation proposals, policy recommendations, and research proposals, and (4) stakeholder cooperation on particular proposals (Vergragt 2000).

Table 6.1 *Overview of the SusHouse project (Vergragt 2000, Vergragt et al.: no date)*

Subproject	Countries
Nutrition (Shopping, Cooking and Eating)	Hungary, Netherlands, UK
Clothing Care	Germany, Italy, Netherlands
Shelter	Germany, Italy, UK
Methodology development and evaluation:	
> *Overall methodology and project design*	Netherlands (Delft University of Technology)
> *Stakeholder analysis*	Italy (Avanzi)
> *Workshop organisation*	Netherlands (Delft University of Technology)
> *Scenario construction*	Italy (Politecnico di Milano)
> *Environmental assessment*	Netherlands (Delft University of Technology)
> *Economic analysis*	UK (UMIST School of Management)
> *Consumer acceptance research*	Germany (University of Hanover)

6.1.2 Participatory backcasting

The project's overall methodology was as follows. For each household function studied in a country, a process of stakeholder identification was performed, covering stakeholders on the demand side, the supply side, research bodies, government and public interest groups. Invited stakeholders participated in creativity workshops that aimed at generating ideas for sustainable function fulfilment in the future and clustered these ideas into proto-scenarios. These proto-scenarios and other workshop results were used for further elaboration of the scenarios, which was done by the research teams, Next, these normative scenarios were assessed in terms of environmental gain, consumer acceptance (by means of focus groups) and economic credibility. They were also used for a scenario-specific second round of stakeholder identification. Subsequently, already involved and newly identified stakeholders were invited to a second workshop in which the scenarios and assessment results were discussed, after which implementation proposals, research agendas and policy recommendations for achieving the scenarios were developed. Backcasting techniques were applied in both series of workshop, as well as during scenario construction.

Figure 6.1 shows the overall SusHouse methodology existing of six steps including possible iterations.

The seventh step 'Realisation of implementation' was officially not part of the SusHouse project. This was due to the fact that the EU did not approve realisation of follow-up as a core objective for a research project and did not fund this type of activities (Vergragt interview 2005). However, the focus on implementation of follow-up was clearly present in internal discussions in the project and the aims of the second set of workshops in the project. A more extensive elaboration on the overall SusHouse methodology has been provided elsewhere by Vergragt (2000), Quist *et al.* (2000), Quist *et al.* (2001b), and Green and Vergragt (2002).

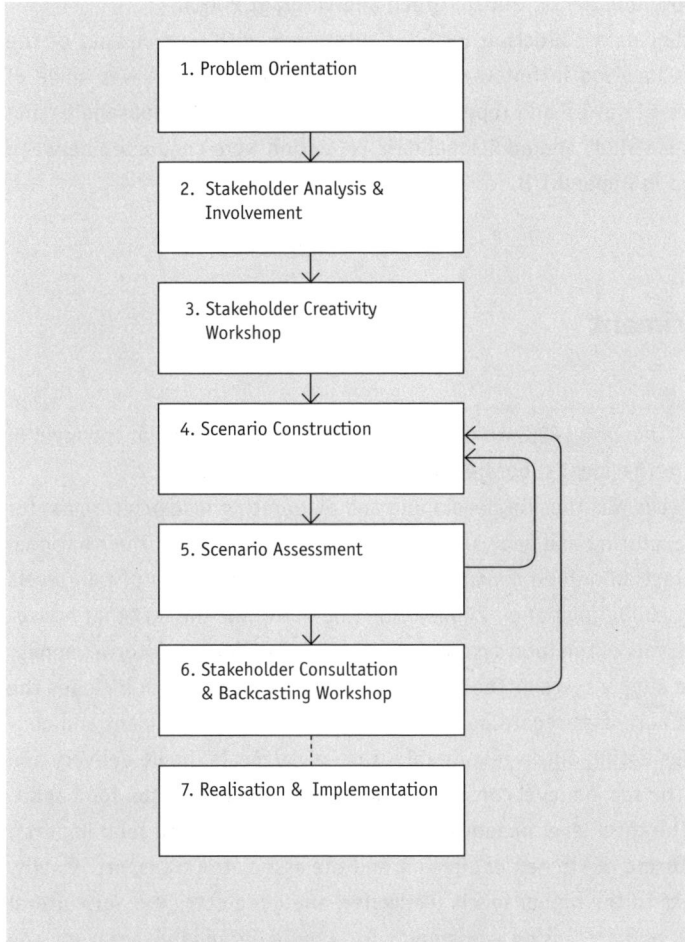

Figure 6.1 The SusHouse backcasting methodology

6.1.3 Nutrition from a household perspective

Shortly before the project was to start officially in January 1998, nutrition was selected as one of the household functions to be studied in the SusHouse project. Based on the available data, estimates had been made of the environmental impact of all household functions before selecting the three functions (Vergragt interview 2005). A range of studies has pointed both to sustainability problems in nutrition and food production (e.g. Heidemij *et al.* 1997, Slob *et al.* 1996, Kramer and Mol 1995, Uitdenbogerd *et al.* 1998, Van Gaasbeek *et al.* 2000, Kramer 2000, Lindeijer 2000, Van der Pijl and Krutwagen 2000, De Vries and Te Riele 2005), as well as to the environmental burden of households and product use (RMB 1996, Slob *et al.*

1996, Noorman and Schoot Uiterkamp 1998, Moll and Groot-Marcus 2002, Schmidt and Postma 1999, Vringer *et al.* 2001, SER 2003). The environmental burden cannot be separated from the way diet and consumption patterns evolved in the Netherlands (e.g. Van Otterloo 1990, Van Otterloo 2000).

The backcasting experiment on household nutrition in the Netherlands was reported in a case study report (Quist 2000a) and a report in Dutch (Quist 2001a), as well as in a number of background reports (these are listed in Quist 2000a). This backcasting experiment was part of the international nutrition subproject (Green and Young 2000, Young *et al.* 2001, Quist *et al.* 2002a) of the SusHouse project that took place between January 1998 and mid 2000 (Vergragt 2000, Quist *et al.* 2001a, Green and Vergragt 2002).

Building on the available data, further data collection included interviews with participants of the backcasting experiment and stakeholders involved in follow-up attempts. Furthermore, use was made of the internet, project documents, academic literature and reports dealing with sustainable households and household nutrition. Interviews for this case study and additional data collection were conducted between March and July 2005. Interviews are listed in Appendix B.

6.2 The SHN backcasting experiment

6.2.1 Process description

The backcasting experiment started in January 1998. An overview of the activities, both at the level of this backcasting experiment and at the international subproject level, is provided in Table 6.2.

In the early stage of the project the focus was thus on developing and elaborating research formats for the various research tasks, as well as on exploring and analysing the household functions in their national contexts. With respect to the nutrition function, a food consumption and production system of four levels was defined (Green 1998, Green and Young 2000, Quist *et al.* 2002a). Starting point was the fact that household decisions have a direct impact on all levels of the food system, while acknowledging the interwovenness and mutual influences of demand side and supply system. The first level of the defined system includes the core of nutrition in the household and all activities regarding food shopping, storage, treatment and consumption in the household. It also includes eating out in restaurants, take-away meals, home delivery and other food services at work or elsewhere. The second level consists of foods retail and wholesale, food appliances for households and food services. The third level includes agriculture, agricultural and food imports, packaging, food processing, the manufacturing of kitchen equipment and the associated transport. Finally, the fourth level consists of all other inputs to the higher levels like water, energy, electricity, agricultural inputs, other resources and transportation and processing equipment. For a more detailed elaboration, see Green and Young (2000: 4-10), or Quist *et al.* (2002a).

The resulting food system framework and overall methodology served as a starting point for the backcasting experiment on nutrition in the Netherlands. Part of the country-specific orientation was the identification of stakeholders in the national food consumption and supply system. In the Netherlands the stakeholder analysis was used as a starting point for more than thirty interviews in the period July – November 1998 and included companies, government, public interest groups and research bodies. Interviewees were asked about major sustainability problems related to food consumption, developments and trends in food, possible environmental improvements in food consumption and their interest and motivation for participating in the first nutrition workshop. The results of the interviews and desk study were combined in an input document for the first stakeholder workshop (Quist 1998, Quist 2001a).

Table 6.2 *Timeline of the SusHouse project and the nutrition subproject*

	SusHouse 'overall'	Nutrition backcasting experiments
Phase 1 *January - April 1998*	> International project start-up meeting January 1998, Delft > Development formats for all tasks > Project meeting, April 1998, Italy	> Orientation on nutrition function (present fulfilment, stakeholders, unsustainabilities)
Phase 2 *May - July 1998*	> Orientation on each function in its national context.	> Internal test workshop > First round of stakeholder identification > Preparing workshop organisation proposal
Phase 3 *August - December 1998*	> Stakeholder enrolment and interviews > First series of workshops > Project meeting, August 1998, Hungary	> Stakeholder and expert interviews > Stakeholder enrolment > Organisation first workshop, workshop input document > Stakeholder creativity workshop > Distribution workshop report
Phase 4/5 *January - September 1999*	> Construction and assessment (consumer, economic, environmental) of scenarios > Project meeting, February 1999, Germany > Project meeting, September 1999, UK	> Elaboration of three nutrition scenarios > Consumer focus group research > Start environmental and economic assessment of the scenarios > Network activities and stakeholder communication > Scenario specific update of stakeholder identification
Phase 6 October 1999 – February 2000	> Second series of stakeholder workshops focusing on backcasting, follow-up and policy recommendations	> Re-involvement of stakeholders and enrolment of new ones > Questionnaire on scenarios sent to stakeholders > Workshop organisation > Second stakeholder workshop > Completing environmental and economic assessment > Evaluation and distribution workshop report
Phase 7 January - September 2000	> Analysis of workshop results > Policy recommendations and reporting > Project meeting, February, 2000, Spain > EU reporting workshop, September 2000	> Elaboration workshop results into action agenda > Post stakeholder management and elaboration follow-up proposals (partly together with stakeholders) > Writing final report backcasting experiment and background reports

In the Netherlands the first stakeholder workshop took place in November 1998. The workshop involved a broad range of relevant stakeholders and resulted in an exchange of views and opinions among participants. The workshop evolved around a plenary creativity and brainstorm session based on the question 'How can we eat in a sustainable way in 2050?'. This was followed by a session on the clustering of ideas into proto-scenarios in subgroups and a plenary discussion (Quist and Maas 1999, Quist 2001a: Chapter 3). A professional facilitator guided the workshop using an extensive workshop facilitation script (see Quist 1999a: 38-41). The workshop resulted in five proto-scenarios of sustainable nutrition in the household and was evaluated positively by the participants (Maas and Quist 1999, Quist 1999a, Quist 2000a).

Eighteen stakeholders from companies, public interest groups, government and research participated. Attendance included the Dutch Association of Rural Women (NBPV), the Dutch Centre for Nutrition, the environmental organisation Milieu Centraal, the Dutch Board for the Hotel and Catering Industry (BHC). Participants also included several food producers like Boekos and Unilever and several kitchen equipment producers, such as Atag and Philips DAP (Philips' business unit on domestic appliances). Research was represented by Wageningen University, TNO Kathalys, DUT Kathalys and SWOKA consumer research). Finally, government representatives included the Ministry of the Environment and the NRLO, which is the Netherlands Agricultural Research Council.

Further elaboration of the scenarios for nutrition was carried out by the Dutch research team and took place in the first half of 1999. Scenario construction was based on the results of the stakeholder creativity workshop, using the ideas from the brainstorm and the proto-scenarios. Scenario construction followed the Design Orienting Scenario methodology of Manzini and Jégou (1998, 2000) as reported by Young et al. (2001). It resulted in three so-called design-oriented scenarios, which are presented in Table 6.3 (see also Quist 2000a).

Scenario construction was followed by scenario assessment in the second half of 1999. While the environmental assessment and economic analysis of the scenarios were carried out within the internal project, for the consumer evaluation external focus groups were used. The consumer evaluation followed the format developed by Bode (2000) and was aimed at evaluating the acceptability and opportunities of the scenarios to Dutch consumers and to identify consumer adoption profiles. A distinction was drawn between three types of consumers: green consumers with high environmental awareness, dynamic consumers, who can be seen as so-called early adopters and mainstream consumers. Three focus group sessions were held during June and July 1999 (Dekker 1999, Dekker and Quist 2000, Quist 2000a: 51-61). For the organisation of these focus group sessions, earlier contacts with stakeholders were used. This resulted in a green focus group of Eco Team members in Rotterdam, a dynamic focus group of Industrial Design PhD researchers at Delft University of Technology and a mainstream focus group of rural women from the Dutch Association of Rural Woman (NBPV) in the province of North Holland. In addition, the scenarios, together with a questionnaire, were also sent to the participants of the first workshop and newly established contacts.

During the organisation of the second stakeholder workshop a new round of stakeholder identification was conducted. It consisted of a general update and a scenario-specific update. The latter included identifying both driving forces of the scenarios and new stakeholders that may emerge in the scenarios and that may be important for realising the scenarios. When the second workshop was being prepared it turned out that a related project of the STD programme and the National Council of Agricultural Research (NRLO) entitled 'Consumers, Nutrition and Environment' was also planning to present its results by means of a workshop. As both projects were addressing similar subjects and stakeholders, it was decided to organise a joint workshop (LEI/DUT 2000, Fonk interview 2005), in which researchers from Delft University of Technology and the Agricultural-Economic Institute (LEI) cooperated.

The second stakeholder workshop took place in January 2000. It focused on backcasting, implementation (strategies) and stakeholder feedback on the scenarios and assessments. The workshop combined elements from strategy and backcasting approaches. The same professional facilitator was hired and he developed a detailed workshop script (Quist 2000b: 19-22). In all, twenty-two people attended, who on average evaluated the workshop positively (Quist 2000b). Around 60% of the stakeholders present had also been present in the first workshop. The workshop started with a structured evaluation of the three normative scenarios, while in the afternoon participants worked in scenario-specific subgroups on backcasting and elaboration of follow-up proposals. The workshop resulted in an evaluation of the scenarios including possible rebound effects and follow-up proposals and cooperation, while new ideas were also proposed. Furthermore, the unprocessed results contained the 'raw material' for policy recommendations and also for the construction of an action agenda (LEI/DUT 2000).

In the months after the workshop additional activities were carried out, such as further elaboration of follow-up proposals, meetings with interested stakeholders like the Ministry of the Environment and dissemination of the results and preparing reports for the EU. Activities in the Dutch backcasting experiment on nutrition were finished in September 2000, after a concluding workshop in Brussels on the outcomes of the entire SusHouse project.

6.2.2 Results

The results of the Dutch backcasting experiment on nutrition include the workshop results, the constructed scenarios for sustainable fulfilment of the household nutrition function that can be seen as future visions, and scenario assessment results.

The (first) stakeholder creativity workshop resulted in a large number of ideas for future sustainable nutrition from the perspective of the household. Participants clustered ideas and translated them into five proto-scenarios. Both the listed ideas and the proto-scenarios were used for the construction of the normative design-oriented scenarios (Manzini and Jégou 2000, Young *et al.* 2001). These scenarios can be seen as future visions as defined in this research.

Figure 6.2 *Picture of the ICS scenario (drawing by Peter Welleman)*

The scenarios for the fulfilment of household functions in the SusHouse project included a core vision describing the general atmosphere and main characteristics of sustainable function fulfilment. Secondly, scenarios also contained several 'proposals' for product-service systems that enabled sustainable function fulfilment. Thirdly, the scenarios used a storyboard or 'snapshot' description of what it would be like to live in that particular scenario in the year 2050, encompassing all the proposals of that scenario. Scenarios were supported by drawings (see Figure 6.2).

Table 6.3 describes the Dutch nutrition scenarios (Quist 2000a, Quist 2001a). These three scenarios or future visions depict more sustainable alternatives for present, as well as possible future ways of living. They were not meant to identify the most sustainable scenario and develop a strategy to bring everyone towards the most sustainable scenario, but to offer possible alternative sustainable future visions.

Table 6.3 *The three scenarios in the Dutch SHN backcasting experiment*

Intelligent Cooking and Storing (ICS) is about a household that can be characterised by high-tech, convenience, do-it-yourself and a fast way of living. Kitchen and food management is optimised with help of intelligent technology, which also organises ordering (electronically), and delivery with help of a so-called Intelligent Front Door. Water and energy are re-used where possible through cascade usage. Meals are either based on a mixture of sustainable pre-prepared components including vegetarian protein foods, or are ready-made meals containing a microchip communicating cooking instructions with the microwave oven. Packaging is biodegradable and contains a (plastic) microchip with relevant consumer information about origin, treatment and preparation.
Proposals include: (1) Intelligent kitchen; (2) Biodegradable and intelligent packaging; (3) Sustainable ready-made meals and meal components; (4) Food delivery service and intelligent front door, and; (5) Novel Protein Foods from non-animals sources.

Super-Rant (SR) combines elements from the present supermarket and restaurant, but these are shaped into a neighbourhood food centre within a compact city. Here you can go for a meal (e.g. by a subscription to the neighbourhood cook), to shop for food, to purchase a take-away meal or to eat together for different prices. In many households only the microwave oven, a water cooker and a small fridge are left. Waste is collected for local energy production. Food is grown in a sustainable way.
Proposals include: (1) Super-rant; (2) Eating together, but individual catering subscription; (3) Sustainable restaurant, and; (4) Sharing and hiring kitchens and appliances.

In Local and Green (LG) household members grow a considerable share of their foods themselves. They also buy and eat seasonal foods that are locally grown and purchased at local shops, small supermarkets, or that are bought directly from the grower or hobby garden as 'fresh' unprocessed ingredients. Regional specialities are important and are consumed in the region by both inhabitants and tourists. Imported products are still available but expensive, because environmental costs are incorporated into the price. There is also a strong green consumer demand in this scenario.
Proposals include: (1) High-tech hobby garden; (2) Collective table, eating and cooking together; (3) Advanced home composting, and; (4) Regional food specialities and chains.

Table 6.4 *More details of the ICS scenario*

Stakeholder panorama
Key stakeholders in this scenario are consumers, retailers, food processors, packaging producers, kitchen equipment and appliances producers and government.

Environmental profit stems from:
> Sustainably grown ingredients (inclusive new ingredients take over the function of unsustainable ingredients like novel protein foods);
> System optimisation (through integrated approach to the kitchen, waste reduction);
> Re-use of heat and water (cascade usage) in the household;
> Waste composting and biodegradable packaging.

Necessary changes (preliminary backcasting analysis)
> Technological: novel kitchen technology and appliances (including a huge efficiency increase), new ICT for kitchen systems and production chain management, plastic chips, biodegradable packaging, cascade usage for water and energy, sustainable transportation, distribution and delivery systems.
> Cultural/ behavioural: sustainability is taken for granted, further shift towards ready-mades and convenience, acceptance of new technologies, shift towards more sustainable substitutes (e.g. vegetable based Novel Protein Foods in stead of meat), shift towards services.
> Structural/ Organisational: the role of supermarkets will change due to large-scale delivery and a shift towards food management services, kitchen manufacturers deliver complete automated systems that communicate in stead of single kitchens and single appliances, close co-operation and joint management throughout the complete production chain plus making information available to consumers; sustainable food production (regional or efficient large scale production where this can most environmentally efficient).

The scenarios were also different in their driving forces. Intelligent Cooking and Storing assumed environmental gain through high-tech and ICT. The expected environmental gain in the Super-Rant scenario was due to scaling-up at neighbourhood level and a product-to-service switch. The key behind Local and Green was do-it yourself and regional food supply and consumption chains. Table 6.4 uses the Intelligent Cooking and Storing (ICS) scenario as an example to show the stakeholder panorama and the preliminary results of both the environmental improvement estimates and the backcasting analysis. The latter was done by looking backwards from the scenario using the question 'What technological, cultural, and institutional-organisational changes are necessary to realise this scenario?'.

In addition to the backcasting analysis, three more scenario assessments were conducted. The first was an environmental assessment through a system approach using indicators to assess whether the scenarios would achieve a factor 20 reduction in household environmental impacts (Bras-Klapwijk 2000, Bras-Klapwijk and Knot 2001). The economic assessment used a questionnaire to assess each scenario on economic aspects (Young and Simms 2000). Finally, the consumer acceptance analysis[3] used consumer focus groups to evaluate the acceptability of the scenarios to consumers and to identify adopter profiles (Bode 2000). Aggregated results of the three assessments for the nutrition backcasting experiment in the Netherlands are shown in Table 6.5. More detailed results are given elsewhere (Quist 2000a, Quist 2001b, Quist et al. 2002b).

The environmental assessment showed that the Intelligent Cooking and Storing scenario and the Local and Green scenario could reduce the environmental burden considerably. By contrast, with regard to the Super-Rant scenario it was found that using energy data from present restaurants (Oudshoff 1996) would even increase the energy requirements (Quist 2000a, Quist 2000b, Van Gaasbeek et al. 2000). The economic analysis indicated that the Local and Green scenario would require most economic changes, while the other two scenarios would require moderate, yet considerable, changes from an economic viewpoint (Quist 2000a).

Table 6.5 *Aggregated assessment results for the three SHN scenarios*

	Environmental Improvement	Economic Adjustments	Consumer Acceptance
Intelligent Cooking and Storing (ICS)	⬤ (high)	⬤ (medium)	⬤
Local and Green (LG)	⬤	⬤	⬤
Super-Rant (SR)	· (low)	⬤	·

Focus group research showed that consumers were very capable of evaluating scenarios describing situations in the far future and connecting them to their present situation and daily lives. The focus group analysis showed that consumers considered the Local and Green to be the most attractive, while the Super-Rant scenario received most of the criticism. In addition, a vast majority evaluated their own present personal situation as most positive (Dekker 1999, Dekker and Quist 2000, Quist 2000a).

The backcasting and implementation workshop (LEI/DUT 2000, Quist 2000d) resulted in stakeholder feedback, such as:

Consumers could combine scenarios or scenario elements. These options could both decrease and increase the environmental gain compared to a specific scenario.

- Combining scenarios or adding elements could increase the attractiveness of a scenario.
- The Intelligent Cooking and Storing scenario reflects the dominant direction of development. As both the Super-Rant scenario and the Local and Green scenario contain important social values, these may be stimulated by the government.

6.3 Analysis

6.3.1 Stakeholder participation
Stakeholder involvement started with interviews at an early stage of the backcasting experiment and was followed by participation in two one-day workshops. Stakeholder involvement is analysed in terms of stakeholder heterogeneity, degree of involvement, type of involvement and degree of influence on content and process of the backcasting experiment. The indicators can also be used to characterise groups according to their role in the process. Two groups can be distinguished in this backcasting experiment from a process perspective: the stakeholders involved and the research team, who organised the backcasting experiment, involved the stakeholders and conducted the research. Heterogeneity results are shown in Table 6.6. Results for the other indicators are listed in Table 6.7.

Table 6.6 *Stakeholder heterogeneity in the SHN backcasting experiment*

	Interviews	Attendance 1st Workshop	Attendance 2nd Workshop
Companies	11	5	7
Public interest groups	9	3	2
Government	4	3	5
Research organisations	7	6	8
TOTAL	31	17	22

Table 6.6 shows that heterogeneity was good, as all four main societal groups were involved. The table also shows that the number of public interest groups attending the workshops was lower than the other stakeholder categories. This indicates that it is difficult to persuade public interest groups to participate. Such organisations are relatively small and involved in a broad range of issues and activities. The interviews appeared to be important in enrolling stakeholders in the first workshop and becoming acquainted to relevant networks (Quist 2000a). Many of the participants attending the first workshop were also present in the second workshop, which indicates a continuing commitment.

The degree of stakeholder involvement was in fact low. It was limited to the initial interviews, two workshops of a single day, returning a questionnaire on the scenarios between the workshops and distribution of the final reports.

The main type of involvement was participation in the workshops. Funding was provided by the EU, which was around € 100,000 for the backcasting experiment on household nutrition in the Netherlands. However, the EU is not regarded as a participating stakeholder in this backcasting experiment.

The influence of the participating stakeholders on the content was high. Idea generation, proto-scenario development, scenario evaluation and elaboration of follow-up proposals were fully based on contributions by stakeholders in the workshops. In addition, stakeholders could also exert influence on the scenarios by returning a questionnaire between the two workshops. By contrast, stakeholders had little or no influence on the formats and the course of the assessments, as these were developed by researchers in the international methodological subproject. Workshop formats were also developed by the international research team, while detailed scripts were elaborated by the local research team for each workshop. The first workshop was focused deliberately on generating creativity and ideas among the participants, attempting to move away from discussions about problem definitions and boundary issues. As a consequence, stakeholders strongly influenced the content, but they had little influence on the process and who participated (see also Table 6.7).

Table 6.7 *Characteristics of participation in the SHN backcasting experiment*

	Degree of involvement	Type of involvement	Degree of influence	
Stakeholder workshop participants	∘	*Capacity*	●	*(on content)*
			∘	*(on process)*
Research team	●	*Capacity*	∘	*(on content)*
			●	*(on process)*

low ∘   high ●

6.3.2 Vision aspects

Generation of the future visions started with stakeholders brainstorming and clustering ideas about future sustainable fulfilment of the household function of nutrition.

Vision: guidance

Guidance by the visions has been analysed in terms of (1) collective normative projection, (2) synchronisation and alignment of the interactions among stakeholders from various backgrounds and academic disciplines, (3) presence of alternative (system of) rules and institutions.

Collective normative projection

Because the three visions were based on results and ideas from a stakeholder workshop and were evaluated in a second workshop and through a questionnaire, they can be seen as collective normative projections. However, as the visions were the result of a creative process, it is possible that the visions are only congruent to a certain extent. For instance, participants were especially in favour of the vision containing most of their ideas and values. In fact, each future vision had its supporters. Besides, some stakeholders did not commit themselves strongly to a particular vision, but had a general interest in the topic. To these stakeholders the three visions were not collective normative projects, but they used the three visions to compare and learn.

Synchronisation

The three visions provided input for debate and dialogue among stakeholders from different academic disciplines and societal backgrounds, like companies, government, public interest groups. In this way it was possible to look at the visions from different perspectives and disciplines, but in a focused way, which suggests synchronisation among these stakeholders. However, because (i) the degree of involvement was rather low (see 6.3.1) and (ii) no clear commitments were asked for, the level of synchronisation is considered as moderate[4].

Alternative rules

Each future vision embodied alternative rules and rule systems and provided the opportunity to look into a possible future in which present rules and rule systems did not automatically prevail. For instance, environmental benefits were achieved in the Intelligent Cooking and Storing scenario through high-tech and ICT, while it could be combined with a 'fast' lifestyle in combination with taking sustainability for granted (thus not as a deliberate choice or action). The Super-Rant scenario combined outsourcing of cooking and using food services in groups with individual choices at specialised locations in the neighbourhood. The Local and Green scenario combined rural living with do-it-yourself. Each of the future worlds depicted in one of the visions could only work if technological, cultural and structural changes would be realised, thus assuming changes in rules and rule systems.

To conclude, although there was some guidance, it was less convincing than, for instance, in the NPF case in the previous chapter, because of limited congruence and sharing in the goals and moderate synchronisation.

Vision: orientation

The orientation of the future visions can be defined in terms of (1) cognitive activation, (2) mobilisation of actors and resources, and (3) providing stability and daily coordination.

Cognitive activation

Each future vision provided cognitive challenges and facilitated cognitive action towards the vision and reflection on the visions and the assessments as well. For instance, the three visions served as input to the scenario assessments and were put up for discussion in a second workshop. This workshop focused on a further elaboration of concrete follow-up activities including policy recommendations. It also evaluated and compared the three future visions, including their potential impact and side-effects.

Mobilisation

Existing and new stakeholders were mobilised and took part in a second workshop, including stakeholders who had already attended the first workshop. Some participants were not necessarily attracted to one of the future visions, but could also be attracted by the topic of households, nutrition and sustainability. Other participants showed interest in specific issues or follow-up proposals. These stakeholders felt thus more attracted to or saw opportunities in one of the future visions. The visions contained not only product-service proposals that could be attractive to companies. For instance, results of the backcasting experiment and its approach were referred to during internal discussions in several companies and in Kathalys (Silvester interview 2005), the institute for sustainable product development in which TNO and the faculty of Industrial Design of Delft University of Technology cooperate (see, for instance, Brezet *et al.* 2001). Visions also embodied research agendas that were attractive to researchers and policy-oriented proposals that may be attractive to policy-makers or public interest groups. However, limited resources were mobilised, apart from the funding by the EU.

Decentralised coordination

Limited evidence was found that the future visions provided coordination in the daily activities of the stakeholders involved in the backcasting experiment, as the degree of involvement was low[5]. However, several initiatives on developing follow-up proposals were initiated during the backcasting experiment, such as the development of a programme proposal for sustainable integration of food chains and kitchen chains (Boekos *et al.* 2000, see also Chapter 5) and a workshop on appliances for Novel Protein Foods (Goekoop 2000, see also Chapter 5). In these examples, coordination was provided, not only at a daily level, but also at the level of interaction and cooperation among groups of stakeholders.

Competing visions

Existing dominant visions representing business as usual developments in household nutrition and food production, or incremental environmental improvement scenarios can be seen as alternative competing visions. Those were not articulated in the backcasting experiment or during the workshops. By contrast, the three generated visions can be seen as competing alternatives, though it is also possible to consider them as non-competing substitutes that may exist and develop side by side.

6.3.3 Learning

Higher order learning is defined in terms of (1) a shift in framing major problems or perceived solutions by specific actors; (2) a shift in the principal approaches to solving these problems and in shifting priorities by specific actors; and (3) joint learning, a shift in congruence and joint opinions in any of the issues related to the previous shifts. This section focuses on instances of higher order learning.

In addition, a distinction is made in this research between learning about the topic of sustainable nutrition in the household, related to one of the future visions, and learning about the participatory backcasting approach. It is assumed that higher order learning about the topic mainly involves participating stakeholders, while higher order learning about the approach is expected among those organising and managing the backcasting experiment. Considerable results of first order learning can be found in the reports on the backcasting experiment, the future visions and the assessments (Quist 2000a, Quist 2001a).

Higher order learning about the topic

It was reported in the backcasting experiment (Quist 2000) and confirmed in the interviews (Silvester interview 2005) that the environmental impact of food services and eating out is considerably higher than that of eating at home. This implies that there is a huge potential for environmental improvement in food services and eating out, which can be seen as a shift in priorities. It also questions the widely advocated shift from products to services, which is a shift in a perceived solution and in a principal approach. The Intelligent Cooking and Storing scenario can be seen as depicting the dominant solution embodying currently dominant values. However, during the second workshop it was argued that the other two scenarios embody important social values related to sense of community or local sustainability solutions that deserve to be stimulated, for instance by the government (Quist 2000a, LEI/DUT 2000). This can be seen as a shift in a perceived solution, or as a shift in priority. Learning thus also took place by comparing the three future visions, the solutions they embodied and their differences. This indicates that discussing several alternative future visions together may facilitate learning among participants, as is confirmed in the scenario literature (e.g. van der Heijden 1996, Berkhout *et al.* 2002). However, this does not necessarily stimulate commitment or congruence regarding a specific vision.

Higher order learning about the approach

Instances of higher order learning about the approach were also found. Several respondents indicated that the interactive and creative approach to generating future visions and backcasting was inspiring and served as an eye-opener, as did the focus on joint vision development in (sustainable) innovation processes (Kramer interview 2005, Silvester interview 2005, Te Riele interview 2005). This can be seen as a shift in approaches. In addition, a lack of follow-up and implementation was noticed, which resulted in learning about how difficult it is to turn attractive and inspiring visions into activities in the short term or into system innovations further along (Te Riele interview 2005, Silvester interview 2005, Green interview 2005, Vergragt interview 2005, Fonk interview 2005). Having interesting workshops raising enthusiasm among stakeholders who elaborate and discuss alternative future visions is not sufficient to achieve follow-up (Te Riele interview 2005). It points to learning about adjusting the principal backcasting methodology. A project or vision champion did not emerge, while such a person is essential for spin-off and should there-fore be actively courted in backcasting experiments (Fonk interview 2005). Several adjustments have been proposed to deal with the lack of follow-up, which points to shifts in priorities and how to deal with the principal approach. For instance, the backcasting experiment could be extended in such a way that estab-lishing the follow-up activities becomes part of it (Green interview 2005). Another possibility is to increase the intensity and level of involvement, for instance by involving stakeholders between the workshops in additional activities and to define special subprojects that are sufficiently attractive to stakeholders that be (co)funded by them or to be submitted to regular funding programmes (Green interview 2005, Vergragt interview 2005). During the evaluation of the backcasting experiment (Quist *et al.* 2000, Vergragt 2000), a proposal was made to extend the backcasting experiment and the applied methodology with additional activities and meetings around single visions and to focus on the opportunities and possibilities in each single vision with a more focused and dedicated group of stakeholders. It was mentioned that in realising follow-up two levels should be distinguished that are both relevant (Te Riele interview 2005, Diepenmaat and Te Riele 2001). There is a strategic level at which high-level actors look for consensus on problems, solutions and approaches resulting in commitment, and there is a practical level on which more concrete solutions and innovations are developed and implemented. It was argued that the backcasting experiment lacked such a strategic level (Te Riele interview 2005).

6.3.4 Methodological aspects and settings

With regard to the methodological aspects, the proposed backcasting framework from Chapter 2 is used to evaluate the backcasting experiment on (i) inter-disciplinarity, (ii) steps, (iii) use of different methods, and (iv) three types of demands.

Backcasting framework: inter-disciplinarity

The backcasting experiment on household nutrition was clearly multidisciplinary. It included economic aspects, scenario construction, environmental analysis, consumer acceptance studies, stakeholder analysis, workshops and broader stakeholder communication. The different research tasks were integrated in an over-all framework and researchers collected data using the research formats developed by other disciplines. The results on the different research tasks in the nutrition backcasting experiment were integrated at the end and presented to stakeholders in a workshop as input for dialogue and debate.

Backcasting framework: steps

The five steps of the methodological framework proposed in Chapter 2 can be identified in the backcasting methodology as applied in the nutrition backcasting experiment (see also Table 6.8). There is a good match between the five framework stages and the steps of the backcasting methodology that was applied. As the methodology was prescribed to all nine backcasting experiments that took place in the SusHouse project, it is likely to apply for the entire project.

Table 6.8 Comparing the steps of the backcasting framework and the SHN backcasting experiment

Steps for participatory backcasting	SusHouse methodology
(1) Strategic problem orientation	Step 1, step 2
(2) Develop future vision	Step 3, step 4
(3) Backcasting analysis	Step 3, Step 4, Step 6
(4) Elaborate future alternative and define follow-up agenda	Step 4, step 5, step 6
(5) Embed results and agenda, stimulate follow-up and spin-off	Step 6, Step 7

Backcasting framework: applied methods

Table 6.9 shows that methods from all four groups of methods and tools were applied in the backcasting experiment on nutrition. An explanation of the methods can be found elsewhere (Green and Vergragt 2002, Vergragt 2000, Quist 2000a, Quist *et al.* 2000).

Table 6.9 Methods applied in the SHN backcasting experiment

Analysis	Design
> Scenario environmental assessment	> Scenario construction
> Stakeholder analysis	> Elaboration of product-service proposals
> Scenario-specific stakeholder analysis	> Drawings of the scenarios
> Scenario evaluation by consumer focus groups	> Workshop design and workshop script development
> Economic questionnaire	> Research design for each research task
> System and function analysis	
> Backcasting analysis	
Participation/ Interaction	**Management, communication, coordination**
> Stakeholder interviews	> International project meetings
> Stakeholder creativity workshop	> Dissemination of workshop reports and project documents to stakeholders
> Brainstoring and other creativity methods	> Participation in related activities (meetings, workshops, expert contributions)
> Scenario evaluation questionnaire	> Scientific dissemination
> Stakeholder backcasting and implementation workshop	

Backcasting framework: demands

Three types of demands have been distinguished in the methodological framework: process demands, normative demands and knowledge demands. A distinction can be made between the demands that were valid in the entire SusHouse project, thus in all nine backcasting experiments, and the demands that were specific to the three nutrition backcasting experiments.

Normative demands that were articulated in the overall project included the factor 20 environmental improvement by 2050, the assumption that both social and technological innovations are necessary in system innovations towards sustainability, a focus on households and consumers, the use of normative scenarios and broad stakeholder participation.

Process demands included heterogeneity in stakeholder involvement covering the four major groups distinguished, as well as endorsement of the future visions and the assessment results, and stimulating follow-up activities. In addition, there were more specific demands about when and how the workshops should be organised. However, there was also a certain flexibility and freedom both in stakeholder involvement and in the organisation of the stakeholder workshops in order to meet with local customs and expertise. This flexibility is in fact both a normative and a process demand.

No clear knowledge demands could be identified. These were left to the subproject responsible for developing research formats and the researchers in charge of developing specific research formats and evaluating the results for a specific task over all backcasting experiments. The decision to include experts for methodology development and to let junior researchers organise the backcasting experiment and conduct all the research tasks was positive in terms of the inter-disciplinarity, but it may have affected the scientific quality (Vergragt interview 2005) and thus the knowledge demands.

Additional demands with regard to the nutrition backcasting experiment came from the international research team on nutrition. These demands or guidelines included not only taking into account the consumption and supply system from the viewpoint of the household, but also assuming that household decisions are decisive to supply chains and the entire food supply systems and dealing with both eating at home and food services, including eating out.

Settings: institutional protection

Both the overall SusHouse project and the backcasting experiment on household nutrition in the Netherlands took place in an academic setting without institutional protection from top level management. The role of the EU was limited to funding, while universities in general do not provide institutional protection to ongoing externally funded research projects with relatively limited budgets. Top levels at universities may provide such protection, but this was not the case either with the SusHouse project or with the nutrition backcasting experiment.

Settings: vision champion

The backcasting experiment did not have a vision champion. The initiator and international project coordinator was an international expert both in backcasting research and in innovation research for sustainability. Thus, he may be seen as the project champion at the international academic level and inside the international project group. However, he can not be regarded as the project champion of the nutrition backcasting experiment in the Netherlands, or of the international subproject on nutrition. No vision champion thus emerged among the stakeholders involved during the backcasting experiment for any of the three future visions. Although with the follow-up initiatives, specific stakeholders or individuals took the lead, they did not turn into project champions after submitted proposals were rejected.

The researcher[6] responsible for conducting the backcasting experiment had to focus on the process aspects of stakeholder participation and could thus not become engaged to any of the visions. This prevented him from pursuing a role of vision champion. In the design of the backcasting experiment it was not planned to organise vision-specific meetings and bring together interested stakeholders. Such an addition may have stimulated the emergence of a vision champion feeling more strongly committed to one of the future visions and the opportunities they embodied.

Settings: focus and management

Although implementation and follow-up was referred to in the goals of the backcasting experiment, the main focus was on developing and testing methodology. The type of management resembles process management with a focus on keeping stakeholders involved

6.4 Follow-up and spin-off

6.4.1 Business and research domains

The research and business domains are taken together, as several initiatives were jointly developed by cooperating companies and research organisations. At the first stakeholder workshop in November 1998 the manager from TNO Kathalys, the institute for sustainable product development, and the R&D manager from Boekos, then a major producer of vegetarian protein foods (and strongly involved in Novel Protein Foods-related activities, see Chapter 5) got acquainted with each other. They saw joint opportunities, which was the starting point for several activities. Near the end of 1999 this led to a workshop focusing on domestic appliances for treating Novel Protein Foods at home, gathering a number of producers of kitchen equipment, including Philips DAP (Goekoop 2000). Follow-up activities resulted in a proposal that was submitted to the EET[7] programme, but which was rejected (Luiten interview 2005). In addition, the Boekos manager used the SusHouse workshops to build an alliance when developing a programme proposal for the NIDO Leapfrogging Contest in 2000. This proposal focused on matching sustainable food chains and sustainable kitchen chains (Boekos Food Group et al. 2000) and involved TNO Kathalys, Delft University of Technology, WUR and the environmental organisation 'The Small Earth'[8]. However, this proposal did not obtain funding (Luiten interview 2005), as only the winning proposal was funded. Moreover, Boekos ended all NPF-related activities, when it was taken over in 2001 (see Chapter 5).

The Technology Assessment research group at Delft University of Technology, which conducted the SusHouse backcasting experiment on nutrition in the Netherlands, developed several follow-up proposals. The most notable one was a programme proposal for a transition towards sustainability in the eating-out and food-service sector, which was submitted to the annual NIDO Leapfrogging Contest in 2001 (Quist and Silvester 2001). This proposal was a joint initiative by the Technology Assessment Group and DUT Kathalys, which is part of the faculty of Industrial Design and had been involved in the backcasting experiment. It involved several other research organisations including the Agricultural Economic Institute (LEI), TNO Kathalys and the BHC[9], the Dutch Board for the Hotel and Catering Industry in which trade and labour unions cooperate. The willingness to participate in the proposed programme or support scope and ambition can be seen as a positive attitude towards the SusHouse nutrition backcasting experiment, the workshops and the outcomes (Silvester interview 2005). The proposal was nominated and received some funding for further elaboration, but the extended proposal was not approved for funding. The proposed line of research on a system innovation towards sustainability in food services and eating out was not continued, but led to a graduate student's research project. This project involved Deli XL, the business unit of Ahold that delivers to restaurants, and used key elements of the SusHouse methodology (stakeholder workshop, scenarios, and scenario assessments). Results included scenarios and proposals for sustainable eating out (Van der Horst

2002, Quist *et al.* 2003).

The approach and results of the SHN backcasting experiment were referred to in internal documents or discussions in companies that participated in the workshops or in related activities, such as the companies Atag and Deli XL (Silvester interview 2005). While Deli XL was interested in the graduate student project in sustainable food services and eating out (Van der Horst 2002), the interest of Atag was of a strategic nature. During the early stages of the SHN backcasting experiment, Atag was until mid-1999 the leading partner in an EET project in which TNO Kathalys, DUT Kathalys and engineering consultancy Gastec participated. This alliance worked on the development of ecologically sound kitchen equipment. The aims of the EET project matched well with the focus on sustainable household food consumption in the backcasting experiment (Silvester interview 2005). Furthermore, the backcasting experiment might result in ideas for kitchens and equipment that could be of interest to the ongoing EET project. However, after Atag faced strong financial losses in 1999, it was taken over by its competitor Etna. Shortly afterwards, the EET project and participation in the SHN backcasting experiment were finished.

Finally, the collaboration with another participatory project on food consumption and environment led to a joint workshop, which can be seen as a mutual spin-off of both undertakings. The former had been commissioned to LEI, the Agricultural Economic Institute, and after the joint workshop, this paved the way for further collaboration in the programme proposal on sustainable eating out and foodservices. As mentioned above, this proposal was turned down, after which this collaboration was discontinued.

6.4.2 Government domain

In the course of the backcasting experiment, the Ministry of the Environment commissioned a study into the possibilities for consumers to reduce the environmental burden of food consumption. This study included consultation of consumers, stakeholders and experts (Van der Pijl and Krutwagen 2000) and aimed at developing policies for sustainable consumption and production. To some extent the policy recommendations of this study where in accordance with the results of the backcasting experiment, especially regarding food services and eating out. However, a policy-maker from the ministry mentioned that results and insights from the SHN backcasting experiment were not used in the policy analysis and policy agenda, but that they served as a source of inspiration.

6.4.3 Public domain

No follow-up among public interest groups has been identified, although several stakeholders from this domain participated in the workshops. However, environmental data from the nutrition backcasting experiment were used by the Milieu Centraal (Milieu Centraal 2001). This is an independent organisation providing environmental information on products, services and consumption patterns to consumers.

6.4.4 Spin-off of the international subproject on nutrition

Similar backcasting experiments on household nutrition took place as part of the international Nutrition subproject (Green and Young 2000) in the UK (Young 2000) and in Hungary (Tóth *et al.* 2000). This subsection briefly looks into the spin-off related to these backcasting experiments in which the same methodology was applied. These results cannot be connected to the backcasting experiment in the Netherlands, but their spin-off may provide additional insights, for instance by analysing the differences and similarities.

The UK backcasting experiment was conducted by an innovation research group at UMIST's School of Management. As a result several follow-up activities were initiated (Green interview 2005), all taking long-term transitional changes towards sustainability or industrial transformations as a starting point. Firstly, the leading researcher in the UK became involved in developing the food section of the IHDP on Industrial Transformation (Vellinga and Herb 1999: 26-33) and in a study on the global aspects of sustainability in food production and consumption systems (Green interview 2005, Green *et al.* 2003). Secondly, the UMIST group initiated research into sustainable food futures as part of the Sustainable Technologies Programme funded by the UK Economic and Social Research Council. This research focuses in a more detailed way on future food systems at the level of specific food categories, using peas, fish, yoghurt, chicken and potatoes as examples (e.g. Green and Foster 2005). Attempts to involve companies that had participated in the UK backcasting experiment in a light version of the SusHouse backcasting methodology focusing on developing business did not succeed. No other spin-offs from stakeholders involved were reported either, although these were not actively investigated (Green interview 2005). Thus, spin-off mainly consists of follow-up activities at the research group that had conducted the backcasting experiment.

The third backcasting experiment on household nutrition was conducted in Hungary by a group of researchers linked to the Szeged College of Food Industry (Tóth *et al.* 2000). This was the first ever holistic and integral approach to nutrition and food production in Hungary (Tóth interview 2005). In particular stakeholder involvement was seen as something new and was much appreciated by the stakeholders involved. Several proposals were developed based on the scenarios developed in the backcasting experiment (Tóth interview 2005). The approach using normative scenarios and involving stakeholders has also been applied in several other research projects, for instance the future of Hungary in the EU and in projects dealing with sustainable futures for various regions in Hungary (Tóth interview 2005). However, no data were collected about stakeholder participation in these follow-up activities, nor was it clear to what extent these activities involved stakeholders, other than the research group that conducted the backcasting experiment.

6.4.5 Related international activities

Two international activities have been referred to as having links with the SusHouse nutrition backcasting experiment (Silvester interview 2005, Vergragt interview 2005): the HiCS project and Suspronet.

The HiCS (Highly Customised Solutions) project dealt with food (www.hicsproject.org, Jégou and Joore 2004). It involved several individuals and organisations that had been involved in the SusHouse project. The general purpose of the HiCS project was to work on highly customised solutions, for instance solutions that support ill or disabled people, while also offering improved environmental and business performance. This project dealt with food delivery services in several countries (Luiten interview 2005, Silvester interview 2005). Despite the links through various persons and organisations with the Dutch nutrition backcasting experiment, the HiCS project was not considered a spin-off of the SusHouse project (Luiten interview 2005). A possible explanation could be that various individuals and organisations were involved both in the backcasting experiment and the HICS project and added similar ideas and methods to both activities, which suggests spin-off that is not there.

Suspronet was an EU-funded network (www.suspronet.org, Tukker and Tischner 2006) dealing with sustainable product service systems. The network approached this subject as a new way of sustainable innovation for companies and involved individuals and organisations that had been part of the backcasting experiment on nutrition. One of the focus areas in Suspronet was food (Tempelman *et al.* 2004), which involved several research bodies that were also involved in the SusHouse backcasting experiment like TNO Kathalys. However, it was contradicted (Tukker personal communication 2006) that the SHN backcasting experiment exerted influence on the Suspronet food area, as the Suspronet activities focused on developing sustainable innovations at companies on the short term.

6.5 Further analysis

6.5.1 Network formation

Activities that can be related to the Dutch backcasting experiment are analysed in terms of network formation using the indicators (1) activities, (2) actors and (3) resources. The analysis takes the activities as starting points to map the actors involved and the resources mobilised (especially funding). Only activities associated with the SHN backcasting experiment in the Netherlands are included in this analysis, as only they can be related to the future visions generated in the Dutch backcasting experiment.

Table 6.10 *Network formation in the SNH case*

	Activities	Actors	Resources
Research and Business domain	*Workshop followed by development of R&D proposal on domestic appliances for NPFs (Goekoop 2000)*	*TNO Kathalys, Philips DAP, Boekos (development proposal) and several other workshop participants*	*Proposal was submitted to EET but not approved for funding*
	Development of programme proposal on chain matching of kitchen and food chains (Boekos Food group et al 2000)	*Boekos, TNO Kathalys, Delft University of Technology, WUR, Kleine Aarde (Small Earth) and several others*	*Proposal of €450,000 was submitted to NIDO but not approved for funding*
	Development of programme proposal on sustainable foods services and eating out (Quist and Silvester 2001)	*Delft University of Technology together with TNO Kathalys, LEI, BHC*	*Proposal of € 450,000 was submitted to NIDO and nominated for further elaboration, but eventually not funded*

In Table 6.10 the results for the combined research and business domain are summarised. In the government and public domains no spin-off activities were identified, therefore these domains are left out. Table 6.10 shows three activities that can be seen as spin-off or follow-up of the backcasting experiment. In each activity a group of stakeholders was involved. Mobilising resources was a major problem; the activities consisted of building stakeholders alliances to develop and submit research proposals to funding programmes. As no funding was awarded, these lines of activities were not continued. This does not mean that no resources were mobilised at all. Mobilising stakeholders and bringing them together in a potential alliance and developing a proposal require resources as well. These were raised internally and were invested to make more resources available. However, these initial efforts did not attract additional resources. The alliances and proposals were thus not capable of mobilising new resources either through internal or external funding.

6.5.2 Vision aspects

The next question is whether, and if so to what extent, the three generated visions have offered guidance and orientation to the listed spin-off activities, and if this relates to possible alternative competing visions. Each of the three visions generated in the backcasting experiment is dealt with separately, as each one had a particular core and embodied different ways of more sustainable future fulfilment of nutrition in households. Each vision also had a different set of supporters and was supported by a different set of 'enabling' proposals that could be either product-service systems, or policy-oriented proposals. Thus, when analysing guidance and orientation, this must be done for all three visions.

Vision: guidance

Guidance has been analysed in terms of (1) collective normative projection, (2) synchronisation and alignment across different domains and scientific disciplines, and (3) provision of alternative (system of) rules.

Collective normative projection

Both the development of the chain-matching proposal (Boekos Food Group *et al.* 2000) and the NPF domestic appliances proposal related to the Intelligent Cooking and Storing vision. The chain-matching proposal had a similar broad scope as the entire vision, while the NPF domestic appliances proposal related to a specific element of the vision and to a distinct enabling proposal that had been formulated. The latter activity can also be seen as connected to the NPF vision (and has been included in Chapter 5) and is thus related to two different visions. These two visions are complementary and non-conflicting, but raise the question whether one of them has prevailed over the other, or that both visions have provided guidance.

The Super-Rant vision was clearly reflected in the sustainable eating out and food-services programme proposal, though the community aspects were largely left out. The focus in the spin-off activities was shifted to a sustainable food service and eating out sector, thus the vision was partly adjusted. The adjusted vision functioned as a collective projection to the initiators and their potential partners. However, when the proposals were not approved, stakeholders moved on and the vision as a collective projection disappeared. No follow-up was found in line with the third Local and Green vision, which suggests that this vision did not last as a collective normative projection among a group of engaged stakeholders. Consequently, the Local and Green vision is not further analysed here. In conclusion, the Intelligent Cooking and Storing vision and the Super-Rant vision partly provided some collective projection, but these were abandoned after proposals could not be put in practice.

Synchronisation

The next question is whether one of the visions increased alignment and synchronisation among stakeholders from different domains and different knowledge disciplines. All three activities involved a range of stakeholders from different domains and different academic disciplines, which suggests synchronisation and increasing alignment. However, due to a lack of access to more substantial resources, the synchronisation and alignment only existed during the time proposals were being developed and alliances emerged. By contrast, it was suggested that the lack of success could be related to fragmentation and limited coherence and consistence within proposals, which suggests (too) limited synchronisation (Luiten interview 2005).

Alternative rules

In the analysis of the backcasting experiment (see 6.3), it was concluded that each of the three future visions contained alternative rules and rule systems. These can also be found in the adjusted visions that play a role in the development of follow-up proposals. However, as these have not been executed, these alternative rules and rule systems were not further articulated, nor was further research carried out with regard to their possibilities and consequences.

Vision: orientation

Orientation has been analysed in terms of (1) cognitive activation, (2) mobilisation of actors and (3) daily decentralised coordination in the emerging network.

Cognitive activation

Both the Intelligent Cooking and Storing vision and the adjusted Super-Rant vision to some extent resulted in cognitive activation, as the visions inspired the development of follow-up proposals. However, when the proposals were not approved for funding, cognitive activation diminished and has fully disappeared.

Mobilisation

The willingness to participate in developing the follow-up proposals, or to join when the proposals would be funded and carried out, indicates attractiveness to stakeholders and the potential to mobilise them through offering interesting opportunities. However, the problem was that this did not include attractiveness to funding parties. Furthermore, those stakeholders interested and attracted were not capable of or willing to fund the proposals themselves. Thus the capability to attract stakeholders and resources was to a certain extent present, but it was too small to be successful.

Decentralised coordination

There was some day-to-day coordination and stability during the development of the proposals, but this disappeared when the activities were discontinued.

The final outcome was that no stability was achieved in the heterogeneous set of actors in any of the three attempts to develop proposals. This also implies that no stability or closure was achieved in any of the three future visions.

Competing visions

As each of the three visions faded away sooner or later, they also stopped functioning as alternative visions. It can be argued that other, totally different, visions were more successful in mobilising resources and actors, while regular dominant visions constrained mobilising additional internal funding. However, this has not been investigated in this research.

6.5.3 Institutionalisation

As spin-off proposals were not executed, institutionalisation did not occur and no institutional resistance emerged. No evidence was found in this research that institutional resistance was the major reason the proposals were not approved.

6.5.4 External factors

Are there possible external factors that may (partly) explain the results presented above? This is important as successful spin-off may have benefited from other developments in the relevant socio-technical system, or even from developments abroad. External factors may also be constraining, which may explain the lack of spin-off in this case.

Table 6.11 lists several developments and events in the socio-technical system of food consumption and production that are potential external factors. For instance, (Te Riele interview 2005) reference was made to several developments and events that relate to what was proposed in the SusHouse sustainable household nutrition visions, but that seem to have emerged fully independently of the backcasting experiment under scrutiny. For instance, the intelligent front door, which allows deliveries of foods when nobody is at home, was put on the market by Siemens, but without connection with sustainability. Community centres in rural areas increasingly offer help and services to local inhabitants, especially elderly and disabled. This is not stimulated from an environmental perspective but from a social perspective, by ministries dealing with social affairs or public health and wellbeing, or by local or regional authorities. Furthermore, local and regional markets for organic foods have emerged. Although several related international research activities have been mentioned, they were not considered spin-off of the backcasting experiment. Thus, Table 6.11 contains several potentially enabling factors and developments have been found, but they have not resulted in an increase in spin-off activities.

Table 6.11 *Potential external factors in the SHN case*

Shocks and crises
> Several outbreaks of livestock and bird diseases

Market and demand side changes
> Gradual growth in consumption of organic foods in the Netherlands, annually with 10-15%, although its share is still below 5%
> Supermarket 'war' starting late 2003 and leading to lower prices, reinforcing the trend towards own brands by supermarkets
> Steady growth in catering services, and to a lesser extent in eating out
> Emergence of slow food movement and organic food markets
> Ongoing growth of the market share of convenience products and ready-mades

Entries and exits
> Transfer of sustainable nutrition and food consumption from the Ministry of the Environment to the Ministry of Agriculture
> Exit of Atag and Boekos due to take-overs resulting in a withdrawal from sustainability-related activities
> Siemens brings intelligent front door on the market, though not in an eco-efficient variety
> Emergence of health and nutrition as an issue of societal concern and policy issue

New practices
> Shift in priorities at the Ministry of the Environment towards transitions and climate change
> Experiments in bottom-up activities in neighbourhood centres
> Enhancement of citizen participation in policy-making in general and in environmental policy especially

International
> Several related international research activities like HiCs and Suspronet

A constraining factor clearly was the exit by the two focal companies Atag and Boekos. These two firms had taken a leading role in environmental innovations and would have been capable of proceeding proposals and ideas that were attractive to them. However, after they were taken over, sustainability was no longer a priority,

Another possibly constraining development is the shift in policy priorities at the Ministry of the Environment (VROM 2001) towards transitions and climate change. This coincided with a reduction in the number of people involved sustainable products and sustainable consumption and the transfer the subject of nutrition and foods to the Ministry of Agriculture. There, the focus has been shifted to food supply chains away from food consumption and food consumption patterns. As a consequence, a successful project focusing on sustainable consumption like the Perspective project (Schmidt and Postma 1999) and a developed policy agenda on sustainable food consumption (Van der Pijl and Krutwagen 2001) did not result in lasting follow-up activities at the Ministry of the Environment.

By contrast, the Socio-Economic Council (SER 2003) made a plea to pay more attention to the opportunities for consumers and citizens in realizing a more sustainable consumption and environmental product policies, while the Netherlands Environmental Assessment Agency MNP has pointed in the same direction (see Vringer *et al.* 2001). Recently, a new interest in involving citizens in environmental policy has emerged (LeBlansch *et al.* 2003, VROM 2005), but it is still uncertain whether this will focus on existing policies or on developing new policy activities stimulating sustainable consumption.

6.6 Conclusions

6.6.1 Backcasting experiment

The SHN backcasting experiment has resulted in the involvement of a broad range of participants from different societal groups and in the generation of three future visions depicting sustainable household nutrition in the future. Stakeholder participation has been characterised by a low degree of involvement and by a considerable degree of influence on the content, but not on the process. The future visions were based on stakeholder workshops and were analysed with respect to their environmental gain, attractiveness to consumers and economic aspects. The three visions provided guidance and orientation during the course of the backcasting experiment, although several stakeholders were more attracted by topic and scope than by specific future visions. This implies that guidance and orientation were moderate.

The applied backcasting methodology matched the proposed methodological backcasting frame-work. Methods from all four groups have been identified in the backcasting experiment, as well as process demands, normative demands and to a lesser extent knowledge demands. No vision champion emerged in this backcasting experiment. Higher order learning has been identified, although the emphasis was not related to the topic of sustainable nutrition in the household, but to the methodology, or to comparing the three future visions. Learning about the methodology resulted in recommendations how the applied methodology could be improved. Insights included the degree of involvement, the generation of endorsement of visions and results, and adding an additional vision-specific step at the end. This additional step should focus on elaborating specific future vision or follow-up activities gathering stakeholders more strongly interested in these.

6.6.2 Follow-up and spin-off

The SHN backcasting experiment in the Netherlands has not led to lasting follow-up and spin-off. From this viewpoint this backcasting experiment cannot be seen as a great success and it has certainly not become the starting point of any regime shift of system innovation towards sustainability. The developed future visions were appreciated by the stakeholders, but have not become guiding images attracting old and new stakeholders and new resources.

Participating stakeholders saw opportunities for new activities and new cooperation and were willing to elaborate follow-up proposals. However, these failed to attract funding. Thus, the results show that the SHN backcasting experiment was successful in involving stakeholders who had the capacity to build alliances and were also able to interest their own organisation in the opportunities for these alliances and proposals. Apparently, there was no interest in funding the proposals internally, so the interest they did have did not mean they gave it priority. Besides, this study has not looked into individual factors that can be at play when developing and submitting proposals, like individual capabilities, sufficient capacity and the quality of proposals.

The development of proposals also illustrates the fact that spin-off is carried forward by individuals who participated in the backcasting experiment. A possible mechanism could be that participants see (strategic) opportunities for their organisation, both regarding ideas and potential partners. They then translate this into action and convince their own organisation of the opportunity, or are in the position that they can develop such initiatives themselves, while also looking for additional relevant partners for bringing such an opportunity to reality. This illustrates the need for participatory approaches in sustainability and system innovations towards sustainability. Ideas and elements of the nutrition backcasting experiment have been found in other projects and initiatives that were not considered as spin-off from the backcasting experiment. It suggests that ideas originate from participants, but that they can also bring it into other projects later on.

External factors also influenced the attempts and limited spin-off. For instance, two companies with a high motivation to participate and continue collaborating with regard to sustainable foods and environmentally sound kitchen equipment finished their activities, due to take-overs. The Ministry of the Environment reduced its activities in sustainable products and consumption, while the food-related activities were transferred to the Ministry of Agriculture. There, this topic got a food chain focus, and shifted away from a consumption focus. Several enabling developments could be identified too. However, these did not stimulate spin-off, even despite overlap in individuals and stakeholder organisations.

One other issue that could explain limited spin-off is that the backcasting experiment included the consumption and production system at the same time and addressed it in an integral way, while also combining products, consumption patterns and full food supply chains. As a consequence, there was no distinct stakeholder or stakeholder group that could be focused on. Furthermore, the focus on households and consumption as a starting point is an interesting one, although the question is who would take the lead first? When combining sustainable consumption and sustainable production chains, there is no stakeholder to take the principal lead, with the possible exception of the government.

6.6.3 Wider effects

The backcasting experiment on household nutrition in the Netherlands was part of a larger EU-funded research project 'Strategies towards the sustainable household'. This included an international subproject on nutrition in which backcasting experiments on household nutrition were also conducted in Hungary and UK. These backcasting experiments have briefly been evaluated. They also showed some follow-up, but largely in the research domain and only initiated by researchers who had been involved in the international project group on nutrition. Thus, a similar pattern emerges as in the Dutch SHN backcasting experiment. Spin-off is limited and what has been achieved is initiated by the 'organising' researchers. Although by no means irrelevant, it again begs the question how one can stimulate spin-off among (external) stakeholders.

The methodological spin-off of the SusHouse project has been more lasting. The developed methodology was applied in several other academic research and design projects. For instance, Partidario has applied it to study the future prospects for sustainability in paint chains in the Netherlands and Portugal (Partidario 2002, Partidario and Vergragt 2002), while it has also been applied to support the creation of an ecodesign network in a developing country (Vergragt et al. 2001).

Finally, the approach developed in the SusHouse project has been adapted for academic education. For instance, the Open University in the UK uses scenario material from the SusHouse project to stimulate creativity in a long-distance course on sustainable innovation (Green interview 2005, Vergragt interview 2005, Roy personal communication 2005). This not so much indicates a direct follow-up from the UK backcasting experiment on nutrition, but rather is the result of academic dissemination on the level of the entire project.

In addition, the methodology has been used in the graduate and postgraduate course programmes at the faculty of Industrial Design of Delft University of Technology (Silvester interview 2005, Vergragt interview 2005), while it has also influenced the key course in the university-wide specialisation in sustainability and engineering at Delft University of Technology (Quist *et al.* 2006).

Notes

1   Nutrition was renamed 'Shopping, Cooking and Eating' to emphasise the focus on households and consumption. However, in this chapter the term nutrition is used.

2   These groups were Technology Assessment Group, Delft University of Technology, also providing overall project design and coordination (the Netherlands); Szeged College of Food Industry (Hungary); Department of Industrial Design, Politecnico di Milano (Italy), Avanzi (Milano, Italy), Manchester School of Management, UMIST (UK), Lehrstuhl Markt und Konsum, University of Hannover (Germany).

3   Results from the consumer focus groups must not be seen as representative, but as indicative due to the low number of respondents involved. Due to the diversity of the groups (mainstream, green, dynamic) a wide range of views and opinions was collected, which makes it possible to improve the scenarios (Bode 2000).

4   The future visions were also input to the international multidisciplinary project group and the international subgroup on nutrition in the SusHouse project. Thus, the future visions also provided synchronisation between different disciplines present in the international project group.

5   Such daily coordination was achieved in the Dutch assessments of the future visions and to a certain extent also in the international research team on nutrition that had decided to have both country specific visions and joint visions in the three parallel backcasting experiments. This appeared feasible after evaluation of all three creativity workshops on nutrition. For instance, the service and outsourcing scenario was only developed in the Netherlands, while the intelligent and high-tech Intelligent Cooking and Storing scenario and the self-sufficient and rural Local and Green scenario had similar ones in Hungary and UK.

6   I am referring to myself here, as I conducted the SHN backcasting experiment in the Netherlands.

7   EET stands for Ecology, Economy and Technology. It is a governmental funding programme for radical environmental innovations. In 2004 this programme was succeeded by a programme funding cooperation in innovation in a broader sense, which was changed again in 2006.

8   The Dutch name is 'de Kleine Aarde'.

9   In Dutch BHC stands for 'Bedrijfschap Horeca and Catering´ (www.bedr-horeca.nl).

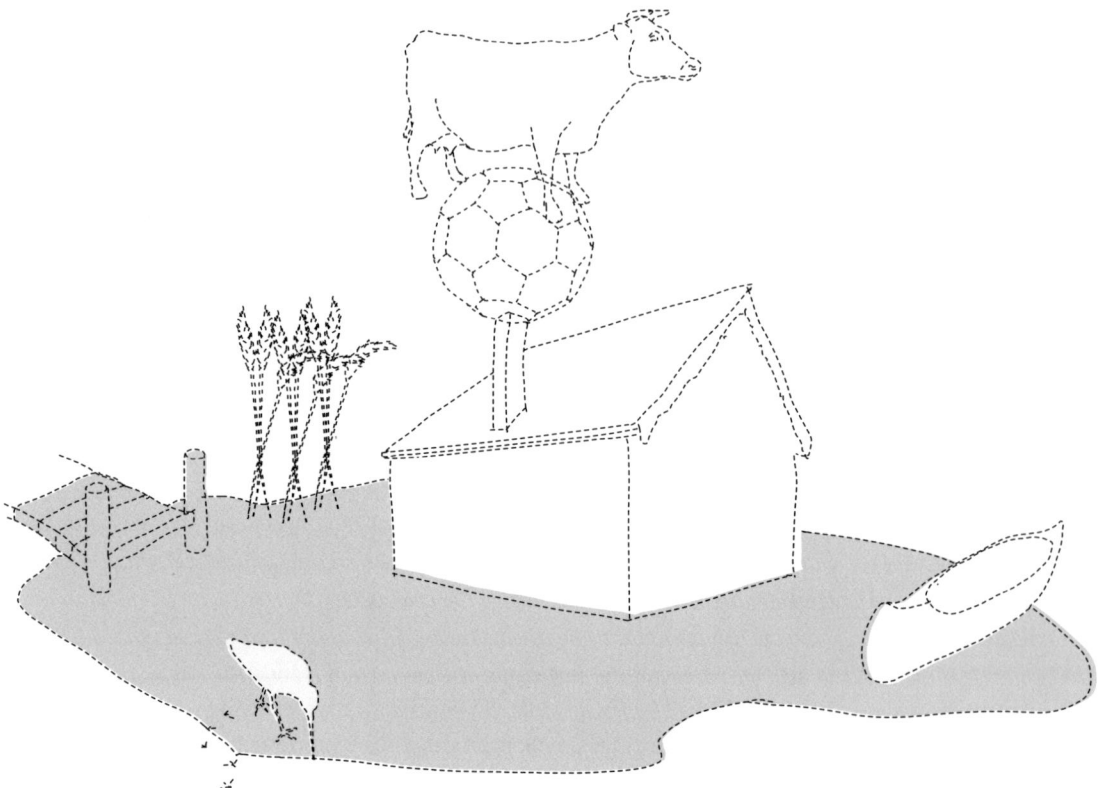

BACKCASTING FOR A SUSTAINABLE FUTURE: THE IMPACT AFTER 10 YEARS

7

BACKCASTING FOR MULTIPLE SUSTAINABLE LAND-USE IN RURAL AREAS

This chapter describes the Multiple Sustainable Land-use (MSL) case, which consists of the backcasting experiment on MSL that ran between 1993 and 1997 and its follow-up and spin-off after eight years. MSL stands for integration of spatial functions like agriculture, nature, recreation and water management, while meeting sustainability demands. This chapter subsequently introduces the topic of MSL (7.1), describes the MSL backcasting experiment (7.2), followed by an analysis (7.3). The chapter also describes follow-up and spin-off (7.4), followed by a further analysis (7.5) and conclusions (7.6).

7.1 Introduction

7.1.1 Agriculture and rural areas

For centuries land-use in rural areas has been dominated by agriculture, but nowadays agriculture is loosing its economic dominance in rural areas. Despite this development, present agriculture in the Netherlands is widely known for its dairy farming, intensive livestock production, intensive arable crop husbandry and large, fossil fuel-driven greenhouse horticulture. Strongly facilitated by Dutch and European agricultural policies[1], it developed into a vast and complex agro-food system[2]. This system not only includes primary agriculture, horticulture and auction hubs, but also food processing industries and suppliers to primary agriculture. The entire Dutch agro-food system still contributes significantly to the national economy and has a strong position on international markets. By contrast, the share of primary agriculture has fallen below 4% of the Dutch GDP in the late 1990s. Nevertheless, the total Dutch agro-food system, including the food industry, is widely considered a strong and competitive sectoral innovation system (Porter 2001).

The success of the Dutch agro-food system has been the result of a process of modernisation of more than a century of scaling-up of production, increasing productivity, rationalisation and mechanisation (Stuiver and Wiskerke 2004). This has contributed to the large surpluses of food on the European market, as well as to low food prices, but has also resulted in considerable side effects, both in environmental terms and in social terms. Focussing on the Netherlands, it has led to the deterioration of the existing rural landscapes, intensive use of energy and pesticides, a strong reduction of jobs in agriculture, a loss of biodiversity, emissions from manure, pesticides and a surplus of fertiliser use[3]. In the last decades intensive livestock production has also raised public concerns with respect to animal welfare, epidemic animal diseases like foot and mouth disease and swine fever, as well as food safety (due to mad cow disease (BSE), its human variety the disease of Creutzfeld-Jacob, bird flu and fodder contamination).

These developments have all contributed to huge changes in rural areas[4]. Primary agriculture (arable land farming, livestock production, horticulture, flower growing, etc.) is no longer the socio-economic carrier in many rural areas. This is enhanced by the gradual elimination of both EU market protection and the EU system of guaranteed prices for agricultural products. In addition, intensive livestock reconstruction policies and cattle epidemic diseases have strongly affected intensive livestock production.

In response to the above-mentioned developments, farmers have started other activities on the farm, or earn part of their income elsewhere. Small-scale 'niche' farming, raising horses, on-farm nature and biodiversity conservation, recreation and retired people settling in the countryside, as well as other activities have become important in many rural areas. As a result, there is a growing tension between different functions.

In the last decades the environmental problems in agriculture as well as the socio-economic problems have resulted in calls for an environmentally sound, or sustainable agriculture and in policy development focusing on rural areas (e.g. Driessen *et al.* 1995, Boonstra 2004: 47-72). However, achieving environmental improvement has appeared to be tough and has proceeded at a slow pace for various reasons. Agricultural actors have strongly defended existing interests and values. This was possible by a strong institutionalisation and the significant political influence of the so-called agricultural policy community, which was exerted through the Agriculture Board, a corporatist structure that was granted the privilege of influencing public policy-making in exchange for cooperation and support. For a long time the actors defending agricultural interests have formed a 'closed shop' or 'closed network'. As a result they successfully managed to resist most kinds of criticism with regard to environmental and ethical problems until the mid 1990s (Stuiver and Wiskerke 2004: 122-123, Grin 2004). EU policies were another major cause for the resistance against changes in agriculture. Finally, problems in rural areas were approached in a fragmented rather than integral way.

7.1.2 Multiple Sustainable Land-use

By the mid-1990s, various responses to the socio-economic and ecological problems in agriculture could be distinguished. The frontrunners of mainstream agriculture advocated a further scaling up and rationalisation of agriculture through the reduction of costs, physical inputs and emissions. For instance, re-structuring policies in intensive livestock production (especially pigs) not only aimed at moving move this type of farms from the vicinity of nature areas to areas of lower ecological value, but also at enhancing the concentration and scaling up of intensive livestock production. This was widely supported by actors involved in other functions, as it resulted in enhanced spatial separation of different functions and more space for other functions. For instance, more space was allocated for nature through the establishment of the National Ecological Network (EHS)[5]. An example of turning agriculture into nature is the island 'Tien Gemeten' that was bought by the nature organisation Natuurmonumenten (VNM) in the 1990s and is being transformed into a nature park after the farmers left the island.

Another major response aimed at dealing with the environmental problems in agriculture was advocated by the environmental movement, which advocated the de-intensification of agriculture and a shift towards organic arable farming, dairy farming and livestock production. This solution to a large extent neglects socio-economic aspects and it assumes significant changes in consumer awareness and consumption patterns. This solution has also been rejected by a vast majority of the agricultural sector and its representatives in national arenas. However, a growing minority has shifted towards organic arable farming and dairy farming, supported by the government and public interest groups. By 2006, organic agriculture is still steadily growing, while it has become institutionalised and has developed influence in the agricultural policy domain. The market for organic foods is small but growing towards 5%, partly because retailers have become increasingly interested in this share of the market.

A relatively recent response is Multiple Sustainable Land-use (MSL) in rural areas. MSL can be defined as integrating various spatial or land-use functions in a sustainable way. Although this term is also used for MSL in urban areas, water areas or land-water areas (e.g. Lagendijk and Wissenhof 1999, De Bruijn *et al.* 2004, www.habiforum.nl), I use it here to refer to MSL in rural areas. It must be noted that MSL is not considered the only future for Dutch agriculture. It has especially potential for areas providing various spatial functions with conflicting claims on the available space. In other rural areas the agricultural function could remain more dominant and might be reconciled with sustainable development in a different way.

The key to MSL in rural areas is the integration of the agricultural function with other functions, not only at the regional level, but also at farm and field level, which makes it possible to increase the earnings per hectare. Examples of other functions include drinking water supply, energy supply, nature conservation and management, and recreation. Two issues are important for realising sustainable and multiple land use:

1) Lowering the environmental burden by a transition towards environmentally sustainable agriculture,
2) Integration of functions in the same area, which leads not only to more diverse land use, but also to higher revenues per hectare and spreading the environmental burden over a larger number of functions.

The MSL concept was preceded in the early 1990s by several research and policy programmes in which farming was combined with nature management (De Graaf interview 2005, Korevaar interview 2005). However, these programmes were meant to combine only two functions on the farm scale. MSL also links to the debate on broadening agriculture, but uses a different starting point. Broadening agriculture takes the farm and the agricultural function as a starting point, and focuses on adding other functions, such as a camping, or producing cheese, to the farm. By contrast, MSL takes the regional land-use as a starting point and aims to integrate more functions in the same area, which makes it possible to combine a range of functions per unit of land area. For a discussion of different types of agriculture including multiple land-use, see Vereijken (2002).

7.1.3 Backcasting for MSL at the STD programme

Developing and implementing MSL in rural areas is a very complex issue. It assumes new ways of farming, and combining them with other activities at the level of fields, farms and regions. It also requires new organisations, new structures and institutions. Furthermore, it requires new knowledge, as the present stock of agricultural knowledge has been developed for mono-functional agriculture, producing high volumes and a strong spatial separation of functions. MSL also assumes new competencies, new activities and new ways of working together for both farmers and actors in rural areas. Introducing MSL requires major changes on multiple dimensions, involves many actors in different domains and functions, and affects the interests of these actors. It also includes uncertainty, complexity and possible conflicts of value and interest, leading to resistance from vested interests. It should thus be seen as a system innovation, also when striving for MSL in a specific region.

One occasion where MSL emerged as an attractive and sustainable alternative was in 1994 at the sub-programme on nutrition of the STD programme. This led to a backcasting experiment on MSL, in which the MSL option was explored and elaborated for the Winterswijk region, which is located on the eastern sand soils in the Netherlands. The backcasting experiment was succeeded by the demonstration and development programme MSL Winterswijk, which was coordinated by the province of Gelderland. Numerous other regional stakeholders were also involved, both at the programme level and at the level of particular projects (Stuurgroep MDL 1999, Akkerman *et al.* 2003).

This chapter describes and analyses the MSL backcasting experiment at the STD programme and its follow-up and spin-off. Data were collected through 10 in-depth interviews (see Appendix B) conducted in the period August - December 2005, an earlier case study (Quist and Vergragt 2001), additional telephone contacts, various documents and the Internet.

7.2 The MSL backcasting experiment[6]

7.2.1 The nutrition domain analysis

In September 1993, the sub-programme on Nutrition was initiated at the STD programme by means of an analysis of the nutrition domain, which took place parallel to the feasibility study on Novel Protein Foods (see Chapter 5). The nutrition domain analysis, which included stakeholder and expert interviews and workshops with visionary key persons from the field, was commissioned to a consortium led by the engineering consultancy Heidemij (presently known as Arcadis). At the STD programme Geert van Grootveld, the director of the STD office, and Oskar de Kuijer, the staff member for Nutrition, supervised the activities. The consortium included several research groups from Wageningen University, as well as the research institutes TNO Nutrition (presently part of TNO Life Sciences) and ATO-DLO (presently part of the WUR institute Agro and Food Innovation) to gather expertise on different parts of the nutrition domain. A steering group was established, consisting of the two STD staff members involved and representatives from the funding ministries VROM and LNV.

The researchers first analysed existing food consumption in the Netherlands, after which they determined the environmental impact of different food categories, like dairy, meat, fish, vegetables, etc. They also determined the share in the environmental impact of different stages in the supply chains of these food categories, such as agriculture, processing, retail and consumption. Major sustainability problems that were identified included intensive livestock production for meat and dairy, greenhouses vegetables and primary agriculture in general (STD 1994, Heidemij *et al.* 1997).

The results of the analysis were fed into a workshop in January 1994 (STD 1994) in which twenty key persons participated. Attendance covered academic food and agricultural research, food retail, food industry and several ministries. The workshop resulted both in trend-based future outlooks and desirable sustainable future images. Using backcasting, three subgroups looked into the technologies that would be needed achieve the sustainable future images (Aarts 1997: 36, STD 1994). Key elements in the future images included closed-cycle systems, food production in the vicinity of consumers, and a major reduction in the use of energy and resources. Four possible sustainable system options were identified, based on the three desirable future images, all dealing with different parts of the nutrition domain. The four system options were (i) genetic plant modification, (ii) fully controlled closed-cycle zero emission greenhouses, (iii) integrated crop conversion, and (iv) multiple land-use (MSL). MSL was at first formulated as agricultural systems, also providing non-agricultural functions. Including the MSL option also made it possible to take on board two major existing visions on sustainability in agriculture and foods, which was supported by the STD office. This would prevent being connected to only one of them (Aarts 1997: 37).

The results of foresighting activities on all four options were fed into a second workshop in March 1994 with sessions on MSL and the two other remaining sustainable options[7]. The workshop involved not only the participants of the first workshop, but also additional experts and stakeholders who were relevant to the generated options and included consumer and environmental organisations. Focusing on the MSL option, the March workshop resulted in further identification and a preliminarily ranking of both critical technologies and uncertainties for the MSL option. It also resulted in support from the participating stakeholders for further elaboration of the MSL topic (Aarts 1997: 38, see also STD 1994). Also, a brochure on nutrition was published containing the future vision on MSL (De Kuijer 1995a).

Table 7.1 *Timeline of the MSL backcasting experiment*

Nutrition domain analysis at STD programme (STD 1994)	
September – December 1993	Analysis of nutrition domain including 15 expert interviews for feedback.
25 January 1994	Stakeholder workshop with 20 stakeholder participants developing future images for nutrition resulting in four broad sustainable options including MSL.
February – March 1994	Elaboration by researchers of the MSL option and three other options for sustainable nutrition and food production using additional expert interviews.
30-31 March 1994	Stakeholder workshop on the MSL option.
MSL feasibility study	
May – December 1994	Preparatory activities for the MSL feasibility study.
January – September 1995	Feasibility study on MSL.
May 1995	First round of seven interviews with stakeholders in the regions of Flevoland and Winterswijk about the concept of MSL and interest for a pilot in the region.
September 1995	Second round of nine interviews in the same regions.
MSL project: 1st stage	
Late (October) 1995	Decision by STD steering committee to approve the selection of the Winterswijk area as the pilot area for elaborating the MSL topic.
February –March 1996	First round of interviews with MSL stakeholders in the Winterswijk region asking for present and future goals regarding functions in the region.
April 1996	Second round of stakeholder interviews on the list of stakeholders' aggregated goals, possible participation and requesting co-funding.
July 1996	Final report of the first stage of the MSL project.
MSL project: 2nd stage	
Mid – late 1996	Further elaboration by the research bodies, resulting in the design of three multiple-goal farming systems: (1) nature, (2) country-estate, (3) meat production.
November 1996	Establishment of steering group MSL Winterswijk consisting of 20 stakeholders.
February 1997	Steering group of the MSL project becomes responsible for the MSL project and decides to develop follow-up demonstration and development projects.
February – April 1997	Selection and elaboration of follow-up project proposals.
June 1997	Consultation of steering group members by project manager.
July 1997	Publication of final reports from the MSL project. The steering group supports the proposed follow-up; the province of Gelderland will coordinate the follow-up.
August – September 1997	MSL project team has stakeholders meetings about proposed follow-up projects.
1 December 1997	Official transfer of the MSL demonstration programme from the STD programme to the province of Gelderland.
End 1997	Local information and discussion meetings in Winterswijk area.

7.2.2 MSL feasibility study and MSL project

MSL feasibility study

The MSL feasibility study was meant as a next step towards a larger project on MSL. Again, a broad consortium was established by three organisations that had been involved in the nutrition workshops. These were (1) the research institute in agricultural biology AB-DLO (presently part of the WUR institute PRI), (2) the environmental biology group of Leiden University and (3) Heidemij consultancy. The latter was experienced in dealing with stakeholders in spatial issues. AB-DLO had expertise in combining agriculture with nature and in environmental aspects of agriculture, for instance with regard to nutrient management in dairy farming (e.g. Van der Meer and Spiertz 1992). The environmental biology group had worked on integrated agriculture and had developed a methodology for developing and evaluating this on a regional scale (e.g. De Graaf *et al.* 1999).

As a first step a general future vision for sustainable land-use was developed by the researchers, which was further elaborated for two rural agricultural regions in the Netherlands (STD 1996d: 8-13). The regions were the Flevoland polders and the eastern sand soils, which includes the Winterswijk region. The Flevoland polders are a typical example of recent land reclamation in the Netherlands with land-based arable land and dairy farming. The target area 'eastern sand soils' was narrowed down to the Winterswijk region, a traditional small-scale landscape, where mixed cropping systems of arable crops and dairy farming were combined with intensive livestock production and some forestry. Here, intensive farming systems, especially livestock production had resulted in serious environmental problems like high nitrogen emissions with pollution of air, ground and surface water as a consequence, phosphate accumulation in soils, and lowering of the ground water level. Through backcasting from the two regional future visions, a list of critical technologies was conceived, as well as cultural and structural conditions (STD 1996d: 29-47).

Two rounds of stakeholder interviews were conducted in both regions. Interviews were held with key stakeholders to determine commitment, opportunities and constraints. Respondents included agricultural organisations, agriculture-related industry, regional authorities and stakeholders from other relevant functions like energy supply, water capture and nature conservation. Next, the Winterswijk region was selected to elaborate the MSL concept, because of enthusiasm in the region, the presence of WCL[8] Winterswijk in which stakeholder representatives from various functions were already cooperating, and the high level of ambition to combine different functions (Aarts 1997: 43).

MSL project: 1st stage

In October 1995, a proposal for a larger project on MSL was submitted to the financing ministries and the steering committee of the STD programme (De Kuijer 1995b). Despite initial resistance from the Ministry of Agriculture, the proposal was eventually approved (De Kuijer interview 2005). The STD staff member for the Nutrition sub-programme became the MSL project manager. This was different from most other STD projects, where external key persons with a strong reputation and an extensive network were appointed as project managers (Aarts 1997).

The MSL project consisted of three interrelated lines of research on: (1) technology and multiple-goal systems that would combine at least three functions, conducted by the agro-biological research institute and supported by expertise from other agricultural institutes, if necessary; (2) regional visions and scenarios carried out by the academic environmental biology group; and, (3) stakeholder involvement and external communication. A project team of seven people was established, including representatives from the three contractors, two representatives from the Winterswijk region, in addition to the project manager and a project coordinator at the STD office (Aarts 1997: 45).

Raising awareness about the concept and opportunities of MSL among stakeholders and achieving participation and funding from third parties like companies, utilities and nature organisations were important at this stage. This was dealt with in two rounds of stakeholder interviews in February and April 1996 respectively. Each round consisted of eight interviews. The first series of interviews focused on providing information on the concept of MSL, as well as on the long term views and goals of stakeholders (see De Graaf and Musters 1997: 46-53). Goals included solving regional environmental problems and maintaining the agricultural function. Goals could also be improving the socio-economic potential in the region or the opportunities for recreation and the production of water and energy. The results of the interviews were used to extend the generic vision on MSL with the aggregated regional goals. This resulted in a regional vision on MSL, while backcasting showed that technologies, organisational structures and institutions were lacking. In addition, aggregated goals were processed into land-use maps reflecting future regional opportunities, while taking the physical characteristics of the region into account. The land-use maps showed that it was possible to combine all the goals, and to do so in a number of different ways. A large part of the area allowed combining five to ten sub-functions like pasture, arable land, nature, forest, recreation, water conservation and different types of intensive livestock production and different types of energy production (De Graaf and Musters 1997: 14-15).

The second round of interviews focused on stakeholder feedback on the processed results of the first round. The interest in participating in the MSL project and funding was also gauged. Several stakeholders like the provincial board, the regional energy utility company Nuon, the regional water provider, several nature organisations and the local branch of the Rabobank were willing to contribute financially. Meanwhile, the researchers developed seven possible multiple-goal farming systems, each combining at least three functions. For instance, arable land farming or livestock production was combined with functions like energy production, water conservation, nature management or recreation (STD 1996e).

MSL project 2nd stage

The second stage of the MSL project started in August 1996. As a first step the seven multiple-goal farming systems were further elaborated (see Aarts and De Kuijer 1997a), which required researchers and expertise from various other agricultural (sub)disciplines and other research institutes of DLO. Next, the number of farming systems was reduced to three, which were elaborated in a greater detail, using elements from all seven proposed farming systems. The ones that were selected were (i) nature farming system, (ii) country-estate system and (iii) meat livestock farming system (Aarts and De Kuijer 1997b, De Kuijer et al. 1997a). Building on the three elaborated farming systems, researchers defined technical building bricks[9] for MSL. These were used as input to constructing scenarios for the region in the year 2020 and assessing these on various environmental and economic parameters (De Graaf and Musters 1997). Next the most important bottlenecks with regard to realising MSL in the region were selected, as well as the most crucial and innovative elements. Based on the results of a workshop attended both by researchers and by stakeholders, nine follow-up proposals were selected (De Kuijer et al. 1997a, Beeren et al. 1997). Also, a related research programme was defined (Zwart 1997).

Parallel to the research activities, a steering group of twenty stakeholders was established in November 1996 (Aarts 1997: 49, De Kuijer et al. 1997b: 7). In February 1997, the steering group became formally responsible for the MSL project, although decisions continued to be prepared and strongly influenced by the project manager and the project team (Aarts 1997: 49). In July 1997, the steering group decided to continue the MSL project after the STD programme would end. The province of Gelderland was willing to coordinate the follow-up activities under the condition that the other stakeholders in the steering group would stay involved as well (Aarts 1997: 51), which was indeed the case. The steering group also requested the STD project team to continue further preparations of the proposed follow-up programme and specific projects. This included mobilising further stakeholder commitment and funding, which took place in August and September 1997. The official transfer of the MSL project and its follow-up to the province of Gelderland took place in December 1997.

7.2.3 The results of the MSL backcasting experiment

It was concluded that applying the MSL concept in the Winterswijk region, in combination with the use of new technologies and closing material flows, could reduce the environmental burden up to a factor 10 (De Kuijer *et al.* 1997b). The results of the backcasting experiment included a future vision on MSL in the Winterswijk region. This vision combined the generic MSL concept with the aggregated demands and goals from different functions, in addition to the evidence in maps that this could be realised in the area (De Graaf and Musters 1997). These maps showed that a large part of the region physically enabled combining between 5 and 10 (sub)-functions asked for by the stakeholders. Backcasting in several stages of the MSL backcasting experiments showed that technologies, organisational structures and institutional arrangements to realise MSL in the region were lacking. Backcasting also resulted in the identification of critical technologies, technological bottlenecks and structural and cultural conditions (De Kuijer *et al.* 1997a, STD 1996d).

Two possible scenarios were developed for the Winterswijk region, which fitted in the future vision and both integrated a range of functions in different ways (De Graaf *et al.* 1999, De Graaf and Musters 1997). The first scenario combined nature, agriculture and water production and was based on existing land use. The second scenario combined recreation, agriculture and drinking water production. It required considerably larger changes in the existing land-use and aimed at restoring the traditional landscape, but resulted in lower energy requirements and more employment due to the growing labour demand for recreation. Both scenarios showed considerable environmental improvement and were attractive in terms of economic value and employment compared with a business as usual scenario based on extrapolations. Environmental improvement would be achieved through closing material flows at the regional level as much as possible; it would require technology to upgrade biomass and re-use of manure in the region. In addition, both scenarios assumed regional self-sufficiency in energy supply, a lower number of cows per hectare, a strongly increased capture of water, and both an extension and enhancement of nature.

Three multiple-goal farming systems combining different functions were elaborated (De Kuijer *et al.* 1997a: 19-24, De Kuijer *et al.* 1997b: 18-21). Firstly, the nature farming system, which combines nature development and water production (based on surface water) with recreation and food production, while being self-sufficient in terms of energy. Secondly, the country-estate farming system combines recreation and production of food or biomass with managing small-scale landscapes, while also providing energy and water capture. Thirdly, the meat livestock farming system is a large-scale pig farm, also producing (i) energy derived from the heat produced by the pigs, and (ii) manure-based fertiliser that enables export of nutrients to regions where part of the required fodder would be produced. The MSL backcasting experiment also resulted in nine project proposals for demonstrating multiple land-use and for solving technological bottlenecks (Beeren *et al.* 1997) and a proposal for a related research programme (Zwart 1997). These can be seen as a follow-up agenda.

7.3 Analysis

7.3.1 Stakeholder participation

Stakeholders were involved in all stages of the MSL backcasting experiment. The type and degree of involvement evolved throughout successive stages. The degree of stakeholder influence and what stakeholders had influence on varied too. Therefore, a distinction is drawn between the three stages of the backcasting experiment: the nutrition domain analysis, the MSL feasibility study and the MSL project. For each stage I look at the degree of stakeholder heterogeneity, the degree of involvement (reflecting how much time is spent), the type of involvement (focusing on other types of participation in addition to capacity to participate like providing funding, expertise, or extra capacity) and the degree to which influence was exerted on the content or process of the backcasting experiment.

Stakeholder heterogeneity is shown in Table 7.2. It shows that stakeholder heterogeneity was good in all three stages of the backcasting experiment. The involvement of public interest groups was lower at the first and third stages in comparison to the other three categories distinguished.

Table 7.2 *Stakeholder heterogeneity in the MSL backcasting experiment*

	Research organisations	Companies	Government	Public interest groups
Nutrition domain analysis	● (>10)	●	●	·
MSL feasibility study	· (<5)	·	●	·
MSL project	● (5-10)	●	●	·

Table 7.3 shows results on the other three indicators for major groups of stakeholders at each stage of the backcasting experiment. Here, stakeholder groups are defined according to their role in the process of the backcasting experiment, instead of by their societal domain. For the nutrition domain analysis a distinction can be made between the contractors, the experts who were interviewed, the participants in the two workshops, besides the STD staff members and the ministries involved. Funding was provided by the Ministry of Agriculture and the Ministry of the Environment, who were represented in the steering group, as were the two staff members of the STD office. Apart from the STD staff members, the steering group had a low degree of involvement, but had considerable influence on the selection of the contractors and on the approach, as they asked to revise the submitted proposal. The consortium was also requested to be extended with experts on nutrition and food processing and to take a more process-oriented, inductive and holistic approach through workshops with visionary people (Aarts 1997: 34).

The nutrition workshops involved more than thirty senior level stakeholders, such as academic professors, directors from research institutes, R&D directors from major food and agricultural companies, senior staff members from the relevant ministries and a few representatives from environmental and consumer organisations. The focus was on consultation, asking for opinions, ideas and support. The stakeholders and experts involved in the workshops and the interviews thus shaped the agenda and content of the nutrition sub-programme. Workshop participants also supported the emergence of the MSL sustainable system option and thus exerted considerable influence on the content and the way it was defined. In this way they had a high degree of influence on the content, where as the STD staff members, the consultancy contractor and the steering group had a considerable influence on the process

Table 7.3 *Characterising stakeholder participation in the MSL backcasting experiment*

Stakeholder group		Degree of involvement	Type of involvement	Degree of influence [i]
Nutrition domain analysis	Workshop participants and interviewees	low	Capacity	medium-high
	Contractors/ research team	medium	Capacity	low-medium
	Ministries	low	Funding	low-medium *(on process)*
	STD office	low-medium	Capacity	low-medium / medium-high *(on process)*
MSL feasibility studies	Interviewees in Flevoland and Winterswijk	low	Capacity	medium-high
	Contractors/ researchers	medium	Capacity	low-medium / medium-high *(on process)*
	STD office	medium	Capacity	medium-high
	Ministries	low	Funding, some capacity	low-medium
MSL project	Regional stakeholders	low	Capacity, co-funding	medium-high
	Contractors/ researchers	low-medium	Capacity, co-funding	low-medium
	Project team/ STD office	medium	Capacity	low-medium / medium-high *(on process)*
	Ministries	low	Especially funding	low

low	low-medium	medium	medium-high	high
● (small)	◉	●	◯	⬤

i   Unless otherwise indicated, reference is made to the degree on content

In the MSL feasibility study the contractors and the responsible STD staff member were part of a project team and therefore had a high degree of involvement; they had considerable influence on the elaboration of the topic and on the content of the research. During the feasibility study seven interviews took place with crucial stakeholders in two possible 'test' regions. The interviewees were high-ranking individuals from regional authorities, key companies and regional farmers' unions. These stakeholders had a low degree of involvement, but had a major influence on the decision to focus on the Winterswijk region.

During the MSL project stakeholder contacts in the Winterswijk region were gradually broadened and intensified. For instance, the project team was extended with representatives from the region; one from a nature organisation and one with an agricultural background. A further distinction can be drawn between the contractors, regional stakeholders and ministries involved. Regional stakeholders were involved through two rounds of eight interviews with individuals at a senior or board level. These interviews strongly influenced the content and direction of further elaboration of the MSL vision. Interested stakeholders were asked to contribute financially and could then become a member of a steering group that would be established. They could then participate in research meetings and become involved in possible follow-up projects. Interested stakeholders included farmers' unions, local and regional authorities, nature organisations, a local branch of a national bank active in the agricultural sector, regional agriculture-derived industry, the water sector and an energy company (De Kuijer et al. 1997b). The steering group, which mainly involved regional stakeholders, became formally in charge during the MSL project and could thus exert considerable influence.

Research stakeholders remained important in the MSL project, most notably the different research bodies that were contracted. Next to the academic environmental biology research group, the main stakeholder in the MSL project was the DLO research division of the Ministry of Agriculture, through its institute for agro-bio-logical research (AB-DLO). This institute also coordinated the contributions from researchers from a range of agricultural research institutes for elaborating various land-use systems and follow-up proposals. The degree of involvement of these stakeholders was high, and they could exert moderate influence on the content.

7.3.2 Vision aspects

Vision: inception

At different stages of the MSL backcasting experiment, different types of future vision were developed. In addition, gradual development and elaboration of the future visions took place. For instance, the nutrition domain workshops resulted in a future vision on sustainable food production and enabling technologies, in addition to a trend-based outlook on the future. During the workshops the MSL concept emerged as an interesting 'idea' and it was further elaborated as an attractive sustainable system option. Shortly afterwards, this was elaborated in a more coherent future vision on MSL (De Kuijer 1995a: 12).

During the feasibility study, researchers developed a future vision on MSL, which consisted of a generic vision describing the core idea and assumptions, in combination with elaborations for two regions. After selecting the Winterswijk region, a future vision on MSL for this region was developed, which vision consisted of the MSL concept of function integration and the aggregated future goals derived from interviews with regional stakeholders. In addition, three multiple-goal farming systems were designed, using the regional future vision as a heuristic. These can be seen as more focused and bounded visions that are nested in the regional vision on MSL. At the same time the regional vision is nested in the generic MSL vision. The two scenarios that were developed for 2020 are not truly future visions, but must be seen as two possible futures within the boundaries of the main vision. However, if it had been decided that one of the scenarios should be strived for, it could have become the future vision providing guidance and orientation.

Vision: guidance

Guidance has been analysed in terms of (1) collective normative projection, (2) synchronisation and alignment of the interactions among stakeholders from different backgrounds, (3) alternative (system of) rules and institutions.

Collective normative projection

The future vision of MSL provided a collective normative projection to the future not only at a general and conceptual level, but also as a regional vision for the Winterswijk area. The regional vision was nested in the conceptual vision and both were aligned to one another.

Synchronisation

The future vision for the region provided synchronisation and alignment among the researchers from different disciplinary backgrounds and a range of stakeholders involved. This was important because regular agricultural research strongly linked reductions in the environmental impact of agriculture to further intensification, rationalisation and up-scaling; de-intensification was not considered economically viable. In addition, the vision allowed people not only to look from different perspectives and backgrounds at the same phenomenon, but also to carry out assessments using the same set of (normative) assumptions. It also helped direct the contributions from various agricultural research institutes, which were all needed to design multiple-goal farming systems. The vision thus shaped and structured the activities in the backcasting experiment. However, the vision in turn was itself shaped by these activities, the stakeholders, the researchers and the project team.

Alternative rules

The future vision and the elaborated multiple-goal farming systems assumed a range of alternative rules, which would be needed to make MSL work both at the level of the Winterswijk area and at the level of the defined multiple-goal farming systems. The conclusion in the backcasting experiment that organisational and institutional arrangements to implement MSL were lacking, confirms that new (sets of) rules were necessary to enable MSL. These were clearly part of the regional vision and its elaboration in the multiple-goal farming systems.

Vision: orientation

Orientation has been analysed in terms of (1) cognitive activation, (2) mobilisation of actors and resources, and (3) daily decentralised coordination.

Cognitive activation

The future vision provided substantial cognitive challenges and resulted in activities dealing with these challenges. For instance, these cognitive challenges were translated into design activities and analytical research by the researchers involved. Cognitive activities included developing the vision using the stakeholder inputs, design of multiple-goal farming systems, constructing and assessing scenarios and developing follow-up proposals. These activities included contributions of the stakeholders involved.

Mobilisation

The future vision was used to mobilise and attract actors and resources. For instance, it was important in obtaining approval and funding by the involved ministries. The mobilisation of regional stakeholders, who were willing to join the steering group and to co-fund the MSL project, also illustrates this. A considerable number of stakeholders became interested in joining follow-up proposals. Mobilisation is also illustrated when the steering group collectively decided to continue MSL follow-up activities in the Winterswijk region, after which the province of Gelderland was willing to do the coordination.

Decentralised coordination

There was a strong coordination among those involved during the MSL backcasting experiment, which was to a considerable extent provided by the future vision and its further elaborated versions. Thus, the future vision not only provided daily coordination, but it contributed to a considerable extent to coordination in decentralised activities that took place on different locations in the Netherlands. The future vision also provided coordination between the different stages of the backcasting experiment when the STD had to develop proposals and commitment among relevant ministries and in the steering group at the level of the STD programme. Furthermore, the overall methodology applied in the MSL project provided daily coordination, as did the project structure itself. The latter included research meetings, stakeholder meetings and the way of commissioning research activities, and the applied project management methods.

Competing visions

Two major alternative visions were around when the MSL vision emerged in addition to the dominant vision on agriculture. The first one aimed at reducing the environmental burden of agriculture by further up-scaling, technology development and intensification, which was aligned with the dominant vision in agriculture and the knowledge infrastructure on agriculture and food. The second one comprised decreasing the environmental burden by de-intensification of agriculture and included concepts like organic agriculture or ecological agriculture. It was connected to the development of the niche of organic foods, strongly supported by environmentalists. Because of the influence of both visions and its supporters, it was advocated (and effectuated) at the STD programme that it should also deal with open multi-functional systems, as this would enable linkages with the supporters of both visions.

Another alternative vision emerged in the early 1990s. This was the vision on multifunctional agriculture, which focused on broadening agriculture at farm level. It had led to experiments and policy programmes on combining agriculture with nature management and related research, while farmers started additional activities on the farm that could provide additional earnings. This vision is to a certain extent related to the MSL vision, but takes agriculture and the farm as its starting points. It was also proposed within the MSL backcasting experiment by the agricultural researchers (Spiertz interview 2005, De Graaf interview 2005). However, the emphasis was kept on integrating functions on the same surface area through multiple land-use in which all functions were equally important.

Finally, in an early stage of the backcasting experiment it had also been proposed to look for services with high added value and green industries as new economic carriers in rural areas, but this had not become part of the regional MSL vision (Spiertz interview 2005).

7.3.3 Learning

In this study higher order learning (see Chapter 4) is defined in terms of (1) a shift in framing major problems and of perceived solutions by specific actors; (2) a shift in the principal approaches to solving these problems, as well as in major priorities by specific actors; and (3) joint learning, a shift in congruence and joint opinions in any of the issues related to the previous shifts.

A distinction is drawn between learning about the topic of MSL and learning about the participatory backcasting approach. It is assumed that higher order learning about the topic mainly involves the MSL stakeholders, while higher order learning about the approach is expected among those organising and managing the backcasting experiment. The purpose here is to provide some empirical evidence of higher order learning. First order learning is not included in this evaluation, but it was achieved including, but not limited to, cognitive knowledge and can for instance be found in project documents and reports.

Higher order learning on the topic

The MSL concept was a new concept and a novel way of thinking for many stakeholders and researchers involved in the backcasting experiment. This resulted in higher order learning, not only concerning shifts in framing the problem and principal approaches to dealing with the problem, but also in shifting priorities as well as in joint learning. Regional stakeholders learnt about the concept, the approach and the opportunities of MSL, which is both a shift in perceived solutions and in priorities. This was also the case when the province of Gelderland took the lead in coordinating the MSL activities after the STD programme. When the MSL steering group collectively decided to continue, this is an example of joint learning and a shift in congruence and joint opinions.

Higher order learning about the approach

Learning took place with respect to future orientation and the backcasting approach using normative desirable future visions as a way to do that. To many of the stakeholders involved this was novel in many aspects and was considered very interesting (Spiertz interview 2005, Tiggeloven interview 2005, Korevaar interview 2005, Kiljan and Moolenaar interview 2005).

At the agricultural research institute, involvement led to learning with respect to inter-disciplinary research and involving stakeholders in research activities. This has led to a broadening of research approaches that can be applied. Working inter-disciplinary crossing the borders of research institutes was also very uncommon before, which was also the case for involving stakeholders in research activities and giving them influence in the direction and scope of the research projects (Spiertz interview 2005). Finally, it was regretted that no time was available to explore and debate the MSL options more extensively on a conceptual level before testing it in a specific region (Spiertz interview 2005).

7.3.4 Methodological aspects and settings

Backcasting framework: inter-disciplinarity

To start with, the MSL backcasting experiment was clearly inter-disciplinary. It involved a range of (sub)disciplines from agricultural sciences through researchers from various specialised agricultural research institutes. It also involved an inter-disciplinary academic research group, which combined land-use modelling, monitoring and scenario construction with socio-economic and environmental analyses. The backcasting experiment also combined design activities with multi-disciplinary analyses and integration of the results.

Backcasting framework: steps

The nutrition domain analysis can to a large extent be seen as a strategic problem orientation. A future vision was also constructed in the stakeholder workshops on nutrition. This vision was used to identify possible sustainable technology (or system) options, uncertainties and critical technologies, which refers to backcasting analysis.

The MSL feasibility study did not only explore and define the topic, but also included elements of problem orientation, problem definition and backcasting. The problem was redefined from 'what are the main unsustainabilities in nutrition and what are potential future sustainable technical solutions' to 'what is the sustainability potential of the MSL concept for rural areas'. A new future vision, taking the form of a narrative (STD 1996a: 8-13), was developed by the contracted researchers, both at a general level and further elaborated for two regions in the Netherlands. This future vision was used for a backcasting analysis, in which not only critical technologies were identified (STD 1996d: 29-43), but also several types of bottlenecks, such as cultural, organisational and structural ones (STD 1996d: 45-47).

The MSL project included various activities. It started with stakeholder consultation, vision development for the region and analysis of what was missing (using backcasting). Parallel to this, design and analysis of multiple-goal systems took place. This refers to the fourth step of the backcasting framework, as do constructing and assessing the two scenarios. The elaboration of follow-up proposals for demonstration projects and an overarching research programme also refers to the fourth step in the backcasting framework (of elaboration, analysis and defining agendas). The activities aiming at embedding outcomes and stimulating follow-up activities among stakeholders relate to the fifth step of the backcasting framework. However, raising involvement and embedding (intermediate) results like the MSL concept and the MSL vision among stakeholders was an ongoing activity throughout of the backcasting experiment. As a consequence, various rounds of stakeholder interviews and various workshops took place throughout the MSL backcasting experiment.

Table 7.4 *Backcasting framework steps compared to the MSL backcasting experiment*

Steps for participatory backcasting	Stages for MSL backcasting experiment
(1) Strategic problem orientation	Nutrition analysis, MSL feasibility study, MSL project
(2) Develop future vision	Nutrition analysis, MSL feasibility study, MSL project
(3) Backcasting analysis	Nutrition analysis, MSL feasibility study, MSL project
(4) Elaborate future alternative and define follow-up agenda	MSL project
(5) Embed results and agenda, stimulate follow-up and spin-off	MSL project

In short, all steps of the backcasting framework can be distinguished in the MSL backcasting experiment, as indicated in Table 7.4. However, the framework steps were not applied in a proposed order, but they were used in different varieties at various stages of the MSL backcasting experiment. In addition, iteration of steps 1-3 took place. For instance, vision development was carried out in three stages, but in a more detailed way, with new stakeholders, or for a more narrowly bounded region. This is in line with what has been said in Chapter 2, that the proposed steps in the framework should not be seen in a linear way, but that iterations and moving forward and backward between steps is possible. Finally, stakeholder interaction and raising commitment was not limited to the last stage, but was repeated and can be seen as an ongoing activity.

Backcasting framework: applied methods

Methods from all four categories in the methodological framework were applied in the MSL backcasting experiment. Examples of methods are listed in Table 7.5. No distinction is drawn here between the three stages of backcasting experiment. Descriptions of the methods can be found elsewhere (e.g. STD 1996d, De Kuijer *et al.* 1997a).

Table 7.5 *Methods applied in the MSL backcasting experiment*

Analysis	Design
> Technological (bottleneck) analysis	> Future visioning (in workshops)
> Environmental and socio-economic scenario analysis	> Construction of future land-use maps based on aggregated stakeholder goals
> Trend and domain analysis	> Development of regional scenarios
> Uncertainty analysis and ranking	> Design of multiple-goal farming systems
> Desk study	> Research design and project design
	> Follow-up proposal elaboration
Participation/ Interaction	**Management, communication, coordination**
> Visioning workshops	> Project start-up meetings
> Brainstorming	> Research meetings and workshops
> Stakeholder and expert interviews	> Stakeholder steering group
> Stakeholder workshops	> Dissemination of brochures and reports
> Stakeholder steering group	> Ranking and selection of activities, gradually narrowing of content and options
	> Regular project management methods

In the nutrition domain analysis various analytical methods and foresighting tools were applied, although for the latter no formalised tools or prescriptive methods were used. In addition, expert interviews and stakeholder visioning workshops were conducted. Within the formats of the two workshops, trend-based future outlooks and desirable future images were constructed using brainstorming techniques in subgroups. No formalised methodologies were reported, although hired professional workshop facilitators developed structured workshop scripts and applied specific creativity, ranking and evaluation techniques (Aarts 1997).

In the MSL feasibility study stakeholder consultation was important as well, which was mainly done through interviews. In addition, the vision for MSL was further elaborated, which took the form of narratives without the use of a formalised prescriptive methodology. Backcasting analysis was (implicitly) applied, in addition to several other analytical methods. To evaluate the demands that functions put on each other, an expert workshop was organised by the consultant from Heidemij and the STD staff member for nutrition, which was attended by the contracted agricultural researchers (Aarts 1997: 42).

In the MSL project a future vision was constructed following a prescriptive methodology (De Graaf and Musters 1997, De Graaf *et al.* 1999); it involved aggregating stakeholder goals and evaluating opportunities for function integration, which was summarised in maps (see Figure 7.1). Backcasting analysis was applied to identify bottlenecks and necessary changes, although without the use of prescriptive or formalised methods. Also, multiple land-use systems were designed, elaborated, evaluated and selected. Scenario modelling and construction methods were also applied, as was scenario analysis. Furthermore, stakeholder interaction methods, such as the establishment of a stakeholder steering group, research (exchange) meetings and workshops were applied at this stage.

With regard to design activities and methods, most interviewees did not mention design activities as such, or did not report on prescriptive or formalised design methodologies (Spiertz interview 2005, De Kuijer interview 2005). Nevertheless, design activities were essential at all stages of the backcasting experiment. For instance, in the first nutrition workshop, future images were constructed using a brainstorm, which is thus both interactive and part of a design activity. Between the two nutrition workshops possible options were elaborated and defined, which also assumes design activities. Future visions were further elaborated in the feasibility study, which also includes design activities. In the MSL project multiple-goal farming systems were designed and follow-up project proposals were elaborated, in these design activities no formalised design methodologies were applied. By contrast, the methodology for developing regional maps based on stakeholder and physical opportunities for multiple land-use and the construction of the two scenarios used a prescriptive methodology.

Backcasting framework: demands

A distinction has been made between (i) normative demands, (ii) process demands and (iii) knowledge demands. These demands can, for instance, be found in the terms of reference that were defined at an early stage (STD 1994), or the objectives and success criteria that were defined at a later stage (De Kuijer 1995b: 3).

To start with, the normative demands set at the level of the STD programme were also guiding the MSL backcasting experiment. Normative demands included, in addition to the factor 20 environmental improvement, taking into account cultural and structural aspects and constraints. Other normative demands were a focus on (i) normative future visions, (ii) sustainable need fulfilment at a societal level, (iii) realising stakeholder involvement and stakeholder support, and (iv) realising follow-up. A normative demand that emerged at a later stage was that MSL in Winterswijk should provide new socio-economic opportunities to the region.

Process demands in the MSL backcasting experiment were defined in terms of acceptance and support by stakeholder groups, numbers of stakeholders involved and reached through communication and dissemination activities, the adoption of follow-up activities by stakeholders, the degree of involvement and obtaining co-funding by other parties than the ministries. Thus, although process demands were articulated in a quantitative and goal-oriented manner, demands reflecting process quality like transparency or accessibility were not articulated in project documents. By contrast, the emphasis on follow-up involving stakeholders and the consultation for interests, views and future goals, indicates a more conscious attitude toward process demands than found in project documents. However, even then, the emphasis was more on meeting the demands of selected stakeholders than on process quality demands like transparency and accessibility.

Knowledge demands were not articulated in the sense of quality of knowledge. This indicates that quality demands regarding the knowledge were left to the research contractors involved. Demands were set regarding content in terms of results like future vision, analysis, proposals for follow-up activities, functionalities of MSL systems, etc. These were partly articulated through the formulation of success criteria in internal proposals (e.g. De Kuijer 1995b: 3).

Settings: institutional protection

The STD programme was initiated by five ministries, which also participated in the steering group of the STD programme. At top level these ministries were committed to the STD programme and wanted it to be successful. However, in the case of the MSL backcasting experiment, some temporary resistance emerged from the side of the Ministry of Agriculture, which can be explained in terms of the dominant vision and the vested interests 'fighting back'. The central organisation of the DLO division also provided some institutional protection and favoured an integrated approach and cooperation crossing borders of institutes, although it was not easy to find approval and agreement at the level of specific institutes (De Kuijer interview 2005).

At the end of the MSL backcasting experiment, the established steering group, involving financing parties and important regional stakeholders on a senior management or board level, to some extent provided institutional protection on the regional level.

Settings: vision champion

The project manager De Kuijer acted as an inspired vision champion for the MSL option and the MSL backcasting experiment. Firstly, he stimulated the MSL option during the nutrition domain analysis and successfully argued that broadening with open systems would enforce the Nutrition sub-programme. Secondly, he responded to the resistance by the Ministry of Agriculture and helped resolve the problems. Thirdly, he created and maintained a stakeholder network relevant to the MSL project in the Winterswijk region, which was crucial for testing the MSL concept in Winterswijk. Fourthly, he was the driving force behind the financial participation of non-government parties and the establishment of the regional stakeholder steering group; he also facilitated the transfer of the follow-up coordination to the province of Gelderland.

After the leading role was taken by the province, the Commissioner for agriculture and environment affairs became the vision champion for the follow-up programme.

Settings: focus and management

The backcasting experiment had a strong emphasis on realising follow-up and implementation.

The operational side of the approach that was applied can be described as an advanced project management approach for complex projects in a multi-actor setting. It was considered important to achieve defined milestones and 'products' in time, as limited time was available at the STD programme. By contrast, to deal with stakeholders a process-oriented approach was adopted, although convincing stakeholders about the opportunities of the MSL concept for the region and of specific follow-up proposals was equally important. Furthermore, although stakeholders were consulted and asked about their views and opinions, they had limited interaction among each other, as most contacts were face-to-face with the project manager or during steering group meetings.

7.4 Follow-up and spin-off

7.4.1 MSL Winterswijk programme: a multi-actor initiative

In 1999, the demonstration and development programme on MSL in Winterswijk started (Stuurgroep MDL 1999, Akkerman et al. 2003). Overall coordination at programme level among the stakeholders involved took place in a steering group, chaired by a Commissioner of the province of Gelderland. The programme comprised eight projects, all involving various stakeholders from the region. In addition to four soil-based demonstration projects, there were three technology development projects and a project for communication and dissemination. Research was part of all projects and was brought together in a research programme at WUR (Korevaar et al. 1999). The four soil-based projects are described in Table 7.6. Table 7.7 summarises the technology projects aiming at closing material cycles, and the dissemination and communication project.

MSL programme: inception

After the MSL steering group had decided in 1997 to continue MSL activities in the Winterswijk region based on the results (De Kuijer et al. 1997a) and the follow-up proposals (Beeren et al. 1997) of the backcast-

Table 7.6 *Soil based projects in the MSL programme (Akkerman et al. 2003)*

Water production and multifunctional farming at the Stortelersbeek

The purpose was to combine water capture and natural filtration with drinking water production and multifunctional farming producing crops, grass, fruits, and nuts, while also managing nature and landscape. Starting point was that the water company Vitens, which needed
new water sources due to falling soil water levels, would pay for the rain water collected
and filtered by the farmers. The results included fundamental hydrological knowledge and
an elaborated demonstration plan. The plan was not executed due to uncertainties about
payments by the water company and about the legal ownership of the water, in addition to
hesitations by a farmer on a key location in the area. The planned pilot has been transferred
to the estate of Lankheet in the communality of Haaksbergen.

Agriculture, nature and recreation in the Winterswijkse Poort area

Users, inhabitants and farmers in this area collectively investigated the opportunities to
combine farming with nature and recreational functions. A design sketch was made for the
area, while several farmers had plans made to broaden their farm. However, these plans
were not put in practice.

Multi-functional pastures and arable lands

Fourteen farmers from the region participated and re-introduced traditional land-use on
selected fields allowing a larger range of plant varieties. This made it possible to combine
crops for fodder with a larger and more varied biodiversity in weeds, flowers and animals.
This would also increase the attractiveness for recreation.

Multi-functional plantations

Eight farmers and landowners demonstrated new ways of growing fruit, mushrooms and
nuts in combination with producing high quality wood and grazing grounds for cattle. This
example has been replicated in the province of Overijssel.

Table 7.7 *Technology projects in the MSL programme (Akkerman et al. 2003)*

Upgrading manure and biomass into high-quality compost and energy

The local pig farmer Jan Leemkuil cooperated with the German composting firm AGR and its subsidiary Wisa. They developed and tested a new way of processing pig manure and various types of biomass, producing both biogas and high-quality compost. A larger installation is economically viable when compost can be sold at € 22 per tonne. Such installations are currently operating in Germany. If permits are granted by authorities and it is allowed in the new development plan, a larger installation of 350 kW can be erected in the region requiring an investment of € 1-1.5 million.

Bio-refinery for upgrading biomass into high-quality fodder and potting peat substitute

A pilot plant was constructed at the site of Avebe, the cooperative potato starch producer that was looking for alternative crops for its farmers on the sand soils in the province of Groningen. As the protein-rich fraction from extensively grown grass is too low to turn it into fodder in an economically viable way, it became less interesting for the fodder cooperative firm ABCTA. Therefore, the focus was shifted towards sustain the grass thermally and turn it into a substitute for current peat used in potting. This would make it possible to maintain moors in Ireland, the UK and Eastern Europe, although the problem is that peat from these moor areas can still be exploited for lower prices. A €20 million business plan was made, but this would require free grass from 10-15,000 hectares of nature area. Further up-scaling would then only be possible if grass crops are grown.

Energy producing pig stable

Energy company Nuon and WUR worked together with several companies on the design of a new pig stable that would reduce emissions by 90%, and would reuse the heat produced by the pigs for heating both water and buildings, or for drying manure. A local pig farmer was willing to invest in a pilot stable on his farm, but the Ministry of Agriculture initially rejected the co-funding asked for. Furthermore, when the outbreak of swine fever led to economic problems among pig farmers, Nutreco was no longer willing to invest in such a radical innovation and has built a cheaper and more incremental demonstration stable at its site in Boxmeer. The swine fever epidemic in combination with higher cost prices of € 3-4 per pig made the pig farmer decide not to invest in a new stable on his farm.

Communication and dissemination

This included development of a handbook for the applied backcasting methodology, the development of an interactive computer model for exploring multiple land-use possibilities in the Winterswijk region and a Community of Practice for rural areas. In the latter representatives from Winterswijk and other regions, where multiple land-use was explored, exchanged results and experiences and developed recommendations for different actors involved in multiple land-use at a regional level. The CoP was also part of the Habiforum programme on multiple and innovative use of space.

---

ing experiment, the follow-up programme could not start until late 1999 due to funding problems. Despite the participation of motivated stakeholders and the verbal support by the minister of Agriculture during the take-off symposium in Winterswijk late 1998, this did not result in separate programme funding by the ministry. Instead, funding had to be obtained through submitting project proposals to regular funding or policy programmes of the ministry, such as those executed by Laser[10], or EU-funded programmes. As a consequence, individual projects had more difficulties in proving their novelty and innovativeness. Nevertheless, with considerable delay and additional efforts all proposed projects were funded, although often with smaller budgets than intended. The proposed research programme that had been defined at the DLO pillar of WUR (Korevaar interview 2005) did not receive separate funding from the Ministry of Agriculture either. Although strongly supported by the ministry, budgets had to be reallocated from existing research programmes that were part of existing regular research budgets provided by the ministry (Korevaar interview 2005).

MSL programme: results

The four soil-based projects have been described in Table 7.6. Demonstration in the area has been achieved for multifunctional plantations and for multifunctional pastures and arable lands. Combining agriculture with recreation and nature in the Winterswijkse Poort area and water filtration and production at the Stortelersbeek was investigated, but not implemented.

The three so-called technology projects are described in Table 7.7. These have resulted in new knowledge, designs and implementation plans. Realising pilots was more difficult. However, a pig stable has been demonstrated at the site of Nutreco in Boxmeer, albeit in a considerably less innovative and environmentally friendly version.

The bio-refinery for upgrading biomass into high-quality fodder resulted in a small pilot plant for grass processing at the site of starch producer Avebe in Foxhol in the province of Groningen. This ended after Avebe decided to focus on its core business of potato starch production and left the project. The involved fodder cooperative ABCTA also got less interested in this project, as the protein fraction in grass from nature areas is too small for processing into fodder in an economically viable way. As a result, the focus shifted towards producing a peat substitute for potting, but this cannot be put in practice yet in an economically viable way either.

Upgrading manure and biomass into high-quality fertiliser and biogas has not been demonstrated as part of the MSL programme. However, this has been done in Germany, although not related to the MSL programme (Leemkuil personal communication 2006).

The results are published in news letters, research reports and articles in various media (see Korevaar and Van Loenen 2002) and two overall accounts (Akkerman *et al.* 2003, Korevaar 2006).

7.4.2 Research domain

The research programme on MSL at WUR was a major part of the MSL Winterswijk programme with a budget of around € 2.5 million. Two thirds of this amount was provided by the Ministry of Agriculture; one third came from other sources, like funding programmes and stakeholders involved (Korevaar interview 2005). The opportunity for researchers was to apply an integral multifunctional approach for an entire region for which a master plan could be made (Korevaar interview 2005).

There were concrete plans to double the multifunctional pastures and arable lands to more than 250 hectares in 2005 (Kiljan and Moolenaar interview 2005). Late 2006 a follow-up project has been approved to participating farmers and estate-owners, which includes multifunctional plantations, but it has not been extended in size (Korevaar personal communication 2007). Another research spin-off by LEI and PRI has quantified the additional economic benefits of MSL at the level of the Winterswijk region (Bos and Korevaar 2006).

The backcasting methodology applied in the MSL project (De Graaf and Musters 1997, De Graaf *et al.* 1999) has been further developed in a full methodology (De Kuijer and De Graaf 2001, De Graaf *et al.* 2003) for participatory spatial renewal in regions funded by Habiforum (Verkade interview 2005, De Kuijer interview 2005, De Graaf interview 2005). This was done by KDO Consultancy. This firm has been established by the former manager of the backcasting project on MSL and involves another key researcher from the MSL backcasting experiment. The developed backcasting methodology for MSL has been applied in several other regions (see 7.4.4).

A Community of Practice (CoP) on MSL in rural areas, funded by Habiforum, consisted of exchanging MSL-related experiences among farmers and other stakeholders from the Winterswijk region and various other rural areas (Verkade interview 2005, De Kuijer interview 2005, Tiggeloven interview 2005, see also Londo *et al.* 2002, De Kuijer *et al.* 2002). The CoP resulted in enthusiasm and exchange of practical knowledge and experiences among participants (Verkade interview 2005). Habiforum subsidises CoPs, because it is interested in developing and collecting generalised knowledge and competencies related to MSL and other forms of multiple use of space.

Replication of backcasting can be found in the agricultural research infrastructure within the DLO pillar of WUR. It has been used for programming new future oriented research programmes in animal production (Spoelstra interview 2005, van Kasteren 2001 Spoelstra *et al.* 2003) and in growing horticultural and arable crops (www.syscope.nl, De Wolf *et al.* 2006). These activities are very much inspired by the STD programme and its sub-programme in Nutrition, and are supported by key players in the agricultural research system. However, no evidence has been found that the methodology was taken from the MSL backcasting project (Spoelstra interview 2005, Vogelesang personal communication 2005).

Finally, the National Council for Agricultural Research (NRLO), with its foresighting and agenda setting-oriented mission, has used the critical technologies identified in the MSL feasibility study as a starting point for commissioning several studies. These studies include sensor and monitoring technology (Chehab and Enzing 1998, NRLO 1999a), technology for processing of manure and biomass (Rooijers and Van Soest 1998) and the use of ICT in participatory regional planning and development of regional future visions (Van Twist *et al.* 1998, Van der Cammen and De Lange 1998, NRLO 1999b). These studies at the NRLO had their own series of workshops, but attendance included numerous people that had been involved in the MSL backcasting experiment. The activities were not continued after the NRLO was turned into the Innovation Network for Rural Areas and Agricultural Systems (INGRA) late 1999; the new organisation shifted towards facilitating system innovations in agriculture. Nevertheless, INGRA has commissioned several essays on how to pay for green and blue services provided by farmers. One of these (Londo *et al.* 2005) was based on experiences from the MSL programme and replication of the backcasting methodology in other regions (De Graaf interview 2005).

7.4.3 Business domain

There was substantial participation from both companies and farmers in the MSL Winterswijk programme and its projects. Participating farmers have become more enthusiastic and other farmers have become interested in joining the spin-off on multifunctional farming (Tiggeloven interview 2005, Korevaar interview 2005). However, the economic side of multifunctional farming was neglected (Tiggeloven interview 2005, Koldewey personal communication 2006). For instance, the market and marketing aspects of the nuts, mushrooms and fruit were not included in the multifunctional plantations project that had a research focus. Paying for green services, for instance by establishing a landscape fund, was also not part of the multifunctional farming project. So far, establishing landscape funds has been achieved in other areas with more political willingness (Wytema personal communication 2006). However, early 2007 a group of farmers and estate-owners involved in land-based MSL projects have initiated new activities to realise a fund that enables payments for landscape and water-related services (Korevaar, personal communication 2007).

By contrast to the farmers, most of the larger companies became less interested in the course of the MSL programme (Korevaar interview 2005, Kiljan and Moolenaar interview 2005), for various reasons. For instance, potato starch producer Avebe made a strategic move back to its core business, while Nuon shifted towards an international orientation and did not want to be connected to pig farming because of the swine fever outbreak. The fodder cooperative ABCTA became less interested because the protein content of grass from nature areas was too small. The German composting firm AGR was especially interested in strategic cross-border cooperation. It was less interested in the particular biogas and composting project, as it had extensive activities in this field running in Germany. The regional water company merged with several others to form the water company Vitens that supplies water to the east and the north in the Netherlands. As a result, there was no longer an urgent need to create new water sources in the Winterswijk region. However, this company uses MSL principles when developing new water production areas (Kiljan and Moolenaar interview 2005). There is spin-off at Nutreco in Boxmeer, where a more incremental version of the green pig stable has been built. Testing is completed by the end of 2006 and successfully evaluated modules of the new stable will be introduced on the market in 2007. In addition, the pig farmer involved in the biogas project is planning substantial investments (€ 1-1.5 million) in a large scale composting and biogas installation.

7.4.4 Government domain

The province of Gelderland has played a leading role in the MSL Winterswijk programme and has provided around € 5 million in capacity and funds. When the MSL programme was completed, the province concluded that such a complex and large programme requires a cooperation of national and regional authorities (Kiljan and Moolenaar interview 2005). The province is willing to allocate some funds for MSL-related activities, but stakeholders in the Winterswijk region have to develop project proposals for MSL activities themselves (Kiljan and Moolenaar interview 2005). Knowledge and experiences gained are used in dealing with environmental and socio-economic problems in other rural regions and reconstruction areas in the province, for instance the veal production reconstruction area in the northern part of the Veluwe region (Kiljan and Moolenaar interview 2005).

The regional Water Board 'Rhine and IJssel' has initiated a water filtration and management project in the municipality of Haaksbergen in the province of Overijssel, using some of the knowledge from the MSL programme (Tiggeloven interview 2005, Korevaar interview 2005). At the country estate Lankheet water is filtered from nutrients by growing reed and subsequently fed into the neighbouring nature area. The Water Board expects to continue after the test stage and sees this natural way of filtration as an interesting approach to improving the water quality in its canals and other surface waters (Van der Veen personal communication 2006). The Water Board has also changed its views on multiple land-use and takes this into account in other projects and activities (Kiljan and Moolenaar interview 2005)

The municipality of Winterswijk has developed a reconstruction plan for the eastern part in cooperation with regional authorities, WCL, and the spatial planning committee[11]. As part of this reconstruction planning process, the backcasting methodology for rural areas, as developed and tested during the MSL backcasting experiment, was replicated first for the entire WCL area and later focused on the reconstruction area (De Kuijer interview 2005, De Graaf interview 2005, De Graaf et al. 2003, De Graaf and De Kuijer 2004a). The resulting reconstruction plan (DLG 2001) has resulted in a regulatory framework stimulating MSL, instead of mono-functional de-intensification as is usually done in reconstruction plans (Tiggeloven interview 2005, De Kuijer interview 2005).

The backcasting methodology that was developed for (rural) areas has also been applied in several other areas. For instance, Habiforum involved KDO in developing a regional vision for the IJssel area downstream and upstream of Zwolle (De Graaf and De Kuijer 2004b, Anonymous 2001, Anonymous 2002b, www.ijsselzone. nl, Verkade interview 2005). This was followed by developing and implementing MSL in a smaller region between Zwolle and the IJssel river (Verkade interview 2005, De Graaf interview 2005, De Kuijer interview 2005, Anonymous 2002a, www.buurtschapzwolle.nl). The latter includes a fund to pay for green and blue (water-related) services. This enables payments for joint activities resulting in water quality demanded by the water company Vitens (De Graaf interview 2005, De Kuijer interview 2005).

Another regional vision development study was facilitated by KDO as part of the reconstruction process in the municipality of Wierden near Almelo (De Kuijer interview 2005, De Graaf interview 2005). KDO is also involved in vision development for restructuring areas along rivers allowing temporary water capture, for instance in the Noordwaard along the river Nieuwe Merwede (De Kuijer interview 2005, De Koning *et al.* 2003). This reconstruction is part of 'Space for Rivers' in the Netherlands, for which the concept of MSL is used. Here, implementation is easier because of the political sense of urgency, enacted river safety regulation and considerable budgets from the Ministry of Transport, Public Works and Water Management.

7.4.5 Public domain

No separate activities have been found in the domain of public interest groups and the wider public. However, nature organisation Natuurmonumenten (VNM) and a local nature organisation were involved in the MSL Winterswijk programme. In addition, WCL Winterswijk was a multi-actor platform involving public interest actors, such as nature organisations. Furthermore, broad stakeholder involvement, including public interest groups, local communities and inhabitants, was a key element in all the regional vision development activities listed in 7.4.4.

7.5 Further analysis

7.5.1 Network formation

In the previous section a considerable number of activities has been described that can be related to the MSL backcasting experiment. In this section these activities are further analysed in terms of network formation, using the indicators (1) activities, (2) actors, and (3) resources. The activities are used as a starting point for mapping both the actors and resources mobilised, with an emphasis on funding. The results are shown in Table 7.8 and Table 7.9. Table 7.8 depicts network formation in the MSL Winterswijk programme both on the programme level and at the level of the projects. Table 7.9 summarises the network formation for the other follow-up and spin-off activities that were found, following the four domains distinguished in this study: the research domain, the business domain, the governmental domain, and the public domain.

Both Table 7.8 and Table 7.9 show the activities that were identified, each involving a range of stakeholders from various domains. No separate activities in the domain of public interest groups were found, but actors from this domain were strongly involved in processes around vision development and spatial planning for particular regions. For instance, public interest organisations, such as nature organisations

Table 7.8 *Network formation in the MSL Winterswijk programme*

Activities	Actors	Resources
MSL Winterswijk programme, consisting of 8 projects below	Province[i], GLTO, WUR, water board, NSW, Nuon, LNV, ANWB, Vitens, Rabobank, VROM, WCL, communality of Winterswijk, VNM, ABCTA	Proposal was submitted to EET but not approved for funding
(1) Water conservation and multifunctional land-use Stortelersbeek	Water Board, farmers, Vitens, GLTO, VNM, WUR, DLG, province, local estate owners	Budget of € 350,000 from, co-funding by Water Board and WCL
(2) Multifunctional agriculture, nature and recreation Winterswijkse Poort	VNM, WCL, NSW-country states, ANWB, GLTO, province, WUR	Co-funding by VNM
(3) Multifunctional pastures and arable lands	14 farmers, GLTO, PRI (WUR)	Budget of € 130,000
(4) Multifunctional plantations	8 farmers/ estate owners, Alterra (WUR), WCL	Funding around € 230,000
(5) Upgrading manure and biomass into fertiliser and energy	Local pig farmer, AGR, Wisa, KDO	Funding by WCL, AGR, and province for around € 250,000
(6) Bio-refinery, upgrading biomass into fodder, or potting substrate	Avebe, ABCTA, Agrifirm, Nedalco, NOM, Rabobank, PRI (WUR)	Budget of around € 175,000, this was part of a larger research co-funded1 by NOM[ii] and BioPartner
(7) Pig stable and energy production	Nuon, AFI (WUR), Nutreco, GLTO, province, ABCTA	Budget around € 200,000, co-funding by Nuon

i   These are the actors in the steering group of the MSL Winterswijk programme (steering group MDL 1999: 42).

ii   NOM is the Northern Development Organisation, which stimulates economic investments in the North of the Netherlands. BioPartner subsidises biotechnology innovations.

and inhabitants, were involved in several projects of the MSL programme. The tables also show that in most activities the actors constituting the activity were capable of mobilising sufficient resources. This often requires considerable efforts, as was the case at the MSL programme resulting in delays of projects. However, in most activities considerable budgets were obtained.

Several clusters of activities can be distinguished. Not surprisingly, a major cluster consists of the MSL Winterswijk programme and its projects (listed in Table 7.8). In each project regional stakeholders were involved, in addition to researchers from WUR institutes (formerly DLO), both in the land-use projects and the technology development projects. Many participants in the projects mobilised resources, both financial resources and capacity, for instance through reallocating budgets in favour of projects in the MSL programme, as was the case at WUR institutes. However, raising external resources was difficult. Nevertheless, substantial resources and a considerable number of stakeholders were mobilised for all projects. Furthermore, the networks constituting the activities were not static, but dynamic in terms of participation and commitment. For instance, farmers in the multifunctional pastures and arable lands project became more enthusiastic and have realised follow-up in which they have taken the lead. However, increasing the number of hectares and the number of farmers has not been successful. Secondly, there were exits and entries of business stakeholders in specific projects for various reasons. This could be due to strategic reorientation of for instance companies what happened in case of Nuon and Avebe, but could also be due to research results which were less attractive from the viewpoint of stakeholders involved, as happened in case of the ABCTA fodder company. However, in other projects new stakeholders entered, like the German composting company AGR and its subsidiary Wisa, and temporarily, Nutreco and its subsidiary Hendrix UTD.

A second cluster of activities involves various spin-off activities of the MSL Winterswijk programme. This cluster includes the approved continuation of the project on multifunctional pastures and multifunctional arable lands in which the multifunctional plantations will be integrated. It also includes the water capture and natural filtration project at another location. After demonstration was not accomplished in the Winterswijk area at the Stortelersbeek, it has been moved by the Water Board to the municipality of Haaksbergen as a pilot for natural filtration. Nutreco is still working on a more incremental pig stable programme, and has constructed a pilot stable at its site in Boxmeer.

In addition, several activities in the communication and dissemination project of the MSL programme can also be seen as spin-off activities. For instance, the backcasting methodology for multiple land-use was further elaborated and written down in an externally available manual (De Kuijer and De Graaf 2001). In addition, the methodology was replicated twice. First for the entire WCL area as part of the reconstruction process in the eastern part of the municipality of Winterswijk (DLG 2001); this involved the municipality of Winterswijk and parts of neighbouring municipalities. This was followed by study focusing completely on the reconstruction area in the eastern part of Winterswijk (De Graaf and Tolkamp 2004). The Community of Practice on MSL in rural areas was in fact also a spin-off aimed at exchanging knowledge and experience from all areas in the Netherlands where MSL was an issue. The results from the MSL programme, the MSL backcasting experiment and repeating the backcasting methodology for the entire WCL area have been incorporated in the reconstruction plan (DLG 2001) in such a way that it allows and enables multiple land-use and multiple-goal farming systems. Finally, many stakeholders use the knowledge and experience gained in the MSL programme in their activities and in projects in other areas (e.g. Province of Gelderland, Water Board, Natuurmonumenten, Vitens).

A third cluster of activities comprises several regional vision development projects in rural areas in which the backcasting methodology of the MSL backcasting experiment is replicated in other areas. This has been the case in the IJsselzone and its spin-off Neighbourhood IJsselzone, in the communality of Wierden near

Almelo, and in the Noordwaard along the river Nieuwe Merwede, near the Brabantse Bieschbosch. Although this cluster is to a large extent positioned in the government domain, in all cases it involved stakeholders from all domains and from various spatial functions. These activities are also strongly influenced by other developments and must therefore be seen as spin-off in which elements from the MSL concept have been used. This cluster also includes application of the backcasting approach in strategic research programmes at WUR, although the connection to the MSL backcasting experiment is not so clear.

A fourth cluster of activities also concerns knowledge development. This was found at the NRLO and its successor INGRA that commissioned studies and essays on topics relevant in the MSL backcasting experiment and the MSL Winterswijk programme.

Table 7.9 *Network formation outside the MSL Winterswijk programme*

	Activities	Actors	Resources
Research domain	NRLO studies on several critical technologies for MSL	NRLO, several research groups and research institutes	Funding by NRLO (provided by LNV) for around € 200,000
	WUR research programme on MSL Winterswijk programme	WUR (former DLO pillar), PRI, AFI, Alterra (all WUR institutes), other stakeholders from Winterswijk	Total budget of € 2.5 million; two thirds funded by LNV, one third from other sources
	CoP Rural Areas	Farmers and other parties dealing with MSL in rural areas, KDO, Habiforum	Funded and commissioned by Habiforum
	Follow-up multifunctional pasture, arable lands and plantations	PRI (WUR), GLTO, farmers in the region	Funding granted by EU, DLG and province, proposed budget of € 400,000
	INGRA study on how to pay for green and blue services	KDO, INGRA	Commissioned and funded by INGRA
Business domain	Development investment plan for larger biogas installation	Local pig farmer Leemkuil	Requires € 1-1.5 million
	Pilot pig stable at site in Boxmeer	Nutreco	Investments by Nutreco
Government domain	Reconstruction process Winterswijk East	Municipality of Winterswijk, WCL, DLG, Reconstruction committee, KDO, local land-users	Funding by province of Gelderland, Municipality of Winterswijk, LNV
	Demonstration of natural water filtration at Lankheet near Haaksbergen	Water board 'Rhine and IJssel', estate owner, nature organisation, WUR (PRI, Alterra), Province of Overijssel	€ 4 million from EU (50%), LNV, Water Board, Living with Water programme
	Reconstruction plan and vision development Water Land Wierden (near Almelo)	Vitens, Water Board, local group, SBB, Reconstruction Committees, Overijssel province, DLG, GLTO, municipalities	Several authorities
	Vision development IJsselzone	ANWB, DLG, GLTO, Habiforum, KDO, ministries, municipalities, province of Overijssel, Vitens, Water Board, Alterra, Delft Hydraulics	Total funding of around € 200,000 by Habiforum, several ministries, Province of Overijssel, city of Zwolle
	Neighbourhood IJsselzone	Similar as in vision IJsselzone, but more local stakeholders and community groups	Habiforum and INGRA. Plan costs around € 0.5 million, various millions for implementation
	Space for Rivers, Noordwaard region	RWS, local farmers and inhabitants, KDO, various other stakeholders	Budget around € 300 million for development and implementation

7.5.2 Vision aspects

The next question is to what extent, the normative future vision generated in the MSL backcasting experiment has offered guidance and orientation during follow-up activities and their emergence. In addition, how has the vision evolved after completion of the MSL backcasting experiment?

<u>Vision: guidance</u>

Guidance of the future vision is analysed in terms of (1) collective normative projection, (2) synchronisation and alignment of the interaction among stakeholders from different backgrounds and scientific disciplines, (3) alternative (system of) rules.

Collective normative projection

The MSL Winterswijk programme was based on the future vision on MSL for the region, consisting of (i) the general level of the MSL concept and (ii) the elaboration for the region. The collective normative projection continued to provide guidance to the activities of the MSL Winterswijk programme. This vision was also used to make a more operational vision for the year 2015 as part of a reconstruction process for the eastern part of the municipality of Winterswijk, which can also be seen as a normative collective projection. The projects in the MSL programme ran fairly independently from one another. This has to some extent resulted in the emergence of collective projections at the level of projects, but this aspect has not been extensively studied. These (sub)visions at the level of projects are still related to the overall regional vision on MSL and can be seen as nested in the regional vision. Furthermore, the MSL topic and the backcasting methodology was transferred to other areas, where the generic vision on MSL and the MSL concept were used as input to other processes of regional vision development leading to collective projections. These visions can be seen as nested in the generic (conceptual) vision on MSL in a similar way as the vision for MSL in the Winterswijk region. In short, diffusion and adjustment of the vision took place, while stability concerned different parts of the vision.

Synchronisation

Synchronisation among stakeholders participating in specific projects in the MSL programme varied from moderate to high, as exchange and interaction between researchers and the non-research stakeholders was essential to achieve not only design and development, but also realising demonstration. Synchronisation at the level of the steering group was reduced during the course of the MSL programme, as several major stakeholders lost their initial high level of interest in the MSL programme. In addition, synchronisation and alignment between the top level of organisations involved in the steering group and operational levels was sometimes limited; embedding of individuals involved in the MSL projects in their organisations could be improved too. At the same time synchronisation across organisations involved in specific projects was high. It has been mentioned that in other areas the same approach was applied to develop future visions and activities, which must have resulted in synchronisation among the stakeholders involved.

Alternative rules

The future vision developed in the MSL backcasting experiment assumed various (sets of) interrelated and interdependent new rules. In the elaboration of the projects, these rules were further detailed in more practical settings. Visions developed in other regions also contained alternative rules.

Vision: orientation

Orientation of the future vision can be defined in terms of (1) cognitive activation, (2) mobilisation of actors and resources, and (3) decentralised daily coordination.

Cognitive activation

The future vision on MSL for the Winterswijk region provided substantial cognitive challenges that were dealt with in both technology and demonstration projects, as well as in spin-off activities of the MSL programme. Cognitive activation was not only related to research, but also to design and implementation and how to achieve further embedding in the region and institutionalisation. Cognitive activation has also been found in other vision development activities in other regions.

Mobilisation

The future vision was used to attract and mobilise to actors and resources, both in the MSL programme and other activities. In most cases mobilising resources required considerable efforts and resulted in delays. New stakeholders entered the projects in the MSL programme, although others left projects for various reasons. The MSL Winterswijk programme did not succeed in attracting and mobilising relevant national networks on topics like rural areas and multiple land-use, but it was very successful in mobilising relevant regional networks. Other related activities in other regions mobilised other actors, although there was a considerable overlap of actors as is indicated by Tables 7.6 and 7.7.

Decentralised coordination

There was daily coordination in the decentralised projects in the MSL programme, both due to the vision at the level of the programme and the structure provided by the projects. In addition, elements of the MSL vision and the MSL concept were found in other vision development activities that had no formal relationships with the MSL programme. This suggests decentralised coordination by the vision to different networks.

Competing visions

The dominant vision continues to focus on enhancing mono-functionality and a further up-scaling, increase of productivity and rationalisation of agriculture in the Netherlands. However, it is increasingly advocated that further up-scaling must be reserved for a few agricultural production areas on clay soils. In addition, the vision of broadening agriculture at the level of farms in rural areas that are less suited for further rationalisation and up-scaling is getting more important as well. The vision of organic agriculture in combination with de-intensification is also gaining strength; this vision seems to have come closer to the MSL vision and vision on broadening agriculture at the level of the farm as well.

7.5.3 Institutionalisation and institutional resistance

A distinction is drawn between institutionalisation and institutional resistance. Institutionalisation has been defined in Chapter 4 as the process by which a practice becomes socially accepted as 'right' or 'proper', or comes to be viewed as the only conceivable reality. Institutional resistance has been defined as the resistance to changes by existing institutions and the actors backing these institutions (see also Chapter 4).

Institutionalisation

The case results show various instances of institutionalisation, for instance in the Winterswijk region, at research institutes of WUR, in other regions and at the national level. An instance of institutionalisation in the region is the incorporation of MSL assumptions in the reconstruction process and reconstruction plan in the eastern part of Winterswijk. This regulation facilitates and enables further development of MSL in the Winterswijk region. Another instance is that MSL has also become a well-known concept in the Winterswijk region and its stakeholders, especially among the stakeholders participating in the MSL programme. This is widely considered a major impact (e.g. Tiggeloven interview 2005). In addition, various stakeholders in different domains use the experience and insights gained in the MSL programme in other activities. For instance, the province uses them in developing region oriented policies and in other reconstruction areas, while thinking in farming systems has been increased too, instead of thinking on, for instance nature targets and goals that must be met. Other organisations have developed more positive views on combining functions, and use their experience in other activities in other areas. These include the regional Water Board, Natuurmonumenten and drinking water company Vitens (Kiljan and Moolenaar 2005). Before, the focus of these organisations was on enhancing and conserving the function of which they were part. Follow-up and spin-off of projects in the MSL programme has stimulated institutionalisation as well. This includes the pilot project on natural water filtration near Haaksbergen conducted by the regional Water Board and the continuation of multifunctional arable lands, pastures and plantations.

Instances of institutionalisation can also be found at the research institutes of WUR. For instance, multi-disciplinary research, integral design and involving stakeholders in research has become a much more common practice at research institute PRI (former AB-DLO) at WUR and others. This has been stimulated by the MSL programme and the MSL backcasting experiment. Backcasting has been replicated at the DLO organisation to the future of crop based farming, horticulture and livestock farming. Although this is not directly related to the MSL backcasting experiment of the MSL programme, it has been supported by top level management at the ministry and the DLO organisation who were positive about the backcasting approach at the STD programme and its results. The research institute AB-DLO had considerable experience in more environmentally friendly dairy farming systems, but not in involving stakeholders in research activities. As a result of its involvement in the MSL activities this research practice has been added to its repertoire; it also has enhanced multidisciplinary cooperation with other agricultural disciplines (Spiertz interview 2005).

Another example is the replication of the backcasting methodology for MSL in rural areas, which is leading to institutionalisation of the MSL concept at a local and regional level in the regions, where it has been replicated. Furthermore, the MSL programme has contributed to the institutionalisation of MSL at the Ministry of Agriculture, partly because leading researchers at the former DLO organisation who were both familiar with and positive about the MSL programme and the MSL backcasting experiment were appointed at the top management level of the Ministry of Agriculture (e.g. Kiljan and Moolenaar interview 2005). However, the increasing interest at the ministry in integrating functions and broadening agriculture at farm level is the result of a much larger change process and ongoing debates on the future of agriculture in the Netherlands.

At the same time, the impact and institutionalisation of the MSL Winterswijk programme in relevant national networks on agriculture and multiple space use, such as around INGRA and Habiforum, has been limited (Hillebrands interview 2005, Verkade interview 2005, Kiljan and Moolenaar interview 2005, Smeets interview 2005). In these networks the MSL programme is not only considered research and technology driven, but also as a regional endeavour and as too much imposed from above on the Winterswijk region (Hillebrands interview 2005, Smeets interview 2005). This is not an example of institutional resistance, but of limited institutionalisation.

Institutional resistance

Institutional resistance to the MSL vision can be found in the results too. For instance, existing regional development plans in general take spatial separation of functions in rural areas as a starting point, thus preventing specific combinations of functions. Furthermore, during in the manure upgrading project, it was found that existing regulation prohibits the processing of biomass at locations where also crops for foods are handled. Also, the hesitance among farmers to change their way of farming, for instance in the Winterswijkse Poort area or at the Stortelersbeek, can be seen as expressions of institutional resistance. It concerns novel practices that are not socially accepted among farmers and require new rules and institutional arrangements.

Despite the steady growth in activities on multifunctional agriculture at PRI, it still is a minor part. Most studies still deal with mono-functional agricultural systems and regular improvements of crops and crop growing. This also relates to institutional resistance: in the competition for research funds regular plant-oriented research prevails over research into multifunctional farming systems.

In the MSL programme the focus was on raising commitment at top levels of relevant stakeholders who delegated the operational activities in general to (staff) members involved in strategy or developing new activities (e.g. Vaessen 2003). This resulted in limited commitment at the level the regular operations, which is also a form of institutional resistance. Something similar was the case between the top of the farmer organisations and the farmers in the region that they represented. While the farmers' unions felt positive about the MSL concept and the MSL programme, initially many farmers did not. Farmers became involved much later than the so-called key stakeholders and needed time to evaluate and discuss the new options and perspectives. By the end of the MSL programme this had changed and more farmers became interested in joining the follow-up on multifunctional farming. Thus, initial institutional resistance seems in a process of shifting. However, it must be realised that increasing the range of individuals involved needs to be done as early as possible and requires sufficient time and capacity.

7.5.4 External factors

External factors have been defined as events and developments in the context of the MSL backcasting experiment and its spin-off that have constrained or enabled the emergence of follow-up and spin-off activities. Firstly, external developments and events were looked for, using interviews and written sources (e.g. Enzing et al. 2005). These are listed in Table 7.10. Next, it is analysed which developments and events have influenced or co-shaped spin-off activities and broader effects that have been associated with the MSL backcasting experiment, and can be regarded as external factors in this research.

An external factor with a negative influence on the implementation of the MSL programme was the swine fever late 1990s. This led to further delay in soil-bound activities and reduced the economic possibilities for parties in the pig breeding sector to invest capacity and funds for an innovative stable concept as was designed in the related project. The swine fever also resulted in decreasing interest at Nuon for the energy producing pig stable. It was feared that participation could negatively affect their image. It also limited the interest and possibilities of both Nutreco to work on a radical environmentally friendly pig stable and of a local pig farmer to build such a stable.

Another external factor was that earlier outbreaks of swine fever in the 1990s had already led to the enactment of the Reconstruction Act targeting those five provinces having areas with intensive livestock production. Although the purpose was originally to concentrate intensive livestock farming in order to make the related system more separated and better controllable, it had evolved in a broader policy programme.

Table 7.10 *Potential external factors in the MSL case*

Shocks and crises
> Several outbreaks of livestock and chicken diseases
> Ongoing economic problems in agriculture

Market and demand side changes
> Gradual growth in consumption of organic foods, annually with 10-15%, though market share is still below 5%
> Emergence of slow food movement and organic food markets
> Growing importance of recreation in many rural areas

Entries and exits
> Large regional companies develop a growing international orientation
> Up-scaling and merging of (semi-)governmental organisations
> Second term for Habiforum dealing with innovative use of space and the establishment of Transforum, which deals with the transition towards sustainability in agriculture and includes MSL in rural areas

New practices
> Increasing research in multiple-goal farming systems (multifunctional agriculture) and organic agriculture
> Growing relevance of environmental issues and sustainability in agriculture and agricultural policy-making
> Decreasing level of facilities in rural areas, both public and private services
> Up-scaling by mergers in companies and (semi-)governmental organisations (e.g. energy, water)
> Agriculture becomes less important in rural areas, resulting in tensions between different functions
> Experiments in combining agriculture and nature management
> Ongoing trend of additional external earnings at farms as well as in broadening farm activities
> The enactment of the intensive livestock reconstruction law, which has led to decreasing intensive livestock farming in the selected reconstruction areas
> Development of area focused policies by VROM and LNV resulting in growth of experiments in vision development and multiple land-use in specific regions
  Related are the experiments in new institutional and organisational arrangement for managing landscape and paying for so-called green and blue services
> Two tracks development in agricultural policy: further scaling-up in production areas and a shift towards multiple land-use and multifunctional farming in more diverse regions
> Enhancement of citizen participation in policy-making including environmental policy-making

International
> Shift of volume-based agricultural price policies towards support of landscape and water management
> Increasing importance of supra-national organisations like EU and WTO
> Decreasing financial support by EU for agricultural production
> EU regulation regarding liberalisation and level playing fields of companies in markets affects current and future support of green and blue services

This broader policy programme included improving regional quality and protecting other functions in rural areas, especially nature, and conserving traditional landscapes.

Apart from the Reconstruction Act and the regular development plan policies, the national government has developed various other region oriented policies since the early 1990s, such as for WCL areas, national parks, national landscapes and the development of regional visions (for an overview see, Driessen *et al.* 1995, Boonstra 2004). One recent policy programme concerns the investments for rural areas, which allocates budgets to regional development under supervision of the provinces. Another example is the programme for region-oriented policies by the Ministry of Agriculture (LNV), the Ministry of the Environment (VROM) and the Ministry of Transport, Public Works and Water Management. In addition, at EU-level there is also increased attention to the level of regions. On average, I would say, this external factor has been a positive factor for the MSL Winterswijk programme and the replication of the backcasting methodology for vision development and integrating functions in other areas. This has been further enhanced by the increasing attention to river security among politicians and policymakers. Making sufficiently more space for rivers in the future is currently considered impossible without allowing selected regions to be flooded annually or once in a few years. This can be realised through multiple land-use comprising nature or ways of farming that allow occasional flooding. By contrast with the MSL Winterswijk programme, river security involves huge budgets for realisation, which is a strong driver. River security thus seems a positive factor for MSL. However, existing spatial policies and regulations also constrain multiple land-use for instance, when these enhance spatial separation of functions.

Another relevant factor is the development of multifunctional farming and reconciling different functions as policy issues at the Ministry of Agriculture and other national agricultural actors. Whereas originally the emphasis was on combining two functions like agriculture with nature management, or agriculture with recreation, so-called stapling of functions is increasingly getting attention. Reconciling different functions and discussing the related problems was for instance a major issue in many so-called WCL areas. Growing interest in multiple space use including in non-rural areas has also resulted in Habiforum, which has been an important actor in funding major parts of the dissemination and communication project in the MSL programme, as well as in transferring the develop backcasting methodology to other areas.

Finally, external dynamics and developments have resulted in the exit or reduced interest of several companies and must thus be seen as an external factor. For instance, Nuon developed international ambitions, which reduced their interest in activities with a regional focus, while Avebe decided to focus on its core business of potato starch production.

7.6 Conclusions

7.6.1 Backcasting experiment

The MSL backcasting experiment resulted in the construction of a normative future vision and in involvement and commitment of a wide range of relevant stakeholders, which also led to mobilising resources from various sources. The degree of involvement, the type of involvement and the degree of influence varied among different groups of stakeholders involved. A high degree of influence was found among rather small, but changing groups of key stakeholders that were consulted during various limited periods in each stage of the MSL backcasting experiment. The first time was at the two stakeholder workshops in the nutrition domain analysis. At these workshops, the MSL option emerged, was partly shaped and received support from attending stakeholders. The second time was during the MSL feasibility study, when after a limited set of interviews with crucial stakeholders, it was decided that the MSL vision would be elaborated and tested in the Winterswijk region. The third time was during the stakeholder interviews in the Winterswijk region in which the views and articulated future goals of important stakeholders were drawn up. The collective decision of the MSL steering group in June 1996 to continue MSL activities and to initiate the MSL Winterswijk programme can be seen as a fourth moment, where a selected group of stakeholders had decisive influence.

The future vision on MSL was gradually developed during the course of the backcasting experiment. A generic vision and a region specific vision can be distinguished; the regional vision can be seen as nested in the generic one. The MSL future vision fulfilled different functions at the same time, such as a tool for integration and analysis, or as a shared vision providing guidance and orientation to the project and its activities, while also structuring these. The vision offered guidance in the sense that it served as a collective normative projection to the future, enabled synchronisation and alignment among researchers from different disciplines and among participating stakeholders from different domains, while also stimulating exchange and debate. Alternative rules could also be derived from the future vision guiding activities and discussions. The future vision provided orientation that facilitated and stimulated cognitive activities on both research and non-research topics. The vision was also used to mobilise (financial) resources and to attract stakeholders who could enter the backcasting experiment on several moments. The vision, together with the project structure and the available resources, also provided daily coordination to decentralised activities among the temporary network.

The backcasting experiment led to higher order learning among participating stakeholders concerning the problems, solutions, principal approaches and priorities. Higher order learning took place at the level of individual actors, as well as among groups of actors leading to consensus or congruence. Among the organisers and some of the participants of the backcasting experiment higher learning took place on the backcasting approach.

Considering the methodological aspects, the MSL backcasting experiment did not follow the exact order of the steps as proposed in the methodological backcasting framework. However, all steps could be identified in the course of the backcasting experiment and its activities. Furthermore, methods from all four categories in the backcasting framework could be identified, just as the three different types of demands. Institutional protection was identified. Two persons acted as vision champions, although in different stages of the backcasting experiment and in different networks.

7.6.2 Follow-up and spin-off

A considerable number of activities have been found that can be seen as follow-up and spin-off of the MSL backcasting experiment, while parties of all four domains (research, business, government and public domain) participated in these activities.

Four clusters of activities have been found. The first cluster consists of activities and projects as part of the MSL Winterswijk programme, which was initiated as a direct follow-up of the MSL backcasting experiment. The second cluster consists of various spin-off activities of the MSL Winterswijk programme and its projects. These included incorporation of MSL principles in the ongoing reconstruction process, continuation of activities at WCL Winterswijk, continuation of the projects on multifunctional farming and multifunctional plantations, and transfer of natural filtration and water conservation to another area. The third cluster consists of replication of the backcasting methodology in several other regions where multiple land-use was considered an attractive solution direction. This could be because of the conflicting spatial claims of different functions, planned reconstruction processes, or due to various other region-oriented policies. In addition, the backcasting methodology has been replicated at DLO in system-oriented strategic research programmes focusing on the future of livestock production and various types of plant production. The fourth cluster consists of studies related to relevant topics of MSL in the Winterswijk region at the NRLO and to a considerably lesser extent its successor INGRA.

The vision provided guidance and orientation throughout the MSL Winterswijk programme, although mobilising external resources appeared to be difficult, as well as getting exposure in relevant national networks. The vision for MSL in the Winterswijk region remained relatively stable, while nested project visions developed sometimes at the level of projects. The visions that were developed for other regions by replicating the participatory backcasting methodology can also be seen as nested in the generic vision on MSL and should be seen as at the same level as the MSL vision for Winterswijk. These visions for other regions also provide guidance and orientation to activities in those regions, based on consensus among and involvement of major stakeholders. In the case of backcasting for future sustainable agricultural systems at DLO also visions were developed but these are not related to any of the MSL visions.

Both examples of institutionalisation and institutional resistance have been found. A certain level of institutionalisation has been found in research practices at WUR, in the reconstruction plan in the Winterswijk region, in the visibility and support in the region and in changes in practices and approaches in major stakeholders involved in the MSL programme, such as the province, the water board, Natuurmonumenten and the regional water company. The follow-up and spin-off of the MSL backcasting experiment has contributed to the process of the embedding and institutionalisation of multiple land-use and multifunctional agriculture at the Ministry of Agriculture. Institutional resistance has been found at operational levels in the organisations involved, at the level of members of farmer organisations and at the national and EU level, for instance regarding regulation prohibiting payments for landscape and water management services, especially when provided by groups of farmers.

External factors and developments exerted influence on the decision-making process of stakeholders and in the further diffusion of MSL and the backcasting approach. On average, ongoing developments and external factors have been supportive; at the level of specific actors and specific activities various constraining factors and developments have also been identified.

7.6.3 Wider effects and system innovation aspects

This chapter has shown that the MSL backcasting experiment has evolved into various clusters of activities. The number of actors involved, as well as the amount of resources that were mobilised, has grown substantially. Instances of institutionalisation have been found, as well as diffusion and adjustment of the vision on MSL. From the viewpoint of system innovations the impact after nine years should be characterised as a niche on MSL or as a set of related niches on MSL.

In addition, various related initiatives can be found in rural areas in the Netherlands (www.neder-landmooi.nl). Many of these focus on landscape conservation, offering additional perspectives to existing farmers, how to pay for collectively produced green (landscape and nature) services as well as blue (water) related services and function integration. The topic of multiple land-use is also gaining importance at different levels of authorities and is dealt with at other initiatives with a national scope like Transforum, Habiforum and INGRA.

Nevertheless, external factors seem to be of huge influence for the envisioned system innovation towards MSL in the Winterswijk region or elsewhere. In the Winterswijk region additional steps are taken, like incorporation of MSL principles in the reconstruction process and the regular spatial development planning process, as well as various other spin-off activities of the MSL programme. Further realisation of MSL in this region or in other regions may require the availability of payment possibilities for green and blue services and substantial investments to be made. The latter can be illustrated by the activities as part of 'Space for Rivers' in the Netherlands. Giving more space to rivers in many cases requires turning agricultural areas into areas combining agriculture with occasional flooding or water capture. Because of major security aspects, substantial budgets are being made available by the national government. For instance, for the Noordwaard, where also a future vision development process has taken place using some elements and principles of the backcasting experiment in the Winterswijk region, around € 300 million will become available to realise the system innovation from a purely agricultural area to an area combining agriculture, nature and occasional flooding.

Notes

1   *For a popular account of these developments and the related changes in Dutch agriculture, see Wester-man (1999). The fixed price policies of the EU did not include all agricultural activities and products. For instance, horticulture, pig breeding, flower growing and potato growing were not included.*

2   *The Dutch agro-food system includes all activities that are necessary for the full chain of production, processing and distribution of agricultural products (both food and non-food) from both domestic and foreign sources. It includes not farmers in rural areas, but also food processors, retailers, horticulturists, growers of flowers, and their suppliers and service providers. Farmers still manage 60% of the Dutch surface area. In 1999 the agro-food system had a share of 10% (over 600,000 full time jobs) in the national employment, as well as of 10% (around € 32.5 billion) in the GDP (LNV 2000: 7-9).*

3   *This view has been criticised as it would neglect that the economic and ecological efficiencies per unit produced in agriculture have been raised strongly over the last decades.*

4   *For popular accounts of this transformation, see Mak (1996) and Westerman (1999).*

5   *In Dutch this is known as Ecologische Hoofdstructuur (EHS). The aim is to realise nearly 730,500 hectares of nature by 2018.*

6   *A major source for this section was, in addition to the conducted interviews and several documents from the backcasting experiment, the process description by Wilma Aarts (Aarts 1997: 33-70).*

7   *Genetic plant modification was left before the second workshop, as expert interviews indicated a limited potential for ecological improvement (which was estimated to be only a factor 2) and substantial ongoing research activities concerning this topic.*

8   *In Dutch WCL is the abbreviation of Waardevol Cultuurlandschap, which means Valuable 'Man-made Land-scape'. In the early 1990s various WCL regions were proposed as part of the fourth white paper on spatial planning, while budgets were made available for WCL areas. In the Winterswijk region representatives of different functions started cooperating and discussing how to bridge the gap between the interests of agriculture and other functions.*

9   *Building bricks or technical building bricks related to actual land-use, such as arable land, pasture, nature, forest. Other building bricks were the production of energy, water, fodder, fertiliser, as well as non land-use activities like livestock production, recreation, storage and distribution. The building bricks were defined based on the characteristics of the area and the goals and interests of the stakeholders who were consulted. They are related too, but not the same as the sub-functions mentioned earlier. Building bricks can be interdependent; for instance, water conservation is only possible in combination with arable land farming or pasture lands. Building bricks can also exclude others, for instance wood and arable land (De Graaf and Musters 1997, Aarts and De Kuijer 1997b).*

10 Laser is the agency of the Ministry of Agriculture that carries out government agricultural innovation programmes and other policy programmes. It used to be a division of the ministry, but is currently an independent organisation.

11 In Dutch this is called a Landinrichtingscommissie.

8

COMPARING CASES AND TESTING PROPOSITIONS

This chapter compares the three cases, starting with the backcasting experiments (8.2) followed by their related follow-up and spin-off (8.3). Next, the propositions are evaluated (8.4), internal and external factors are identified (8.5) before the research methodology and applied indicators are evaluated (8.6).

8.1 Introduction

In this research I have studied three cases, each consisting of (1) a backcasting experiment and (2) its related follow-up and spin-off after five to ten years in the research, business, government and public domains. In each backcasting experiment one or several system innovations towards sustainability have been explored. The first is the Novel Protein Foods (NPF) case, which comprises a backcasting experiment on sustainable meat alternatives. In this backcasting experiment a system innovation was envisaged in which a substantial share of the production and consumption of meat and meat products is substituted by that of protein foods from non-animal sources. The second is the Sustainable Household Nutrition (SHN) case, which was part of the international 'Strategies towards the Sustainable Household (SusHouse)' project between 1998 and 2000. The third case is the Multiple Sustainable Land-use (MSL) case, which is based on a backcasting experiment that dealt with function integration in rural areas involving agriculture and other functions related to landscape, nature, recreation, water production and water management. The first and third cases include backcasting experiments that were conducted between 1993 and 1997 at the Sustainable Technology Programme (STD).

While all three cases relate to a substantial part of the food production and consumption system in the Netherlands, each case focused on a different socio-technical system with different characteristics. The NPF case focused on a system of production and consumption. The SHN case looked at nutrition and consumption and production from the point of view of households and consumers. The MSL case focused on a regional system in which the agricultural function was integrated with other functions and as a consequence spatial planning aspects were important. The cases and types of systems are summarised in Table 4.8 in Chapter 4.

8.2 Comparing the backcasting experiments

This section compares the results of the three backcasting experiments as described in the case chapters. To start with, all three backcasting experiments resulted in one or several future visions, broad stakeholder involvement, higher order learning, proposals for follow-up and a follow-up agenda. In these terms all three backcasting experiments can be characterised as successful.

At the same time, considerable differences were found between the three backcasting experiments. These are analysed below by comparing the three backcasting experiments with regard to the four groups of indicators developed in Chapter 4: (1) the indicators regarding stakeholder participation, (2) the indicators covering the future vision aspects, (3) the indicators concerning higher-order learning, and finally, (4) the indicators on methodological aspects and settings. When appropriate and possible, indicators are aggregated. This is discussed when it occurs.

Stakeholder participation

Table 8.1 summarises the results of the three cases with respect to stakeholder participation. To start with, the number of participating stakeholders and individuals was considerably higher in both the MSL case and the NPF case than in the SHN case. With regard to stakeholder participation, the degree of heterogeneity was high in all three cases. This means that in each backcasting experiment stakeholders from all four distinguished societal groups were actively involved. The four societal groups that were distinguished were research, business, government, and public interest groups and the wider public.

Table 8.1 shows that the degree of involvement was low in the SHN case, while it varied across different groups in both the NPF case and the MSL case. In the latter two cases especially research parties, to which part of the research in the latter two backcasting experiments was commissioned, showed high degrees of involvement. As far as the types of involvement are concerned, stakeholders involved in the SHN case provided only the capacity to attend two stakeholder workshops. In the MSL case and the NPF case various stakeholders were willing to provide co-funding and substantial capacity for contributing to the activities (more than meetings, workshops, etc), including additional research capacity.

Table 8.1 *Stakeholder participation in the three backcasting experiments*

	NPF case	SHN case	MSL case
Participation:			
> Number of individuals	*100 - 150*	*40 - 50*	*100 -150*
> Number of stakeholders	*50 - 60*	*20 - 30*	*50 - 60*
> Heterogeneity	*High*	*High*	*High*
> Degree of involvement	*Varying*	*Low*	*Varying*
> Type of involvement	*Funding & capacity*	*Limited capacity*	*Funding & capacity*
> Degree of influence (on content)	*Moderate-Low, small group High*	*High*	*Moderate-Low, small groups High*

In all three backcasting experiments stakeholders had no or very limited influence on the process. In all three cases the process was designed and managed by the organisers of the backcasting experiment. With regard to the influence the stakeholders had on the content, there is again a difference between the MSL and the NPF case, on the one hand, and the SHN case on the other hand. In the latter case stakeholders had a high degree of influence. To a large extent this influence was exerted through two stakeholder workshops in which content was generated for scenario construction, scenario assessments, backcasting analysis, as well as for defining and elaborating concrete follow-up proposals and a follow-up action agenda. A majority of participating stakeholders in the MSL case and the NPF case had only low to moderate degrees of influence on the content. In general, they had to work with what was decided and done before they came on board.

However, at particular moments in the NPF case and the MSL case a limited and carefully selected group of stakeholders could exert a high degree of influence. In the NPF backcasting experiment this was the case in the initial six interviews and the related meeting that involved technological experts and R&D directors. After this there were no major changes with regard to the course and scope until the end of the backcasting experiment on NPFs.

In the MSL backcasting experiment four such key moments can be identified. The first one was during the workshops in the nutrition domain analysis, in which the MSL option emerged and received support from the attending stakeholders. The second was when a limited number of stakeholders was interviewed, after which it was decided at the STD office to select the Winterswijk region for a pilot. The third was when two rounds of interviews took place with selected stakeholders from the Winterswijk region, whose feedback was used to elaborate the future vision for the region. The fourth was when the stakeholders in the steering group of the MSL backcasting experiment decided to continue, which was the starting point of the MSL Winterswijk programme.

Vision aspects

The results for the vision indicators in the three cases are shown in Table 8.2, partly in an aggregated way. In all three backcasting experiments one or several future visions were generated. In the SHN case three future visions were constructed, which were viewed as a set of three future alternatives that were kept and analysed together. In the NPF case vision development started with the idea of novel sustainable meat alternatives that would be as attractive to consumers as meat and meat products have been. The MSL backcasting experiment started with a broad orientation on the nutrition domain, its major sustainability problems and possible future alternatives. After the MSL concept was approved by the stakeholders involved, a vision was elaborated.

Table 8.2 *Vision indicators in the three backcasting experiments*

	NPF case	SHN case	MSL case
Number of visions	One, but nested	Three	One, but nested
Degree of guidance	High	Moderate	High
> Collective projection	High	Moderate	High
> Synchronisation	High	Moderate	High
> Alternative rules	High	High	High
Degree of orientation	High	Low	High
> Cognitive activation	High	Low – Moderate	High
> Mobilisation	High	Low – Moderate	High
> Coordination	High	Low	High
Competing visions	Various	Various	Various

There were differences in the ways the future visions evolved in the three backcasting experiments. In the SHN case, the future visions were constructed in one period, after which no further elaboration or adjustment took place. In the MSL case and the NPF case, the visions developed more gradually; different 'versions' showed nested relationships. For instance, in the MSL case, after initial idea articulation a generic vision on MSL was generated. At a later stage a vision was developed for the pilot in the Winterswijk region, which combined the generic assumptions on MSL with a region-specific elaboration, which was nested in the generic one. Two scenarios that were later developed for the Winterswijk region were not considered future visions providing guidance and orientation, but possible scenarios. These showed more detailed pictures of the future vision and were assessed in ecological and economic terms. Something similar took place in the NPF case; after the ideas had been articulated and tested, a fairly generic vision was constructed, while in a later stage of the backcasting experiment a more detailed version was elaborated that was nested in the more generic one. The later version, which assumed a 40% substitution of meat, was less widely supported by stakeholders and in this respect resembled the two possible scenarios elaborated in the MSL backcasting experiment.

Comparison of the guidance and orientation of the visions in the backcasting experiments is based on the average of the underlying indicators. As Table 8.2 shows, in both the MSL and the NPF backcasting experiment guidance and orientation were evaluated as high. By contrast, all visions in the SHN case were considered to provide moderate degrees of guidance and low degrees of orientation.

When looking for guidance at the level of the indicators, the MSL vision and the NPF vision scored high on (i) collective normative projections, (ii) synchronisation, and (iii) alternative rules. The three visions in the SHN case scored moderately on the indicators collective projection and synchronisation, and high on alternative rules. In the SHN case the collective projections were seen as moderate, because there was no full consensus on particular future visions; the consensus was that there was an interesting set of future visions. The degree of synchronisation was regarded as moderate, because of the low degree of stakeholder involvement. The alternative rules in the visions were evaluated as high in all three future visions in this backcasting experiment.

Table 8.2 also shows that the MSL and NPF visions scored high on the indicators (i) cognitive activation, (ii) mobilisation of stakeholders and resources, and (iii) decentralised coordination. The visions in the SHN case were considered moderate for the indicator cognitive activation. Though cognitive activation has been considered high among the stakeholders during the workshops, it was evaluated as moderate to even low because of the low degree of involvement. Mobilisation of stakeholders took place to a moderate degree, but was limited to attending the workshops, and developing and submitting proposals. Decentralised coordination was to some degree found among the stakeholders working on a few follow-up proposals, but not on a larger scale among stakeholders involved in the SHN case.

Finally, competing visions were found in all three cases, although differences can be noted. In the SHN case the three generated future visions 'acted' as alternatives to one another. In the NPF case, although there was the dominant vision of regular meat production strongly present in the socio-technical system, the vested interests hardly saw the NPF vision as a seriously competing alternative. As a consequence, they did not really 'fight back'. In the MSL case various competing visions could be found.

Learning

Higher order learning has been analysed with respect to three types of shifts (that are interrelated). These are (1) shifts in framing major problems and of perceived solutions by specific actors; (2) shifts in the principal approaches to solving these problems and in shifting priorities by specific actors; and (3) joint learning, shifts in congruence and joint opinions in any of the issues related to the previous shifts or shifts in the relationships among the participants. In this comparison I combine the first and second types of shifts, as they both concern higher order learning at the level of individual actors. By contrast, the third type of shift reflects higher order learning at the group level.

Table 8.3 *Higher order learning in the three backcasting experiments*

	NPF case	SHN case	MSL case
On the topic			
> Individual actors	Yes	Yes	Yes
> Group level	Yes	No	Yes
On the approach			
> Individual actors	Yes	Yes	Yes
> Group level	?	?	?

The results are summarised in Table 8.3. In all three cases higher order learning on the topic and higher order learning on the approach could be observed. Higher order learning on the topic has been more substantial in the MSL case and the NPF case than in the SHN case in terms of instances and the range of aspects. In addition, the SHN case showed no higher order learning at the group level, whereas both the MSL case and NPF case show instances of learning at the group level. Learning in the SHN case was also of a different nature; in the SHN case higher order learning was (partly) the result of comparing the three future visions. In the other two cases, the emphasis was on higher order learning with regard to the proposed solution (reflected in the envisaged sustainable system innovation), as well as reframing the sustainability problem underlying the proposed solution and the principal approach to the sustainability problem.

In addition, higher order learning took place on the backcasting approach, especially among the organisers of the backcasting experiments. This led to reflections as well as recommendations on how to improve the approach. In all cases learning about the approach was observed among the stakeholders as well, but to a lesser extent than among organisers of the backcasting experiment. It has not been possible to determine whether learning with regard to the approach included joint or congruent learning on a group level.

Methodological aspects and settings

With regard to methodological aspects there are strong similarities between the three cases in terms of the applied indicators. Results are summarised in Table 8.4. To start with all three experiments were interdisciplinary in the sense that knowledge and methods from a range of disciplines was combined to explore the sustainable system innovation envisaged in the future visions.

In addition, in all three cases the five steps of the proposed methodological framework for participatory backcasting could be identified. However, the steps were not necessarily applied in the proposed linear order, while iteration of steps (or elements of steps) also took place. Iteration took place in the sense that, (parts of the) initial steps on problem orientation, future vision construction and backcasting were iterated and replicated in later stages of the backcasting experiment. This was found to a larger extent in the MSL case and the NPF case, which (1) consisted of various stages that partly involved different stakeholders, while (2) in subsequent stages the topic was further focused, for instance by selecting a test region (MSL case) or by narrowing down the number of NPF options to be analysed. In addition, it has been found that the embedding of outcomes and follow-up action agenda, as well as raising commitment among stakeholders, was not limited to the final step. This was part of stakeholder communication throughout the backcasting experiment and was intensified in the final stage of these backcasting experiments when results and follow-up agenda had become available. In the SHN case no iteration took place, but the applied backcasting methodology allowed for adjusting the future vision after assessments and stakeholder consultation.

In all three backcasting experiments the four distinguished groups of methods could be identified, as could the three different types of demands. Thus, in all backcasting experiments design methods, analytical methods, participatory methods, as well as methods for communication, coordination and management were applied. In addition, knowledge demands were found, although mainly articulated in terms of 'content' deliverables and not explicitly in terms of quality aspects of knowledge. Process demands were to a large extent articulated in terms of numbers of 'reached', 'informed', 'co-funding', or 'involved' parties, but hardly in terms of process quality like transparency or stakeholder influence. Regarding normative demands, they were partially introduced as initial conditions in all three cases, for instance the factor 20 environmental improvement, and they were partly the outcome of stakeholder consensus in the backcasting experiment.

Table 8.4 *Methodological aspects and settings in the three backcasting experiments*

	NPF case	SHN case	MSL case
Backcasting framework:			
> Inter-disciplinarity	Yes	Yes	Yes
> Framework steps	Yes, but iteration	Yes, but iteration	Yes, but iteration
> Four types of methods	Yes	Yes	Yes
> Three types of demands	Yes	Yes	Yes
Settings:			
> Mobilised budget	€ 2 Million	€ 200,000	€ 2 Million
> Institutional protection	Yes	No	Yes
> Vision champion	Yes (2)	No	Yes (2)
> Main focus	Implementation & follow-up	Academic achievements, methodology development	Implementation & follow-up
> Type of management	Project management	Process management	Project management & process management

With regard to the settings, Table 8.4 shows that the SHN case concerns a relatively low budget backcasting experiment of around € 200,000, while the other two backcasting experiments had budgets of around € 2 million. In addition, in both the MSL case and the NPF case there was institutional protection from the top level of the founding ministries of the STD programme as well as from the top management levels of the co-funding stakeholders. By contrast, no institutional protection could be identified in the SHN case, possibly due to the fact that it was conducted in an international academic setting. In the SHN case no vision champion emerged, while in both the MSL case and the NPF case two vision champions were identified who acted in a complementary way.

With regard to focus, the backcasting experiments on NPF and MSL had a strong focus on follow-up and implementation. By contrast, the SHN backcasting experiment had a strong focus on academic achievements and methodology development and only to a limited extent on follow-up and implementation.

With regard to the type of management, in the NPF backcasting experiment a project management approach prevailed. In the SHN backcasting experiment a process approach prevailed. In the MSL backcasting experiment a project management approach prevailed in the research part; as far as the external key stakeholders are concerned, the focus was on raising consultation and raising commitment (and interest).

8.3 Comparing follow-up and spin-off

This section compares the follow-up, spin-off and broader effects of the three backcasting experiments, as well as the external factors that exerted influence. Follow-up relates to activities that can be seen as directly related to the backcasting experiment and partially involving the same stakeholders and individuals. Spin-off refers to activities that have to some extent been influenced by the backcasting experiment, for instance because these activities contain parts that were derived from the backcasting experiment. Broader or wider effects refer to institutionalisation and institutional resistance.

Briefly, the SHN case showed very limited follow-up and spin-off, while the NPF case and the MSL case both showed substantial follow-up and spin-off. Cases are compared on the three groups of indicators developed in Chapter 4. These are network formation, vision aspects, institutionalisation and institutional resistance.In addition, external factors are compared too. Sometimes, indicators are aggregated if this helps make a useful comparison.

Network formation

Network formation has been analysed using three indicators: activities, actors and resources; the emphasis of the resource indicator was on funding. Both the NPF case and the MSL case have resulted in a considerable number of networks involving all four distinguished domains, while all networks included (focal) activities, actors and resources. By contrast, in the SHN case only a few networks have been identified. These networks were still in an early stage, with actors working on mobilising resources through developing and submitting proposals. These networks fell apart when proposals were rejected.

A practical way to compare the networks and focal activities is at the level of clusters that have been distinguished in the case chapters. The clusters of activities make up a larger network of various (focal) activities, actors and resources. The results for the three cases at the level of clusters are given in Table 8.5, which shows for each case (i) the identified clusters of activities, (ii) the number of actors involved, as well as (iii) an estimate of the total amount of financial resources mobilised in the cluster. In both the NPF case and the MSL case five clusters have been identified. It is also possible to define two tiny clusters of activities in the SHN case, but these are not further discussed here.

The main follow-up initiatives in the MSL case and the NPF case have been defined as separate clusters of activities. These were the MSL Winterswijk programme and the Profetas programme respectively. The numbers of actors involved is higher in the MSL programme than in the Profetas programme. This can partly be explained by (1) the issue of function integration in MSL, which requires the involvement of actors from various functions, and (2) the fact that demonstration of multifunctional farming at the level of individual fields required the participation of individual farmers. In other words, this difference may relate to the different types of socio-technical systems involved: a production and consumption system in the NPF case versus a regional system involving a range of spatial functions in the MSL case.

Table 8.5 *Network aspects per cluster of activities in the three cases*

	Cluster of activities	Number of actors	Amount of resources	Domain
NPF case	Profetas programme	20 - 50 [i]	€ 1.5 Million [ii]	Research domain
	NPF Business R&D and supply chain development	10 - 20	€ 5 - 10 Million	Business domain
	SME cluster	10 - 20	Regular investments	
	Ministry of Environment, targeting food actors & public interest groups	10 - 20 [iv]	< € 500,000 [iii]	Government domain
	Public interest groups	10 - 20 [iv]	< € 500,000 [iii]	Public domain
SHN case	Programme proposals submitted to NIDO contest	10 - 20	< € 100,000 [v]	Research domain
	NPF domestic appliances proposal	< 10	< € 100,000 [v]	Business domain
MSL case	MSL research programme	20 - 50	€ 2.5 Million [vii]	Research domain
	Studies for NRLO & INGRA	10 - 20	< € 500,000	
	Replication regional vision development	20 - 50 each	€ 1 - 2 Million [viii]	Governmental domain
	MSL Winterswijk programme	50 - 100 [i]	€ 5 - 10 Million	Multi-actor initiative covering all domains
	Spin-off of the MSL Winterswijk programme	20 - 50	€ 2 - 5 Million [vi]	Covering all domains

i   This is without symposium participants and informed stakeholders.

ii This is the total external funding, which does not include internal resources of universities and additional capacity of the stakeholders involved.

iii Budgets for public interest groups were largely provided by VROM; the figure presented here is an estimate of the total amount.

iv This figure includes only actors actively involved and does not include visitors of organised public events.

v   This figure reflects largely invested capacity.

vi This estimate accounts for the follow-up of multifunctional farming fields, multifunctional plantations and natural water filtration at the Lankheet estate. It does not include the pig stable pilot at Nutreco, as this is hardly influenced by the MSL programme. However, this pilot and the business development activities are likely to involve several millions of Euros.

vii This research budget is also included in the total funding of the MSL Winterswijk programme.

viii This reflects only vision development, plan development. Implementation involves much higher budgets ranging from € 2 million up to around € 300 million when it is part of 'Space for Rivers'.

Table 8.5 also shows that in the overall budget of the MSL Winterswijk programme was higher than that of the Profetas programme, although research budgets were about the same. This difference is due to the fact that in the MSL Winterswijk programme considerable additional funding was mobilised for the non-research elements, such as demonstration in several projects, the coordination by the province, and communication and dissemination. Some similar non-research activities can also be found in the NPF case, but these were initiated independent of the Profetas programme like the cluster of activities in the domain of the public interest groups. Another difference is that the MSL Winterswijk programme has led to a cluster of follow-up and spin-off activities, while the Profetas programme has not led to new spin-off and follow-up yet by late 2006. However, it can be argued again that such activities are conducted independent from the Profetas programme and involved various companies that did not participate in the Profetas programme.

When comparing the total follow-up and spin-off the MSL case and the NPF case, some additional observations can be made. For instance, the NPF case shows a separate cluster of activities in the public domain. In the MSL case no such cluster was found, though nature organisations were involved in some of the projects of the MSL programme. In addition, the NPF case shows more and more substantial activities in the business domain. In the MSL case companies were involved in some clusters, but to a lesser extent than in the NPF case. While in the NPF case product development, market introduction and investments in production facilities has been found, in the MSL programme mainly R&D activities have been found.

Other observations take a more general approach to the overall results in Table 8.5. Firstly, the case chapters have shown that government financial resources and funding are important for follow-up and spin-off in all domains. This includes funding of academic and applied research, as well as the funding of innovation activities at business through innovation funding programmes. Activities like policy-making and policy implementation by the government, not only at the national level but also at the regional and local levels, also require allocation of government resources. In addition, many activities by community groups or public interest groups very often involve some kind of government funding. The case chapters suggest that most of these activities would not have come about without government funding, or only to a very limited extent. In a way this is confirmed by the SHN case, where unsuccessful attempts were made to mobilise resources from different funding programmes the budgets of which were provided by the government. In the SHN case activities were not continued after proposals were turned down. This suggests that government funding and allocating internal budgets are highly important to stimulate sustainable system innovations in their early stages. This also points to the relevance of financial aspects of policy instruments addressing system innovations towards sustainability.

Secondly, the NPF case shows a considerable growth in mobilised resources in the business domain, after companies decided on product development, production investments and market introduction. NPF related product development, market introduction and supply chain development at DSM Food Specialties, Campina and the alliance of Sodexho and Schouten Europe, respectively, resulted in the mobilisation of substantial resources. This included staff, product testing, investments in production facilities and investments in marketing campaigns. This has resulted in multi-annual investments of at least several millions of Euros. In the MSL case company involvement did not go beyond research activities and was finished before follow-up steps and investments would have commenced. This points not only to the relevance of contributions by companies in system innovations towards sustainability, but also to the stepwise increase in investments and mobilised resources, due to well known business investment cycles.

Thirdly, a large share of the mobilised resources concerns research and generating new knowledge. This is in line with the widely acknowledge relevance of research and new knowledge to both sustainable development in general, and system innovations towards sustainability in particular.

A fourth observation concerns the possible relevance of activities by public interest groups and bot-tom-up community groups. Despite the low share in mobilised financial resources, this type of activities in the domain of the wider public and public interest groups can have a significant direct and indirect impact. For instance, organising public events and relating topics to other broader societal concerns among a larger group of citizens may influence public opinion as well as the way social values evolve. In addition, public interest groups may promote certain opinions, products and changes to their members or supporters, or to a wider audience. Such instances were found for topics like animal welfare, vegetarian lifestyles and world protein security in case of NPFs and meat alternatives, and for topics like liveability, organic foods and landscape concerns in MSL case. In this way public interest groups may contribute significantly to raising awareness and acceptance among the wider public, or larger groups of citizens. This also points to a relevant contribution by public interest groups in system innovations towards sustainability.

Vision aspects

In the SHN case the three future visions 'faded away' after the submitted follow-up proposals were not approved for funding. This means that they did not provide any guidance and orientation. In the Profetas programme, which was the major direct follow-up of the NPF backcasting experiment, the vision and problem definition shifted slightly. This was the result of the entrance of new parties in the network who influenced the vision and the activity in such a way that it matched the preferences and mission of their organisation more closely. In other follow-up and spin-off activities the vision was adjusted as well, in a way that it matched better with the views, preferences and missions of the actors in the related network.

In the MSL case the vision showed strong stability in the MSL Winterswijk programme, whereas flexibility dominated when replication of the backcasting approach took place in vision development experiments in other regions. In the latter spin-off activities new visions were generated, which used the same MSL con-cept and similar basic assumptions as in the MSL backcasting experiment. This resulted in various regional MSL visions that were different, because they were adjusted to the goals and views of stakeholders in the region as well as to the physical characteristics of the regions. These new regional visions are found at the same level as the MSL vision for the Winterswijk region, while all of them are nested in the generic vision on the MSL concept.

Table 8.6 Vision aspects and institutional effects in the three cases.

	NPF case	SHN case	MSL case
Visions:	Adjustment per cluster of activities	Faded away	Replication leads to new visions all nested in the generic MSL vision
> Guidance	Yes	No	Yes
> Orientation	Yes	No	Yes
> Competing visions	Yes	?	Yes
Institutional effects:			
> Institutionalisation	In knowledge domain & in business domain	No	Especially regionally, contribution to the national level
> Institutional resistance	Decreasing support from LNV Hesitance on further investments by companies	No	Conflicting EU market policies

The results for vision aspects are shown in Table 8.6, using the average of the underlying indicators. Table 8.6 shows that at an aggregated level the evolving and adjusted visions provided high degrees of guidance and orientation in the NPF case and in the MSL case, while no guidance and orientation was provided in the SHN case. Another observation is not reflected in Table 8.6, but it is worth mentioning here. This concerns the changed nature of the provided decentralised coordination by the vision in both the NPF case and the MSL case. Within particular activities as well as the preceding backcasting experiment, such coordination is partly provided by project structures, contracts and agreements. Such coordination is no longer provided across (clusters of) activities and domains. However, such implicit coordination seems to be provided by the vision, despite decentralised adjustments to the vision.

Institutionalisation and institutional resistance

Wider effects or institutional effects, as dealt with in this research, concerns institutionalisation and institutional resistance. In Chapter 4 institutionalisation has been defined as *"the process by which a practice becomes socially accepted as 'right' or 'proper', or comes to be viewed as the only conceivable reality"*. In my research, institutionalisation covers the entrenchment of the follow-up and spin-off of a backcasting experiment, as well as resulting changes in institutions and rules. Institutional resistance concerns the rejecting reactions from the vested interests, the existing socio-technical system, or other existing institutions. In general it is put into action by actors in the socio-technical system in which the follow-up and the spin-off are taking place. In fact, these reactions can be seen as wider effects too, albeit as 'negative' ones.

In the SHN case neither institutionalisation nor institutional resistance has been found. Theoretically, it is possible that institutional resistance was so strong that it has prevented follow-up and spin-off, but no such evidence has been found. It is very unlikely that such degrees of institutional resistance would not have been uncovered by in-depth interviews of stakeholders involved.

In the NPF case and the MSL case instances have been found of institutionalisation. For instance, in the NPF case institutionalisation relates to the generated knowledge base in the Profetas programme that has become part of the food innovation system in the Netherlands. The development and successful market introduction by Campina, as well as the development and pilot activities and continuing these by the catering company Sodexho and producer of eat alternatives Schouten Europe, are also clear examples of institutionalisation.

With respect to the MSL case, examples of institutionalisation include adjusted practices at participating stakeholders, such as the Province of Gelderland, the water board and the water company. In most cases this is not only due to the MSL backcasting experiment or the MSL Winterswijk programme. In general it is the result of various activities among the stakeholders where this has taken place to which the spin-off of the MSL backcasting experiment has contributed. Another example is the incorporation of MSL principles in the reconstruction plan and regional spatial development plan for parts of the Winterswijk region.

Examples of institutional resistance have been found in both cases as well. For instance, the Ministry of Agriculture has ended its active support and funding of NPF activities, partly because it considers the meat sector economically important. There is also hesitance to move towards further product development or market introduction among major food companies in the Netherlands, which can partly be seen as driven by vested institutions. With regard to the MSL case, institutional resistance originates from existing regulation.

External factors

In each case study external events and wider developments in the socio-technical system and its context have been identified. In all three cases I evaluated which events and developments in the socio-technical system had exerted an enabling or constraining influence on the emergence of follow-up and spin-off. If this was the case, then events and developments were regarded as external factors. Besides, I investigated whether external factors had emerged in the context of the socio-technical system. As a result, all three cases showed various enabling (+ in Table 8.7) and constraining (- in Table 8.7) external factors that could originate from within the socio-technical system, or from the context of the defined socio-technical systems. Results are shown in Table 8.7.

External factors exerting huge influence that emerged from the context were in the NPF case the policy change at the General Board of NWO, the growing interest of foreign food multinationals in meat alternatives and the EU enacted Novel Food regulation. In the SHN case it was the emergence of international projects on similar topics involving some of the stakeholders and individuals that had been involved in the SHN case and the competition with other proposals for funding. In the MSL case external factors in the context are mainly due to developments and events at EU-level. The MSL case showed that an external factor could be both enabling and constraining at the same time. For instance, the livestock reconstruction reinforced the opportunities for multiple land-use in regions, but at the same time it enhances spatial separation of different functions like agriculture and nature.

Table 8.7 External factors in the three cases

	NPF case	SHN case	MSL case
From socio-technical system (+)	> Increasing market share of vegetarian foods > Shift towards consumer oriented technology development at WUR > New entries (e.g. Campina)	> Emergence of initiatives on related and similar topics without any relationship to backcasting experiment or stakeholders involved	> Increasing policy interest in MSL and area oriented policies > Livestock reconstruction act > Habiforum funding various follow-up and spin-off activities
From socio-technical system (-)	> Supermarket war 2002-2005 > Exit of stakeholders (due to take-overs)	> Decreasing policy interest in sustainable consumption > Ext of major stakeholders	> Pork plague outbreak > Limited support by relevant national networks > Exit of stakeholders involved > Livestock reconstruction act
From context (+)	> Policy change by general board of NWO > Foreign food multinationals invest in meat alternatives	(none)	> Entries by German companies > Decrease of agricultural support by EU
From context (-)	> EU enacted Novel Food regulation in 1990s	> Competition by other proposals > Rejecting of submitted follow-up proposals	> Conflicting EU market policies > Uncertainties about new EU policies for green and blue services

8.4 Evaluating the propositions

In Chapter 4 a set of propositions has been developed that relates the degree of follow-up and spin-off to aspects of the backcasting experiment, which may point to enabling and constraining factors. In this section the propositions are evaluated for which especially the results of the cross-case analysis in the preceding sections is used.

Propositions relating participation to learning or follow-up

The first group of four propositions takes aspects of stakeholder participation as an input and has been divided into two subgroups. Propositions P1A, P1B and P2 relate aspects of stakeholder participation to the degree of follow-up and spin-off.

P1A: High degrees of stakeholder influence on the content result in a higher degree of follow-up and spin-off.
P1B: High degrees of stakeholder influence on the process result in a higher degree of follow-up and spin-off.

With regard to Proposition P1A, the case results vary; the proposition is neither confirmed for stakeholder influence on the content, nor rejected. Stakeholders in the SHN case had large influence on the content of the generated future visions and the elaboration of follow-up proposals and a follow-up agenda during two stakeholder workshops, but no significant follow-up and spin-off has been observed. By contrast, the backcasting experiments in the NPF case and the MSL case resulted in considerable follow-up and spin-off, while only during limited periods selected, rather small, groups of stakeholders could exert considerable influence. In the NPF backcasting experiment this happened at a very early stage of idea articulation and testing. In the MSL backcasting experiment this happened on four different moments in the backcasting experiment involving different groups of stakeholders. In all, two cases show that moderate degrees of stakeholder influence on the topic can lead to considerable extents of follow-up and spin-off; the third SHN case shows a high degree of stakeholder influence on the content in combination with very limited follow-up and spin-off. This proposition is thus not confirmed for stakeholder influence on the content, but not rejected either.

The evaluation of proposition P1A suggests that high degrees of influence may be limited to small selected groups of key stakeholders in combination with a broader group of stakeholders for the purposes of dissemination, elaboration, and testing the preferences and ideas generated by the group of key stakeholders. If this is indeed the case this has implications for who to select and how to design and manage a backcasting experiment.

Regarding Proposition P1B, which deals with stakeholder influence on the process, all three cases show low degrees of stakeholder influence on the process. Process organisation and process management was in all three cases fully under control of the organisers of the backcasting experiments. Proposition 1B is thus rejected; the cases show that a low degree of stakeholder influence on the process does not automatically lead to low extents of follow-up and spin-off.

P2: A greater variety in types of involvement by participating stakeholders results in a higher degree of follow-up and spin-off.

Proposition P2 assumes there is a relationship between the variety in types of involvement by stakeholders and the degree of follow-up and spin-off. This proposition has been confirmed by the case study results. Stakeholders involved in the SHN case, which showed very limited follow-up, did not provide types of involvement other than a (limited) capacity to participate. In the MSL case and the NPF case, each showing significant follow-up, various stakeholders provided co-funding and substantial additional (free) capacity for research, expert contributions and knowledge development.

P3: *A high degree of stakeholder heterogeneity results in more instances of higher order learning on the topic among larger groups of stakeholders.*

Proposition P3 assumes there is a relationship between stakeholder heterogeneity and higher order learning. The case studies show that all three backcasting experiments had a high degree of stakeholder heterogeneity; stakeholders from all four distinguished domains were involved in all cases. All three cases have also shown instances of higher order learning on the topic, although the extent of higher order learning in the SHN case has been evaluated as moderate and no instances of congruent or joint learning have been found in this case. Thus, high extents of stakeholder heterogeneity have been found with varying numbers of instances of higher order learning. As a consequence, P3 is not confirmed, nor rejected by the case results. A high extent of stakeholder heterogeneity may result in more instances of higher order learning, but because none of the cases showed low extents of stakeholder heterogeneity, there is no confirmation that low stakeholder heterogeneity results in less instances or no higher order learning.

P4: *A high degree of stakeholder involvement results in more instances of higher order learning on the topic among larger groups of stakeholders.*

Proposition P4 assumes there is a relationship between the degree of stakeholder involvement and the extent of higher order learning. Both the MSL case and the NPF case show both a high degree of stakeholder involvement for several groups of stakeholders, especially research bodies, as well as co-funding stakeholders. Both cases also show considerable follow-up and spin-off, as well as broader effects. The SHN case shows a low degree of stakeholder involvement in combination with a low to moderate number of instances of higher order learning, as well as no instances of joint or congruent learning. Thus, Proposition P4 has been confirmed; high degrees of involvement lead to more instances of higher order learning. In addition, high degrees of involvement also come along with instances of congruent and joint learning.

Propositions relating learning to follow-up

The second group consists of two propositions that relate higher order learning to the degree of follow-up and spin-off:

P5: *More instances of higher order learning on the topic lead to a higher degree of follow-up and spin-off.*

Proposition P5 assumes that higher order learning influences the degree of follow-up and spin-off. The MSL case and NPF case show more instances of higher order learning, as well as high degrees of follow-up and spin-off. The SHN case combines a low degree of follow-up and spin-off with a moderate number of instances of higher order learning. These results would appear to support this proposition, but to my opinion not yet in a sufficiently convincing way. Instances of higher order learning have also occurred in the SHN case and it can easily be argued that higher order learning does not always lead to higher degrees of follow-up and spin-off. However, P5 is also not rejected. As higher order learning is widely considered a condition for achieving change, more detailed analyses of learning results and learning process may be required to be conclusive on this relationship.

P6: *More instances of joint or congruent learning on the subject lead to a higher degree of follow-up and spin-off.*

One way to gain greater insight into this issue is to focus within higher order learning, which exactly Proposition 6 does. It assumes there is a relationship between the number of instances of joint and congruent higher order learning and the degree of follow-up and spin-off. This type of higher order learning concerns shifts leading to congruence and consensus among groups of stakeholders on problem definitions, perceived solutions, principal approaches to solving the problem, or changing priorities in competing yet desirable objectives (see Chapter 4). The MSL case and the NPF case both show instances of joint and congruent learning, while in the SHN case no such learning has been found. This refinement thus increases the insight and understanding about what type of higher order learning relates the most to the degree of follow-up and spin-off. More instances of congruent and joint (higher order) learning go together with higher degrees of follow-up and spin-off. Proposition P6 is thus confirmed.

<u>Propositions concerning visions</u>

The third group of propositions deals with relationships between vision aspects and the degree of follow-up and spin-off.

P7: *Visions providing high degrees of guidance and orientation result in a higher degree of follow-up and spin-off.*

Regarding Proposition P7, the NPF case and the MSL case contain visions with high degrees of both guidance and orientation, in combination with high degrees of follow-up and spin-off. The SHN case shows a moderate degree of guidance and a low degree of orientation in combination with very limited follow-up and spin-off. Thus the cases confirm proposition P7 in two ways.

P8: *Multiple visions backcasting experiments lead to a less significant degree of follow-up and spin-off than single vision backcasting experiments.*

With respect to Proposition P8 the case studies show that the NPF case and the MSL case concern single vision backcasting experiments, while the SHN case concerns a triple vision backcasting experiment. When combined with the various degrees of follow-up and spin-off, this confirms P8. It needs to be mentioned that at the first stage of the MSL backcasting experiment three other visions were generated, but these were dealt with in separate backcasting experiments with different sets of stakeholders parallel to the MSL backcasting experiment. Apparently, generating multiple visions together at an early stage of the backcasting experiment is not the deciding factor; the distinction is that elaboration and analysis of a single vision in a backcasting experiment enables a higher degree of follow-up and spin-off. This may be due to a more focused group of stakeholders, while 'domestication' of the vision by stakeholders or groups of stakeholders may also be stimulated in a single vision backcasting experiment.

P9A: *Visions with high degrees of stability over longer periods of time result in a higher degree of follow-up and spin-off.*

P9B: *Visions with high degrees of flexibility over longer periods of time result in a higher degree of follow-up and spin-off.*

Proposition P9A assumes there is a relationship between the degree of follow-up and spin-off and the degree of closure or stability of the vision, while Proposition P9B assumes there is a relationship between the degree of follow-up and spin-off for the degree of flexibility of the vision. These propositions cannot be evaluated based on the time period of the backcasting experiment, because the vision is then still being developed. It requires taking into account longer periods of time. Therefore, the SHN case is not used to evaluate this proposition because the three visions faded away too soon after the end of the backcasting experiment. The question underlying this proposition is, whether follow-up and spin-off require closure and stability, or flexibility in the content of the vision.

The cases provide mixed results with respect to these propositions. The NPF case shows adjustments to the vision that roughly follow the five identified clusters of activities (and their networks), which confirms flexibility. At the same time the core of the NPF vision has remained very stable across various clusters of follow-up and spin-off activities. This stable core consists of the idea that environmentally sound and highly improved meat alternatives (or protein foods) from non-animal protein sources are important in future sustainable protein food consumption and production. This confirms stability or closure of the vision. The MSL case shows a similar pattern; there are adjustments in the vision largely at the level of clusters, but there is also a stable core. One difference with the NPF case is that the vision at the MSL Winterswijk programme hardly changed at all compared to vision when the MSL backcasting experiment was completed. However, in other clusters of activities in the MSL case considerable adjustments of the vision were found. Like in the NPF case, there is a stable core in the vision and adjustments in specific clusters and different regions. Thus, in the MSL case both flexibility and stability are found as well. As a consequence, both Proposition 9A and Proposition 9B are confirmed. The two cases illustrate that both propositions are not conflicting and do not point in opposite directions: both flexibility and stability of future visions are required. It may be possible to distinguish between a deeper core of the vision that should not change and other parts that can easily be changed, but this has not been analysed systematically in this research. Another possible explanation may relate to core convictions or world views in specific domains, in specific networks or of particular stakeholders; it assumes that adjustments of the vision are driven by the world views and core convictions of specific actors or groups of actors.

Propositions concerning settings and methodological aspects

The last group of three propositions relates the degree of follow-up and spin-off to methodological aspects and aspects related to settings.

P10: Institutional protection of the backcasting experiment leads to a higher degree of follow-up and spin-off.

Proposition P10 assumes there is a relationship between institutional protection of the backcasting experiment and the degree of follow-up and spin-off. Both the MSL case and the NPF case show that the backcasting experiments had institutional protection; in the SHN case no institutional protection of the backcasting experiment has been found. Thus Proposition P10 has been confirmed.

P11: The emergence of a vision champion in the backcasting experiment results in a higher degree of follow-up and spin-off.

Proposition P11 assumes there is a relationship between the emergence of a vision champion in the backcasting experiment and the degree of follow-up and spin-off. Both the MSL case and the NPF case have shown individuals that acted as vision champions, while no vision champion has been found in the SHN case. In both the NPF case and the MSL case two vision champions emerged who acted in a complementary way, for instance over time or across different domains. Proposition P11 is thus confirmed.

P12: If the conducted backcasting experiment has a better match with the methodological framework, there is a higher degree of follow-up and spin-off.

Proposition P12 relates the extent to which the conducted backcasting experiments matches the developed methodological framework for backcasting to the degree of follow-up and spin-off. In all three cases the methodological framework matched the way the backcasting experiment had been conducted. However, in the SHN case this does come along with considerable follow-up and spin-off, thus Proposition P12 is not confirmed. However, P12 is also not rejected by the case results. It may be that P12 is conditional, meaning that the match with the methodological framework is a necessary, but not sufficient condition for the emergence of substantial follow-up and spin-off. Settling this issue requires further research.

Finally, Table 8.8 summarises the status of the propositions after evaluation; it also gives some implications.

Table 8.8 *Evaluation of the propositions and possible Implications*

	Status	**Implication**
P1A	*Not confirmed* *Not rejected*	High degrees of influence on the content by all stakeholders involved are not necessary for considerable follow-up and spin-off; high degrees of influence by small groups of selected key stakeholders on the topic can be important for realising considerable follow-up and spin-off.
P1B	*Rejected*	High degrees of stakeholder influence on the process are not necessary for considerable follow-up and spin-off.
P2	*Confirmed*	Other types of stakeholder involvement in addition to the capacity to participate may stimulate considerable follow-up and spin-off.
P3	*Not confirmed* *Not rejected*	A high degree of stakeholder heterogeneity does not guarantee a high degree of higher order learning (on the topic).
P4	*Confirmed*	A higher degree of stakeholder involvement leads to more instances of higher order learning on the topic.
P5	*Not confirmed* *Not rejected*	More and wider higher order learning does not guarantee a high degree of follow-up and spin-off.
P6	*Confirmed*	More instances of joint and congruent learning among groups of stakeholders may stimulate considerable follow-up and spin-off.
P7	*Confirmed*	Visions providing high degrees of guidance and orientation lead to more substantial follow-up and spin-off.
P8	*Confirmed*	Single visions backcasting experiments may lead to substantial follow-up and spin-off; multiple visions do not.
P9A *P9B*	*Confirmed*	For providing high degrees of guidance and orientation and substantial follow-up and spin-off visions need to have both stability and flexibility.
P10	*Confirmed*	Institutional protection is important for substantial follow-up and spin-off.
P11	*Confirmed*	Vision champions help realise substantial follow-up and spin-off.
P12	*Not confirmed* *Not rejected*	A good match of the backcasting experiment with the methodological framework for participatory backcasting does not guarantee substantial follow-up, but can be a necessary condition.

8.5 Internal and external factors

Although the case chapters have provided a broad picture of the cases from a multitude of sources, they have not focused on the internal factors that may constrain or enable the process involved in raising follow-up and spin-off. In addition, external factors in the different cases have only been compared to a limited extent so far. This section therefore takes a closer look at the internal and external factors, for which I use the case comparisons made in Sections 8.2 and 8.3, as well as the evaluation of the propositions in Section 8.4.

The NPF case and the MSL case showed considerable follow-up and spin-off, while the SHN case did not. By comparing the differences between, on the one hand, the NPF case and the MSL case and, on the other hand, the SHN case, factors may be found that affect the degree of follow-up and spin-off and are characteristics or internal factors of the backcasting experiment. The differences that were identified are listed in Table 8.9. Please note that I did not take into account that some of these differences and the underlying factors may be related and interdependent.

Table 8.9 *Differences between cases with and without significant follow-up*

MSL case & NPF case	SHN case
High degrees of stakeholder involvement among some groups of stakeholders	Only a low degree of stakeholder involvement
Various types of stakeholder participation including co-funding and substantial capacity	Only one type of stakeholder participation, limited capacity for workshop attendance
Limited (selected) groups of stakeholders have high levels of influence	All participating stakeholders have high levels of influence
Single vision	Multiple visions
High degrees of guidance and orientation by the vision	Moderate degree of guidance and low degree of orientation by the visions
Considerable budgets (around € 2 millions)	Limited budget (around € 200,000)
Institutional protection	No institutional protection
Several vision champions	No vision champion
Strong focus on follow-up and implementation	Focus on academic methodology development; little focus on follow-up and implementation
More instances of higher order learning on the topic at the level of specific stakeholders	Moderate instances of higher order learning on the topic at the level of specific stakeholders
Joint and congruent learning among groups of stakeholders	No joint or congruent learning among groups of stakeholders

To summarise, all the characteristics of the backcasting experiments on MSL and NPF listed in the left column of Table 8.9 are potentially enabling factors. The characteristics of the SHN backcasting experiment listed in the right column of Table 8.9 are potentially constraining factors. It must be noted that the absence of an enabling factor is not automatically a constraining factor, thus each difference must be evaluated on this. The resulting factors are listed in Table 8.10. Sufficient budget is not seen as a factor here, but as a necessary condition, and it is therefore not included in Table 8.10.

Table 8.10 *Internal factors affecting the degree of follow-up and spin-off*

Enabling internal factors	Constraining internal factors
> A high degree of stakeholder involvement > Other types of stakeholder participation in addition to the capacity to participate > Limited, selected groups of stakeholders have high levels of influence > Single vision backcasting experiment > High degrees of guidance and orientation by the vision > Institutional protection > Presence of a vision champion > Focus on follow-up and implementation > More instances of higher order learning on the topic at the level of specific stakeholders > Sufficient joint and congruent learning among groups of stakeholders	> Multiple visions backcasting experiment > Focus on academic methodology development

Table 8.10 shows a substantial number of enabling internal factors and two constraining internal factors. With regard to the low number of constraining factors, it did appear that many of the characteristics of the SHN backcasting experiment relate to the lack of enabling factors. The substantial number of enabling factors also raises the questions (1) if some enabling factors are more important than others, and (2) how factors relate to one another. For instance, different factors may enhance each other's enabling effect, or may be conditional, or may have to come about together. It is even possible that all the enabling factors are required. However, the empirical results from the cases do not allow for ranking the enabling factors identified or for relating them to each other.

Table 8.11 *'Generalised' enabling and constraining external factors*

Enabling external factors	Constraining external factors
> Entries of 'motivated' stakeholders, due to contingent or external factors > Presence of and access to government funding programmes > Contingent factors due to initiatives in the context (e.g. policy change by NWO board)	> Exit of stakeholders due to contingent or external factors > Competition by other visions or proposals > Constraining influence by supranational organisations like EU and WTO

It has already been mentioned that external factors can have a big influence on the emergence of follow-up, spin-off and wider effects. External factors are highly context-dependent and can also be highly contingent, which complicates identifying more generic enabling and constraining factors that may have a wider relevance. Nevertheless, I make a first attempt, using the case comparison earlier in this chapter on the external factors that had exerted influence. Results are shown in Table 8.11.

8.6 Evaluating research methodology and indicators

In Chapter 4 the research methodology for this multiple case study has been developed, which included turning the proposed conceptual framework into an analytical framework consisting of indicators. The results of applying these indicators have been dealt with in the case chapters and the preceding part of this chapter. The empirical part of this research has yielded evidence that the research methodology served the purpose of this research and has yielded interesting results.

Despite this positive overall observation, it is interesting to focus on the level of the indicators as well, and more specifically on those indicators that yielded similar results in all three cases. The latter group of indicators has provided no insights to explain the differences between the cases regarding the different extents of follow-up and spin-off in the three cases.

Indicators providing similar results and thus having limited explanatory power in this research can be found in all building blocks in which the backcasting experiment has been conceptualised. Firstly, stakeholder heterogeneity was high in all three cases and was thus not supportive in explaining the differences in the degree of follow-up and spin-off. It is possible that stakeholder heterogeneity should not be seen as a relevant indicator, but as a condition for backcasting experiments. Evaluation of backcasting experiments having moderate or low stakeholder heterogeneity is necessary to shed more light on this.

Secondly, in the group of indicators on future visions there are two indicators that provided similar results for all backcasting experiments. These are the indicators for alternative (sets of) rules and for decentralised coordination. Regarding the 'rules' indicator it has been found that all visions contained alternative sets of rules. This suggests that presence of alternative (sets of) rules is an inherent characteristic of future visions and not a varying characteristic that provides information about the degree of guidance. The 'rules' indicator has been derived from one of the sub-functions of the Leitbild concept, but this does not conflict with my finding that it is an inherent characteristic. Decentralised daily coordination has also been found in all backcasting experiments, but this was not only due to the structuring and integrating 'power' by the vision. It was also due to other structures that were created, such as contracts and project structures. Thus, although coordination seems an important characteristic, it does not appear to have real explanatory power for the degree of follow-up and spin-off. These two indicators may be left out or need to be adjusted in future research aimed at evaluating backcasting experiments and their follow-up and spin-off. However, the decentralised coordination indicator might be relevant in analysing follow-up and spin-off when separate clusters of activities can be identified.

Thirdly, regarding the three indicators for higher order learning, it has been found that only congruent and joint learning provided results that helped explaining the differences in the degree of follow-up and spin-off. Thus differences in shifts in problem definitions, perceived solutions, principal approaches for dealing with the problems at the level of individual actors or stakeholders could not be related to the degree of follow-up and spin-off; all cases provided instances of higher order learning at the level of individual actors or stakeholders.

Fourthly, with respect to the group of indicators for methodological aspects and settings, the indicators for methodological aspects that have been based on the developed methodological framework for participatory backcasting provided similar results for all three cases and were thus also less useful for explaining the differences in follow-up and spin-off. Like in the case of the stakeholder heterogeneity, it may confirm the relevance of these characteristics when applying the backcasting approach but not for explaining the differences in the backcasting experiments studied in this research. Then, these characteristics and the entire methodological framework should be seen as conditions to 'good' participatory backcasting. As this aspect goes beyond the evaluation of the analytical framework and the indicators, I return to this issue in 9.3, when I reflect on the backcasting approach and the developed methodological framework for participatory backcasting.

Fifthly, the purpose of the three groups of indicators, reflecting the impact after five to ten years, was to map and analyse the follow-up and spin-off. These indicators served this purpose very well, so little more needs be said about this. One observation is the difference of the results on the future vision indicators across the cases and in comparison to phase of the backcasting experiment; regarding follow-up and spin-off the vision indicators provided more explanatory power including the decentralised coordination indicator.

Finally, it can be mentioned that the checklist for external factors has been useful for identifying possible external factors. In addition, emergent external factors have been found that exerted enabling or constraining influence on follow-up and spin-off that were not covered by the proposed checklist for the socio-technical system, but originated from the context of the socio-technical system.

BACKCASTING FOR A SUSTAINABLE FUTURE: THE IMPACT AFTER 10 YEARS

9

CONCLUSIONS, REFLECTIONS AND RECOMMENDATIONS

This chapter provides conclusions based on the empirical findings in the previous chapters (9.1). It also provides conceptual and theoretical reflections (9.2), reflections and improvements related to the backcasting approach (9.3) and recommendations (9.4).

9.1 Conclusions

This research has been guided by two main questions. Central in this section are the first main question and conclusions based on the empirical chapters. The first main question has been formulated as:

A. *What factors determine the impact of backcasting experiments after five to ten years?*

In this research I have investigated this question by evaluating three cases, each consisting of a backcasting experiment and its follow-up, spin-off and broader effects after five to ten years (see Figure 9.1). Selected cases had to show broad stakeholder involvement and had to have resulted in the development of at least one desirable future vision. The first case was the Novel Protein Foods (NPF) case, which focused on sustainable and attractive meat alternatives and had envisioned a system innovation in which a substantial share of meat and meat products is replaced by protein foods from non-animal sources. The second case was the Sustainable Household Nutrition (SHN) case, which had been conducted as part of the international 'Strategies towards the Sustainable Household (SusHouse)' project from 1998 to 2000. The third case was the Multiple Sustainable Land-use (MSL) case, which dealt with function integration in rural areas involving agriculture and other functions related to landscape, nature, recreation, water production and water management. The NPF case and the MSL case were based on backcasting experiments that were conducted in the period 1993 to 1997 at the Dutch governmental Sustainable Technology Programme (STD). While these cases relate to substantial parts of the food production and food consumption system in the Netherlands, each case focuses on different socio-technical systems with different characteristics. The NPF case focused on a production and consumption system of protein foods, which includes meat alternatives, meat and meat products, and in which food companies are the central players. The SHN case approached nutrition and food consumption and production from the viewpoint of households and consumers. The MSL case focused on a regional system in which the agricultural function was integrated with other functions and as a consequence spatial planning aspects were important. Finally, Figure 9.1 shows again the backcasting experiment and its impact after five to ten years also summarising the building blocks for the two phases.

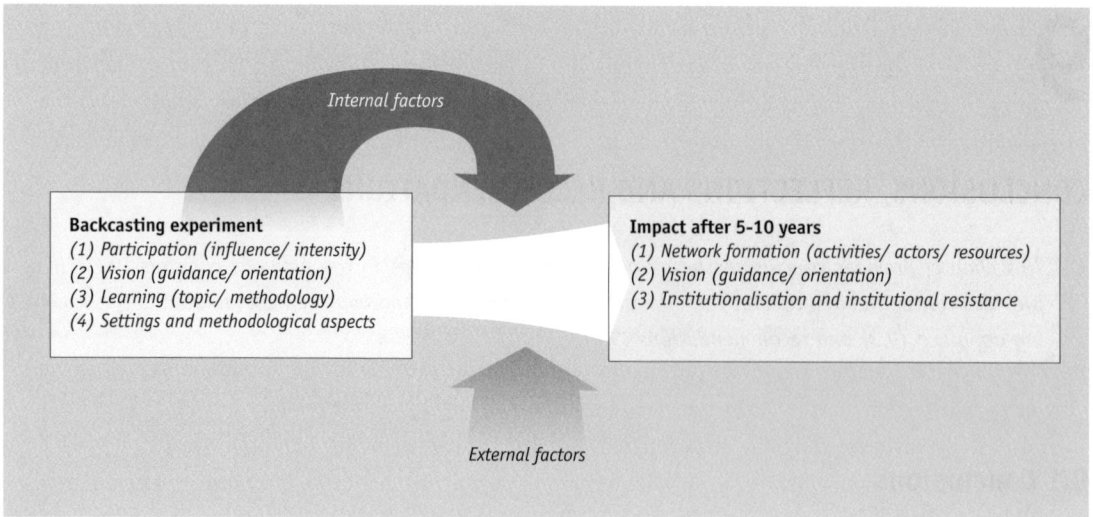

Figure 9.1 *The backcasting experiment and its impact after five to ten years*

Empirical conclusions

The first conclusion is that backcasting experiments involving various stakeholders from different societal domains can result in the development, exploration and analysis of desirable visions of the future that provide guidance (where to go) and orientation (what to do) to some of the stakeholders involved; backcasting experiments can also lead to instances and processes of higher order learning among stakeholders and in the formulation of follow-up agendas.

The second conclusion is that this does not automatically lead to follow-up, spin-off and implementation in line with the vision and the follow-up agenda, but that this depends on various internal and external factors that can be both enabling and constraining. Internal factors are characteristics of the backcasting experiment, whereas external factors originate from the surrounding socio-technical system and its context. Table 9.1 depicts both types of factors that influence the extent of follow-up and spin-off.

Important enabling internal factors, relating to the backcasting experiment, are institutional protection from top level of participating stakeholders, the emergence of vision champions, a high degree of stakeholder involvement, other types of participation in addition to the capacity to participate, for instance, co-funding or substantial 'free' capacity, sufficient budget for the backcasting experiment, a strong focus on follow-up and dissemination as well as communication, a single vision backcasting experiment, high degrees of influence on the part of (at least) some key stakeholders and joint or congruent learning. Constraining internal factors relate to the backcasting experiment include a multiple visions backcasting experiment and a strong focus on academic achievements. In addition, external factors and developments that evolve in the surrounding socio-technical system or its context can be both constraining and enabling, but these are case specific (more details are provided later in this section).

Table 9.1 *internal factors enabling and constraining the extent of follow-up*

Enabling internal factors	Constraining internal factors
High degree of stakeholder involvement	-
Diversity in types of stakeholder involvement	-
Single vision backcasting experiment	Multiple visions backcasting experiment
High degrees of guidance and orientation of the future vision	-
Institutional protection	-
Presence of vision champions	-
Strong focus on follow-up and implementation	Strong focus on academic achievements
Joint and congruent learning	-

The third conclusion is that the desirable vision of the future is important in follow-up and spin-off activities and provides high degrees of guidance (where to go) and orientation (what to do). Follow-up and spin-off activities are constituted by networks of actors that have been successful in mobilising sufficient resources to establish the activities. Future visions at play in follow-up and spin-off show stability as well as flexibility; they co-evolve with networks in the sense that the networks and actors involved in follow-up and spin-off are influenced and inspired by the visions, while they conversely influence and adjust the vision as well. Adjustments can be related to preferences and worldviews of the actors involved or to the characteristics of different societal domains.

The fourth conclusion is that when substantial follow-up and spin-off occur after five to ten years, this is still at the level of niche activities, or concerns a set of niches in the four distinguished domains of research, business, government, and public interest groups and the general public. This follow-up and spin-off comes along with first instances of broader effects like institutionalisation. However, although the niches have 'grown out' of the backcasting experiments and can be characterised as first steps or stepping stones towards system innovations towards sustainability, that does not guarantee that the envisioned system innovation towards sustainability will fully come about in the next 30 to 40 years. The backcasting experiments themselves can be seen as the initial niche for experimentation.

Other empirical findings: backcasting experiments

Backcasting originates from the 1970s and was originally developed as an alternative to traditional forecasting and planning. The original focus was on policy analysis for energy planning, which later shifted to exploring sustainable futures and solutions. Stakeholder participation and achieving implementation, in which the Netherlands was a front runner, became important in the 1990s. A distinction can be made between approaches in which modelling is a key element of constructing future visions and approaches in which modelling is hardly used when generating future visions. Backcasting can be applied at the level of organisations, regions, industrial sectors, socio-technical systems countries, or even on a global scale. While the approach is named after backcasting or backwards looking from the desirable future, this particular element is at the same time the least developed and elaborated part of the approach. Backcasting is not the only approach that uses normative or desirable future visions; participatory backcasting approaches are part of a family of approaches, all combining desirable future visions or normative scenarios with stakeholder participation.

A methodological framework for participatory backcasting has been developed, which consists of five steps and the outline of a toolkit containing four groups of methods and tools: design tools, participatory tools, analytical tools and management, Coordination and communication tools. This framework also distinguishes different types of demands, as well as different types of goals. A further discussion is provided in 9.3. The way the three backcasting experiments were conducted was in accordance with the developed methodological framework for participatory backcasting. Although the five steps could be distinguished in each of the cases, they were not applied in the proposed order in all three cases; iteration of steps took place, especially with respect to the future vision and backcasting. The methodological framework and the case chapters provide suggestions to further improve the approach, as well as to develop (prescriptive) guidelines (see 9.3).

All three backcasting experiments involved a wide range of stakeholders, developed one or several desirable future visions, and proposed follow-up activities and agendas. They also induced higher order learning among the participating stakeholders. The three cases illustrate how backcasting can be a powerful approach to developing alternative sustainable future visions, utilising the expertise and knowledge of a broad range of stakeholders. Higher order learning occurred with regard to the topics under study in the backcasting experiments, as well as the backcasting approach itself. Follow-up agendas included R&D-activities, strategy development, policy recommendations and short-term proposals. Despite these similarities, the three backcasting experiments varied considerably in the degree of stakeholder influence, the degree of stakeholder involvement, whether a vision champion emerged, whether other types of participation than the capacity to participate occurred (e.g. co-funding), and regarding the degree to which the developed future visions provided guidance and orientation.

Other empirical findings: follow-up and spin-off

The case studies showed strongly varying extents of follow-up, spin-off and broader effects after five to ten years. The SHN case showed very limited follow-up and spin-off. By contrast, the MSL case and the NPF case showed substantial follow-up and spin-off, as well as broader effects like early forms of institutionalisation. In the NPF case and the MSL case follow-up and spin-off activities in general involved actors from more than one domain and were found in all four distinguished societal domains.

Follow-up and spin-off can be seen as activities constituted by networks consisting of actors that have successfully mobilised resources and other actors to establish and carry out the activities. The main share of financial resources in follow-up and spin-off activities involved government funding or allocation of internal budgets at the government. It includes funding for academic research, for R&D and innovation at companies, the allocation of budgets for policy implementation (programmes) and policy-making, the funding of experimentation on various levels of government, as well as subsidizing activities of public interest groups or local communities. The second major source of (financial) resources concerned investments by companies. Although companies are often hesitant with respect to larger investments, when they decide to start product development or market introduction and realising production facilities, substantial financial and personnel resources are mobilised as the NPF case shows. In all cases most follow-up and spin-off activities could be grouped in clusters of activities based on shared characteristics. Clusters of activities relate to the emphasis within the future vision or to shared adjustments to the future vision.

In the cases where there was substantial follow-up and spin-off, the future vision provided high degrees of guidance and orientation to processes of activity generation and network formation in line with the future vision. At the same time the future vision showed a stable core, as well as flexibility, which both could vary in different activities and different societal domains. With regard to stability, the core of the original vision was still present in follow-up and spin-off, and was supported by the actors, including new entrants. With regard to flexibility, the visions were adjusted in the direction of new stakeholders' preferences and worldviews, as well as adapted to the societal domains in which these actors are rooted. This can be characterised as a process of alignment, domestication and making the vision more congruent with existing visions by actors or groups of actors.

Visions diffuse to other societal domains and other settings through actors and individuals, in a process of attracting and mobilising actors and resources. This includes processes of re-evaluation and adjustment of the vision, as well as processes of network formation. These processes are mutually interdependent: a vision exerts influence on networks around activities and, conversely, the networks and the actors making up the network influence the vision leading to adjustment or emphasis on specific elements. Thus, the vision and mobilised networks of actors develop in a co-evolutionary way.

External factors may have a positive (enabling) or negative (constraining) influence on the emergence of spin-off and broader effects. Factors that originate outside the socio-technical system are often highly contingent and can be highly important in establishing specific activities. Events and ongoing developments within the socio-technical system have a more diffuse influence, but may also lead to strongly changing actor preferences and decisions to join or leave a specific activity. Anticipations and expectations for the future, which can be actor-specific or shared by a group of actors, can also exert influence. They relate to possible external developments that still have to become important, but may considerably influence actors' strategies.

9.2 Reflections

9.2.1 Conceptual framework

This study has provided evidence that backcasting experiments involving stakeholders can be conducted in various ways leading to stakeholder support, future visions, analytical results and follow-up agendas, as well as to higher order learning among stakeholders involved. This study has also shown that backcasting experiments can lead to substantial follow-up, spin-off and broader effects after five to ten years, but this is not guaranteed. The selected cases suggest that inducing higher order learning among individual stakeholders is very well possible and that the key to achieving follow-up and spin-off is to induce congruent or joint learning at the level of groups of stakeholders.

It must be noted that this finding can relate to the way the cases were selected. Criteria were discussed in 4.4 and included that the backcasting approach had been completed, as well as that broad stakeholder involvement had been achieved and that there had been a focus on system innovations towards sustainability. It may thus be that higher order learning is provoked by characteristics of the backcasting approach, or by the focus on complex sustainability problems (see also 9.3). For instance, Van de Kerkhof (2004) who has evaluated a participatory backcasting experiment focusing on options for mitigating climate change, has also reported considerable higher order learning at the individual level.

The difference between cases with substantial and little follow-up and spin-off points to the importance of joint and congruent learning among groups of stakeholders involved, as well as how to translate this into action, activities and behavioural changes by individuals and organisations. Follow-up and spin-off as found in two cases relates to niche activities or sets of niches, thus to initial stages of system innovations.

How do these observations, as well as other findings and conclusions summarised in the previous section, relate to the conceptual framework and the underlying theories, and what are the possible implications?

The conceptual framework developed in Chapter 4 is shown in Figure 9.2. It covers both the first phase of the backcasting experiment and the second phase of follow-up, spin-off and broader effects after five to

Socio-technical system

Internal factors

Backcasting experiment

(2) Vision

(4) Settings

(1) Participation

(3) Learning

External factors

Context

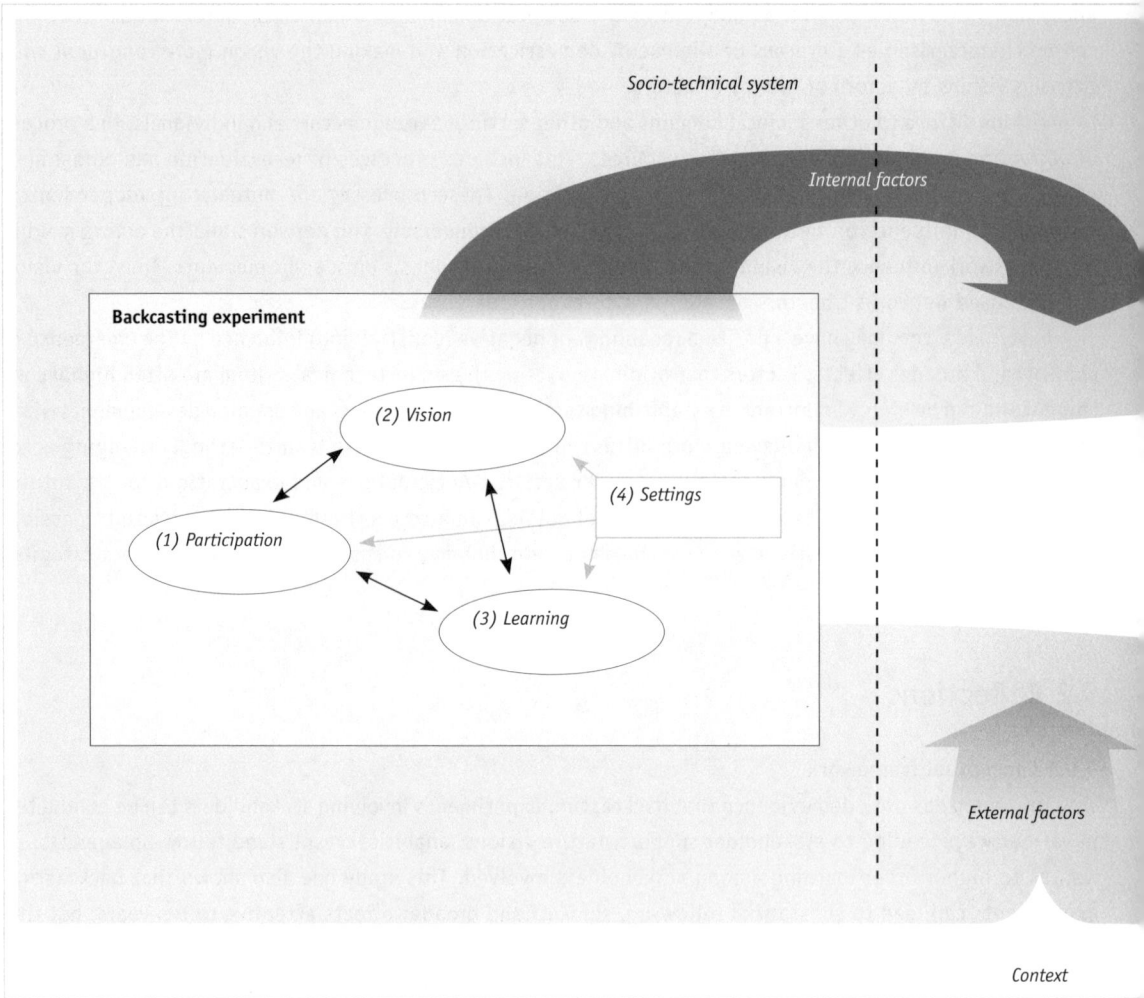

Figure 9.2 *The conceptual framework used in this research (see Chapter 4).*

ten years. In this conceptual framework the backcasting experiment is made up of four building blocks: (1) stakeholder participation; (2) future visions; (3) stakeholder learning; and (4) settings and methodological aspects. The phase of follow-up, spin-off and broader impact consists of three building blocks: (1) network formation, (2) future visions, and (3) institutionalisation. In addition, internal factors and external factors are distinguished that both exert influence on the emergence and the extent of follow-up and spin-off.

Each of the building blocks has been based on the theory for which the theoretical exploration in Chapter 3 has provided the input using insights and concepts from stakeholder theory, actor learning, the sociology of technology, industrial network theory, as well as institutional theory. The conceptual framework as a whole has been relevant and useful to guiding the research and obtaining the results and conclusions. It has guided and structured the conceptualisation of both the first phase of the backcasting experiment and the second phase of follow-up, spin-off and broader effects after five to ten years. The conceptual framework served as a useful starting point for elaborating the research methodology, as well as for developing the propositions. Below, reflections on separate building blocks are made.

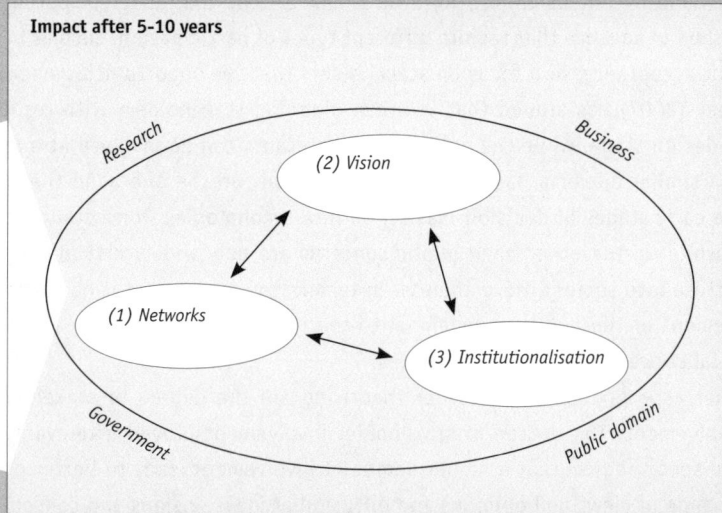

Socio-technical system
(5–10 years later)

Impact after 5-10 years

Research

(2) Vision

Business

activities
getting realised

(1) Networks

(3) Institutionalisation

Government

Public domain

Context

Stakeholder participation

The building block of stakeholder participation has used elements from stakeholder theory in (public) decision-making, science and technology studies and sustainability studies. It consists of the aspects 'stakeholder heterogeneity', 'stakeholder influence', 'degree of stakeholder involvement' and 'types of involvement'. In short, these stakeholder theories argue that more influence and control by stakeholders on both content and process provide higher degrees of acceptance, legitimacy, accountability and commitment, and therefore a higher probability of follow-up and acceptance. By contrast, this study has shown that considerable follow-up and spin-off can be achieved when stakeholders have no or very limited influence on the process of the backcasting experiment. Considerable follow-up and spin-off were also possible when most stakeholders have a moderate degree of influence on content and small groups of key stakeholders have a high degree of influence. This suggests that careful network management and theories of network management may be more relevant than theories of stakeholder participation in (public) decision-making and science. However, stakeholder theories concerning decision-making and science have (partly) been developed to enhance democracy by giving citizens more of a say and taking into account a broader range of social aspects. From this viewpoint stakeholder participation is not only a matter of efficiency or effectiveness, but also of social judgement and democracy. This may point to a dilemma and a normative issue, which may need broader reflection and theorising. For instance, there is an ongoing debate on how to relate representative and deliberative democratic practices when they occur at the same time, as well as in what kind of arrangements they can be combined and how this relates to different types of governance (Driessen *et al.* 1995, Teisman and Edelenbos 2004). However, it may also be that different purposes of participation must be distinguished in advance that require different types of participation, such as broad participation aiming at wide public acceptance, or a focus on stakeholders that see opportunities when aiming at follow-up. For instance, Peek (2007) has argued that in urban planning stakeholders with resources should be involved in the early design stage, while the public and inhabitants can be involved at a later stage. In technology assessment a similar dilemma has not yet been resolved. On the one hand there are pleas to involve the public in the early stages of decision-making on new technologies, for instance through citizen juries and lay panels, while on the other hand public concerns are only widely articulated when new technological artefacts diffuse into society. Nevertheless, in technology assessment it has always been emphasised that early involvement of the public is beneficial to the purpose of acceptance, as well as to articulate social aspects, social concerns and user preferences.

Two other aspects that need further theorising are the degree of stakeholder involvement and the types of involvement. The degree of stakeholder involvement proved a relevant explaining factor in the cases; it also seems logical that a higher degree of involvement leads to better opportunities for learning, debate, exchange of views and opinions and diffusion of ideas, visions and concepts. However, no theoretical base for this has been found in stakeholder theory or in theories of higher order learning. This can be due to the roots of various stakeholder theories focusing on involving citizens and social actors in (public) decision-making and science or improving the external communication of companies, but deserves further theorising. It may be that relevant elaborations can be found in learning theories focusing on education or learning-by-doing in innovation studies.

In addition, the necessity of different types of involvement is acknowledged in economic or industrial network theory, where it is referred to as different types of resources. By contrast, in various stakeholder theories no such concept has been found, while this research points to exploring the relevance of such a concept and further reflection and theorising on this.

Future visions

The future visions building block has been conceptualised using the Leitbild concept of Dierkes *et al.* (1996). This has provided the aspects of guidance (where to go) and orientation (what to do) among groups of actors, which was originally developed for emergent visions around the genesis of new radical technological innovations.

Three extensions of the Leitbild concept have proved applicable in this research. The first was the idea that a Leitbild provides guidance and orientation can also be applied in case of visions with a strong normative or ethical component. The second was that the process of synchronisation guided by the Leitbild would not only apply across various scientific disciplines, but also across various societal domains and stakeholders with very different values and different worldviews in different societal domains. The third assumed that analysing guidance and orientation at the level of actors and groups of actors would suffice.

One direction of further conceptualisation is the nested character of related visions. Nested visions have clearly appeared in two cases, but this phenomenon has not been conceptualised as part of this research. A promising direction is to build on the literature on technological expectations and technological promises, where the nested character of technological promises and expectations has been proposed (Van Lente 1993).

In addition, in Chapter 3 I have pointed to the similarities between promises and visions, while arguing that differences relate to the nature of supporting expectations: technological expectations in which the likelihood of shared anticipations is important, versus what Berkhout (2006) has called normative expectations in which normative principles and assumptions are important. As this comparison has only briefly been touched upon and different types of expectations have not been conceptualised in this building block, this seems another interesting direction for further reflection and theorising; this may also relate to the nested character of both visions and technological promises.

Another direction of further theorising and conceptualisation relates to the balance between flexibility and stability in visions in order to realise maximum guidance and orientation, as well as what such a balance could look like. This may also relate to the nested character of visions.

The aspect of competing visions has been introduced as part of the vision building block, which could be emergent alternative visions, as well as the dominant vision in a socio-technical system. This aspect has been useful in the analysis of the cases, but may also be an issue for further theorising and conceptualising. For instance, the relationship between the future vision generated in the backcasting experiment and other emergent alternative visions seem relevant, as well as how social dynamics around different visions may affect each other. This relates to Berkhout (2006) who sees visions as proposals or 'bids' that are employed by actors in processes of coalition formation and coordination. The mechanism proposed by Berkhout seems related to the dynamics that can be found in policy networks or policy domains. I see Berkhout's point as particularly relevant for the phase after the backcasting experiment is completed and the vision is confronted with the favourite visions and wishes of other actors.

Another direction for further reflection is how the vision in the backcasting experiment and the challenged dominant vision relate, how they may influence each other and how the related social dynamics develop. This is partially related to similar processes mentioned above. The difference is, however, that the existing dominant vision is strongly institutionalised and strongly backed by actors and existing coalitions of actors.

Learning

The conceptualisation of higher order learning taken from Brown et al. (2003) has yielded substantial instances of higher order learning at the level of individual actors in all three cases. This may suggests two different things. First of all, it may suggest that realising higher order learning is less difficult than in general is assumed by numerous authors. Secondly, it may also suggest that backcasting is a good approach to achieving higher order learning, for instance because of the system orientation, the long time horizons and the applicability in case of complex and persistent problems at a societal level. It is also possible that both suggestions are valid. The consequence is then that the decision to analyse on backcasting experiments has led to selecting cases with substantial higher order learning. Independent of which of the two suggestions prevails, the cases suggest that the real challenge is how to turn higher order learning into action, activities and changed behaviour of individuals and organisations, and how to achieve congruent and joint higher order learning among groups of stakeholders.

The conceptualisation provided by Brown et al. (2003) is a recent one and distinguishes between three interrelated shifts or types of higher order learning. The first and second types comprise shifts in problem definitions, perceived solutions, principal approaches to dealing with the problems at the level of individual actors or stakeholders, but were of limited relevance for explaining the differences in follow-up and spin-off in the three cases. The third type, reflecting congruent and joint learning at the level of groups of stakeholders, has provided results that helped explaining the differences in the extent of follow-up and spin-off.

What are the possible implications of the results in this study for the conceptualisation of higher order learning as proposed by Brown *et al.* (2003)? A first direction for further theorising is to make the conceptualisation by Brown *et al.* (2003) more dynamic. Then the first shift and the second shift occurring at the level of individual stakeholders should be seen as inputs to a learning process at group level that may result in the third shift which emphasises joint and congruent learning by several members of the group, or the entire group. I return to this issue below in the discussion on the linking pin between the backcasting experiments and its follow-up and spin-off.

Another direction may be to look further into existing conceptualisations of higher order learning, such as organisational learning (Argyris and Schön 1994), policy-oriented learning (e.g. Sabatier and Jenkins-Smith 1999) and Fischer's (1980, 1995) changes of appreciative systems. It is possible that other conceptualisations of higher order learning place higher demands to instances of learning before determining them as higher order learning, for instance because higher order learning results need to be shared. Such a result would call for adjusting the conceptualisation by Brown *et al.* (2003). However, even if higher order learning in this study would be limited to the third type, the number of instances is still considerably higher than reported in numerous other studies evaluating higher order learning (e.g. Hoogma *et al.* 2002, Van Mierlo 2002). By contrast, it is also possible that no higher demands are set in other learning theories, which would enhance the confirmation of the appropriateness of backcasting for inducing processes of higher order learning. This may also explain why other authors identify much less instances of higher order learning then this study has done.

Brown *et al.* (2003) have also pointed to higher order learning in the context of the BSTE. This has not been studied in this research, because such diffusion processes have been mapped by following the diffusion of visions. Nevertheless, it could be an interesting extension to include this aspect in the developed conceptual framework.

Network formation

The adjusted conceptualisation of Håkansson's (1987, 1989) industrial network perspective that consists of activities, actors and resources and was originally developed for supplier-customer networks of focal companies, appeared helpful in describing and analysing follow-up and spin-off of the backcasting experiments. This study has proven that the adjusted conceptualisation enables to use this network concept in other domains than of companies, as well as with respect to activities that involve actors from various domains. In addition, the use of focal activities, instead of companies as focal actors, as well as the definition of activities at a more aggregated level than Håkansson has proposed, proved a useful starting point for identifying actors involved and mobilised resources in this research. This confirms that network analysis using the conceptualisation of Håkansson does not necessarily have to start with an actor or a group of actors, but can also start with the activity and can probably also use resources as a starting point. Finally, both the adjusted and the original conceptualisation emphasise the mobilisation of resources. While in many network conceptualisations mobilisation of resources is rather neglected, this aspect was relevant with regard to the extent of follow-up and spin-off.

There are four directions for extending this building block. The first is to add the concept of clusters of activities and networks, which emerged from this research as an additional element. This may relate to what to the distinction by Luiten *et al.* (2006) between micro-networks and meso-networks, which they also call technology networks. While these authors (Luiten *et al.* 2006: 1032) have defined the micro-network as *"a group or actors who co-operate in developing a specific industrial energy-efficient technology"*, they have defined the meso-network as *"the total collection of micro-networks around an innovative technology"*. Despite the fact that the latter study focused on a business setting of new technologies, this issue should be further looked in. It also refers to the widely acknowledged notion that networks can be bounded in different ways (e.g. Håkansson 1989, Oerlemans 1996).

The second direction for extending this building block is to distinguish between the various stages of network formation. Network formation starts in an early stage when proposals for concrete activities are developed. These early stages include making the design for an activity, as well as selecting who will or should be involved. A third, more fundamental extension is to include a stronger focus on relationships between actors involved in the network constituting an activity, which is in line with the vast majority of network theories (see Chapter 3). In this research the relevance of relationships has been acknowledged, but relationships have not been analysed. Such an extension would yield a lot of additional insights on a more detailed level of the dynamics within the identified networks. The fourth extension concerns the possible relationships between visions and networks, which is discussed below.

Diffusion of visions

The emergence of follow-up and spin-off can be viewed as the diffusion of a vision. The adjusted Leitbild concept of Dierkes *et al.* (1996) has also been useful for analysing spin-off and other effects after five to ten years. In addition to the reflections on visions mentioned above, one more direction for further elaboration is the relationship between visions and networks. It has been concluded that vision and networks of actors evolve in a co-evolutionary way and mutually influence each other. This is an extension to the conceptualisation offered Dierkes *et al.* (1996), who emphasise strongly that visions exert influence and guide sets of actors at different locations in case of the development and diffusion of radical technological innovations, but do not relate this process to network theory.

Conceptualising the dynamic relationships between visions and networks may also shed more light on the process by which visions are adjusted during the process of defining follow-up (proposals) and mobilising resources and actors and of affecting existing activities leading to spin-off. This may to a certain extent build on known theoretical concepts. For instance, Callon (1995) has focused on both dynamics and more structural aspects of innovation networks, but not on the shared future vision as something that can provide guidance and orientation and may influence network dynamics. Dierkes *et al.* (1996) describe network development as a result of the guidance and orientation of a future vision, but have neither conceptualised this, nor the interaction between vision dynamics and network dynamics, nor do they refer to any kind of network. Looking into both types of dynamics and their relationships and mutual influence may also shed more light on the nested character of related visions, as well as on the issue of stability and flexibility in visions. Furthermore, in policy sciences research has been conducted into the deconstruction of the value systems or belief systems of actors in policy controversies such as environmental deadlocks. For instance, Van Eeten (1999) has used discourse analysis to of such policy controversies to define new alternative agendas that may bridge the gap among stakeholders. Such agendas can be framed as potential alternative future visions, but this type of policy analysis does not include if and how the agendas can become visions providing guidance and orientation.

Institutionalisation and institutional resistance

The applied conceptualisation of institutionalisation and institutional resistance appeared appropriate in this study. Interestingly, both institutionalisation and institutional resistance refer to the structure side of Giddens' duality of structure and action (Giddens 1984), but depict very different aspects. Institutional resistance refers to the existing structures and institutions that guide the actions by actors, whereas institutionalisation as used in this study refers to creating new structures and institution and is the result of the actions of actors 'against' existing structures and institutions. In this research actions are the follow-up and spin-off activities conducted and constituted by the actors making up the network surrounding an activity and refer to the action side of Giddens' duality.

An interesting direction for further theorising is on how institutionalisation and institutional resistance relate, for instance in the cases evaluated in this research. Furthermore, institutional theory, as described by Van den Hoed (2004), offers two concepts which deserve further theoretical exploration with regard to their applicability and added value for my conceptual framework. Firstly, institutional theory distinguishes between (Scott 1995, Scott 2001) regulative, normative and cognitive aspects of institutions (or "pillars" of institutions as Scott actually calls them), which seems an interesting extension to evaluate institutionalisation in a more detailed way.

Secondly, in institutional theory the concept of organisational field has been developed, which is the locus of institutional change and institutionalisation. The organisational field is external to organisations, develops around issues and involves a range of actors. It is within the organisational field that rules, regulations and standards are debated, adjusted and decided upon, and where new practices are rejected or accepted. This implies that institutionalisation resulting from follow-up and spin-off of backcasting experiments needs to be approved within the relevant organisational field. The relevant issues from the viewpoint of my research are (1) how the organisational field for the spin-off and follow-up of the backcasting experiments in this study has emerged, and (2) how the process of approval and decision-making has taken place for the instances of institutionalisation found in the cases of this research.

The linking pin between backcasting experiments and their impact

One element of the conceptual framework that certainly deserves further theorising and conceptualisation is the (change) process through which the follow-up and spin-off 'grow' out of the backcasting experiment. So far, the backcasting experiment and its impact have been conceptualised separately, but the process that connects the two phases has not been conceptualised. This is certainly a direction in which further research is possible using insights from two different fields.

In the first place the multi-layered models of higher order learning from, for instance, Sabatier and Jenkins-Smith (1999), and Fischer (1995), as they offer explanations about processes of higher order learning at individual actors, as well as groups of actors. Secondly, these learning theories need to be connected to theories of organisational behaviour that acknowledge that changes in mental frameworks, appreciative systems or value systems are insufficient to change actor behaviour. Changes in organisational behaviour also relate to aspects like opportunity, available resources, and constraining and enabling structures, as can be found in models of consumer behaviour (Fishbein and Azjen 1975), diffusion of innovations (Rogers 1995) and existing environmental behaviour models (e.g. Hoevenagel *et al.* 1996). Institutional theory may also be useful in explaining the behaviour of organisations in their institutional environment, as it is the institutional environment that enables and constrains organisational behaviour (e.g. Scott 1995, 2001). Eventually this could result in connecting higher order learning among groups of stakeholders to organisational behaviour or organisational decision-making and in achieving insights in the linking pin between the backcasting experiment and its follow-up.

Internal factors

Internal factors have been empirically identified, based on the evaluation of the propositions and the comparison of cases with and without follow-up and spin-off. No attempt has been made to conceptualise these factors theoretically. This may nevertheless be an interesting direction to pursue. In Chapter 8 it has already been mentioned that the substantial number of enabling factors raises questions, such as whether some enabling factors are more important than others, and how factors within the backcasting experiment relate to one another. Theorising and conceptualising on this issue may enable to shed more light on these questions.

9.2.2 System innovation theory

When developing the conceptual framework for this research in Chapter 4 I have decided not to use system innovation theories, as follow-up and spin-off were expected to occur in the early stages of system innovations or transitions (see also Chapter 3). This has been confirmed by the empirical findings. However, it is interesting to relate the results to various system innovation theories, keeping in mind that I have concluded that follow-up and spin-off was at the level of niche activities or sets of related niches. Below I address the Multi-Level Perspective, transition contexts, innovation systems and institutional theory; explanations of these theories have been provided in Chapter 3. Furthermore, I do not deal separately with the Large Technical Systems (LTS) approach (see Chapter 3), although it may be interesting to explore how Hughes and other scholars using this approach have dealt with the early stages of LTS.

Multi-Level Perspective

In Chapter 4 it was shown that the first phase of the backcasting experiment, the second phase of follow-up and spin-off after five to ten years and the third phase of the 'completed' system innovation towards sustainability could also be conceptualised as three separate, but nested levels each with a different scale, involving different numbers of actors and different degrees of institutionalisation. Then, the backcasting experiment should be seen at the lower level. The intermediate level consists of the impact after five to ten years. The higher level includes the long-term impact after 40 to 50 years, after which in some cases the system innovation towards sustainability has resulted in a highly adjusted socio-technical system. This conceptualisation can be related to the Multi-Level Perspective (MLP) of Geels (2005) and others (e.g. Rip and Kemp 1998, Kemp *et al.* 2001). The MLP also comprises three levels: the niche level, the level of socio-technical regimes and the landscape level. The MLP suggests that system innovations or transitions start in niches and can grow out of the niche 'replacing' the dominant socio-technical regime, while the landscape refers to external events and developments that exert influence to the niches and the regimes. In addition, Geels (2005) has proposed four phases of development that are more or less similar to the phases suggested in Transition Management (Rotmans *et al.* 2001). Neither the three phases distinguished in this research, nor their conceptualisation as three levels conflict with the MLP. Backcasting experiments, however, are much 'smaller' than the level of technological or market niches. Follow-up, spin-off and broader effects after five to ten years can be related to the niche level in the MLP, while 'matured' or 'completed' system innovations can be related to level of socio-technological regimes.

Niches and niche experiments

This research has shown that the impact at a niche level after five to ten years are not limited to technological niches, or to niche replication in which demonstration projects are repeated elsewhere, as Van Mierlo has pointed to (2002). In the case of the follow-up and spin-of of backcasting experiments, clusters of activities can be seen as niches, which can be found in all four societal domains distinguished. These findings contain various suggestions for refining the niche concept in the MLP. Firstly, a niche is not necessarily a simple phenomenon like a market niche or a technological niche, but can comprise various types of niche activities in different societal domains that may have some explicit kind of coordination or an implicit decentralised kind of coordination provided by a future vision. There is thus a need to look further into different types of niches and the different types of activities in different domains that may be necessary to make a niche successful. For instance, Verheul and Vergragt (1995) have pointed to social experiments in which niches are emergent and emerge in a bottom-up process in society. In addition, niches may also take the form of policy experiments in which a range of stakeholders or the wider public is involved. Secondly, the results of this research suggest there is a phase before the niche, which can be a R&D lab in industry, a research institute or a backcasting experiment. This phase is neglected in current niche and regime change concepts, but can be the phase in which important decisions are made.

Transition contexts

It has been argued that transitions and regime change starting in niches are not the only plausible mechanism, but one of several possible mechanisms (Berkhout et al. 2004, Smith et al. 2005). These authors have argued in favour of taking into account whether necessary resources are internal or external to the regime, as well as whether the degree of coordination is high or low. An interesting direction to explore is whether the origin of resources and the degree of coordination can be identified in backcasting experiments and their follow-up and spin-off. This may also have explanatory power for the extent of follow-up and spin-off or can account for other characteristics of follow-up and spin-off. Another direction for further research is how guidance, orientation and decentralised coordination by visions that have both flexibility and stability as proposed in this research relate to the concept of purposive transitions (Smith et al. 2005) that resembles transitions in transition management (Rotmans et al. 2001). A major agreement between the idea of visions generated in backcasting experiment guiding system innovations towards sustainability and the purposive transition in line with transition management is that the visions are generated purposefully. A major difference is that I neither assume central coordination nor centralised management by the government or a government organisation. By contrast, I assume decentralised coordination, guidance, and orientation by the vision, as well as flexibility and stability of the vision.

Innovation systems and system innovations

Innovation systems (see Chapter 3) are systems of knowledge production and knowledge use by companies that turn generated knowledge and developed technologies into competitive products and services. A distinction can be drawn between national systems of innovation, sectoral innovation systems and technology focused innovation systems. Because of the system orientation and the focus on knowledge production and knowledge use by companies, and the relevance of generating new knowledge for system innovations towards sustainability, an interesting direction for further theorising is how the theories and concepts of innovation systems relates to my findings on follow-up and spin-of.

In innovation systems the focus is on producing knowledge that can be commercialised and the way this process is facilitated by existing structures and institutions, as well as how this process can be improved. However, any change that requires radically new or modified structures and institutions is not, or to a lower extent facilitated and to a large extent neglected by innovation system scholars. Nevertheless, empirical findings and theorising on early stages of emerging innovation systems, such as in ICT or in modern biotech and life sciences may be interesting to explore and relate to my findings on follow-up and spin-off of backcasting experiments. Something similar can be said about the approach of technology-oriented innovation systems (see Chapter 3), as in this line of innovation system research has been focused on emerging technologies like renewable energy technologies.

However, it must be realised that system innovations and innovation systems may also be conflicting concepts. System innovations towards sustainability assume new knowledge, which may be partly fundamental knowledge, or strong reconfiguration of existing knowledge from several domains and scientific disciplines that are in general not connected. This implies that structures and institutions are lacking, as well as that existing institutions and structures are constraining rather than enabling in nature. Nevertheless, it can be argued that system innovations towards sustainability must be accompanied by either new innovation systems or strong reconfiguration of existing innovation systems, including connecting parts of existing innovation systems. Then, the new or reconfigured innovation system can be thought of as part of the institutionalisation process accompanying the system innovation that goes towards sustainability. As there seems to be hardly any scientific literature on these issues, this may be an interesting direction for further research and theorising. At the same time my results points to some of these issues. For instance, in the NPF case instances of institutionalisation relate to the entrenchment of the NPF knowledge base in the Dutch food innovation system; in the MSL case integrated use of knowledge on different spatial functions was constrained by disciplinary and organisational boundaries. Also, in industrial ecology similar problems have been encountered because innovation systems are in general sectorally structured and industrial ecology includes connecting chains that are part of different industrial sectors.

Institutional theory

Institutional theory views on change in organisations as processes of institutionalisation and de-institutionalisation. These processes are driven by what is agreed and decided on by the actors in the surrounding organisational field. The organisational field is considered to be the locus of institutional change and develops around issues. An interesting direction would be to study the possibility to conceptualise system innovations towards sustainability as processes of institutionalisation and de-institutionalisation, as well as how an organisational field develops and evolves around a system innovation towards sustainability. Also, the extent in which a backcasting experiment can contribute to an organisational field for a certain system innovation towards sustainability is worth exploring.

9.3 Backcasting and methodological reflections

9.3.1 Refining participatory backcasting

This section deals with reflections on the backcasting approach and the developed methodological framework, in particular for exploring and shaping system innovations towards sustainability. This section also suggests methodological improvements and addresses the second main question, which has been formulated in Chapter 1 as follows:

B. *How should participatory backcasting be applied for exploring and shaping system innovations towards sustainability?*

The methodological framework

This section takes as a starting point the methodological framework for participatory backcasting and how this has been evaluated in this research. This framework is depicted in Figure 9.3. It consists of five steps

Three types of demands:

(1) Normative demands

(2) Process demands

(3) Knowledge demands

Different goals:

> Involvement of a wide range of stakeholders

> Future visions and follow-up agendas

> Awareness and learning among stakeholders

> Commitment and follow-up by stakeholders

> ...

Five steps:

STEP 1:	STEP 2:	STEP 3:	STEP 4:	STEP 5:
Strategic problem orientation	*Develop future vision*	*Backcasting analysis*	*Elaborate future alternative & define follow-up agenda*	*Embed results and agenda & stimulate follow-up*

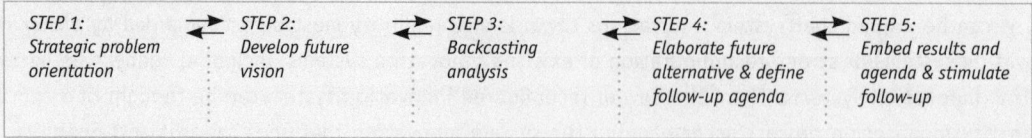

Four groups of tools and methods:

(1) Participatory/ interactive tools and methods

(2) Design tools and methods

(3) Analytical tools and methods

(4) Tools and methods for management, coordination and communication

Figure 9.3: *The methodological framework for participatory backcasting*

and the outline of a toolkit containing four groups of methods and tools, three types of demands, as well as different types of goals. The methodological framework assumes that participatory backcasting is both trans-disciplinary and inter-disciplinary. Backcasting is inter-disciplinary because methods and knowledge from different disciplines are used and combined in such a way that the whole is more than the various parts. Participatory backcasting is also trans-disciplinary, because inter-disciplinarity is combined with stakeholder involvement and stakeholder knowledge and views are used. Stakeholder heterogeneity needs to be high, which can be achieved by involving stakeholders from all four distinguished societal groups. Despite the fact that the steps are presented in a linear fashion, iteration of steps and moving forward and backward between steps is possible and in fact likely to occur.

With regard to the approach and the methodological framework the case studies showed the following. The way the three backcasting experiments were conducted, matched well with the developed methodological framework for participatory backcasting. All three backcasting experiments were both inter-disciplinary and trans-disciplinary. The five steps of the methodological framework could be identified in all the cases, but did not follow the proposed order. The backcasting step was less well elaborated than the other steps in terms of prescriptive methods and tools. All four groups of methods could be identified in the cases. In general design methods and visioning methods were less formalised than analytical tools and methods. Finally, embedding of outcomes and follow-up action agenda, as well as raising commitment among stakeholders was part of stakeholder communication throughout the three backcasting experiments.

What, then, are the possible implications of these findings for the backcasting approach and the methodological framework for participatory backcasting that was tested?

Strategic and generic issues

The key to participatory backcasting can be described as developing and exploring desirable visions and turning them to action and activities in a bottom-up process driven by stakeholders involved. Key elements are stakeholder participation, future visions and stakeholder learning, which are all three mutually dependent on one another and mutually influence each other. Both the settings of backcasting experiments and the way the backcasting approach is applied are also important, but in my opinion these aspects must be seen as conditional and not sufficient to achieve high degrees of follow-up and spin-off. Follow-up and spin-off is driven by stakeholders who are not only attracted or inspired by the vision, but also see opportunities for their organisations.

This research has positioned participatory backcasting as a useful approach to exploring system innovations to sustainability, as well as to facilitating first steps towards envisaged sustainable future visions. In this way alternative solutions in the long-term can be identified for complex sustainability problems in socio-technical systems that cannot be solved adequately on the short term, or by optimisation of the socio-technical system.

Applicability of participatory backcasting does not have to be limited to complex sustainability problems. It can be applied to any complex or unstructured problem at the level of society or socio-technical systems. But, as Dreborg (1996) has argued, backcasting is particularly applicable when there is a need for a major change, when dominant trends are part of the problem, when there are externalities that cannot be satisfactorily solved in markets, and in case of sufficiently long time horizons allowing alternatives that need long development times. So far, backcasting has hardly been applied to complex problems other than sustainability problems, which is thus an interesting direction for further development, as well as for further application of the approach.

So far, participatory backcasting has applied to organisations (e.g. Holmberg 1998) as well as to system innovations towards sustainability and sustainable technologies (see Chapter 2; Quist and Vergragt 2006). Examples of participatory backcasting on a national or international level or involving neighbouring regions from two countries are rarer. As such backcasting studies without stakeholder involvement are reported in literature (see 2.2) and this seems an interesting direction for further application of participatory backcasting. However, it may be that at a national or international level it is more difficult to induce bottom-up processes driven by stakeholders that lead to follow-up and spin-off.

Steps and iteration

The five steps provide a useful starting point for developing or organising participatory backcasting experiments and applying participatory backcasting. There is no need to adjust the number of steps or redefine one or several steps.

In addition, the five proposed steps must not be seen as a prescriptive order, but rather as five key parts that need to be addressed sufficiently in a specific participatory backcasting experiment. This allows replication and iteration of steps; in the cases such replication and iteration has been especially found with respect to generating future visions and the backcasting step. This has been acknowledged by various authors that have proposed backcasting methodologies (e.g. Weaver *et al.* 2000: 76, Quist *et al.* 2001a: 78), but has never been emphasised. However, most authors have only reported the steps applied in the specific backcasting experiments and methodologies that they reported on (e.g. Van de Kerkhof 2004, Holmberg 1998).

It is possible that different steps in a backcasting experiment are organised as different activities or projects, as the NPF case and the MSL case show. Then different stages may involve different sets of stakeholders and have a different focus (e.g. national level versus regional level, or a generic concept versus more elaborated versions). Such a setting enhances the need for replication and iteration, especially of the future vision, the backcasting analysis and other assessments. Two issues are relevant here. Firstly, iteration and replication stimulates that new parties get acquainted with other stakeholders involved, as well as may enhance raising commitment and momentum. This type of dynamics and settings has not been taken into account in the methodological framework, but would allow organising and designing more complex back-casting experiments in line with what has been found in the cases. Secondly, the cases show that iteration and replication lead to gradual development and elaboration of visions, and did not lead to fully new future visions. In addition, the cases with gradual development of vision did also show considerable follow-up and spin-off. As a consequence, this may even point to an internal factor that has a positive influence on the degree of follow-up and spin-off, but this has not been looked into in this research. In addition, it is pos-sible that a backcasting experiment made up of stages or rounds may impose additional process demands, like entry or exit rules and organisational demands like process management or leadership competencies. This could be a topic for further study, as well as for further development of the approach.

This research has shown that the embedding of outcomes and follow-up action agenda, as well as raising commitment among stakeholders, involves more than just the final step. In all three cases this was part of stakeholder communication and stakeholder involvement throughout the backcasting experiment, though 'embedding' was intensified in the final part of the backcasting experiments. This suggests that raising sup-port and commitment should be part of all five steps, and should be pursued in addition to other functions of stakeholder participation. A way for further development of the backcasting approach may be to identify a range of possible functions of stakeholder participation and prioritise these for each step.

The backcasting step

The backcasting step has been less well elaborated in terms of prescriptive methods and tools than the other steps. Often this was done in an intuitive and non-formalised way. This calls for further methodology development regarding this issue, as already mentioned in Chapter 2 when developing the methodological framework. The three cases have provided useful starting points for the development of such tools and methods, at least for the development of a structured list of questions. Furthermore, use can be made of earlier methodological evaluations such as Aarts (2000), Quist *et al.* (2000) and application of the backcast-ing approach in teaching (Quist *et al.* 2006).

Four groups of methods

In all three backcasting experiments the four distinguished types of methods have been identified. Thus, in all backcasting experiments design methods, analytical methods, participatory methods and methods for communication, coordination and management were found. The results do not suggest that certain methods in each group can be prescribed. By contrast, the results suggest that the four groups make up a toolkit of tools and methods that can be used to design a backcasting experiment. To conduct each step the organisers of a backcasting experiment can select methods and tools from each group.

The distinction between the four groups of methods has been useful for the purpose of this research. However, it is possible that methods can be related to more than one group. For instance, brainstorming with stakeholders can be classified as a participatory method, but when it leads to generating visions it can also be classified as a (part of) a design method. Also, developing and using a model includes design steps and analytical steps, while it is possible to apply a model in a participatory setting. This may be a direction for improving the methodological framework.

From the cases it appeared that in general design methods and visioning methods were less formal-ised than analytical tools and methods. While all three backcasting experiments obviously included design activities, those involved in the backcasting experiment were not aware that they were conducting design activities. Design activities were often conducted without using formalised or prescribed design methods. Adding formalised design methods to the toolkit, as well as testing these in backcasting experiments is thus a direction for further improvement of the framework and the approach. In addition, further elaboration of the toolkit with methods in all four groups is a direction for refining the backcasting approach.

Demands

In all three backcasting experiments the three different types of demands have been identified. Thus, in all backcasting experiments knowledge demands, process demands and normative demands have been found. Knowledge demands were mainly articulated in terms of 'content' deliverables and not explicitly in terms of quality aspects of knowledge. Process demands were articulated largely in terms of numbers 'reached', 'informed', 'co-funding', or 'involved' and hardly in terms of process quality, such as transparency, or influence. With regard to normative demands, these were in all three cases partially introduced as initial conditions, for instance the factor 20 environmental improvement, and were partly the outcome of stakeholder consensus in the backcasting experiment.

Goals

Different types of goals relate to content (sustainable option, vision, and assessments), process (stakeholder support, dissemination and follow-up) and to normative preferences. In the methodological framework for participatory backcasting, as well as in the case studies limited attention has been paid to goals of backcasting experiments. Further elaboration may be needed to compensate for this. Use could be made of the goals that In 't Veld (2001) has proposed the TO3 approach (see Chapter 2). Further development may also focus on how different goals relate to different steps in the backcasting process, as well as to normative demands. The latter issue can be illustrated by the question whether the factor 20 should be a normative demand, as argued in this research, or that it should be seen as a goal.

Improvements and guidelines

A major challenge for the organisers of participatory backcasting experiments is to induce higher order learning among groups of stakeholders. Another major challenge is to facilitate and stimulate stakeholders groups to start or join follow-up activities, or to include elements of the future vision in ongoing activities. Although the second challenge is partly a bottom-up and self-organising processes among stakeholders who are attracted and inspired by the vision and pursuit opportunities in the future vision, some guidelines can be provided based on the outcomes of my research. These guidelines for conducting backcasting experiments, aimed at achieving follow-up and spin-off, are summarised in Table 9.2 and can be applied in combination with the methodological framework.

Efficiency and effectiveness

Finally, can something be said about the effectiveness and efficiency of the two backcasting experiments that resulted in considerable follow-up and spin-off, assuming that these allow for generalisation? In the fall of 2006, follow-up and spin-off of the backcasting experiments reported in this thesis were presented during an environmental research-policy exchange meeting, where a policy-maker of the Ministry of the Environment argued that 'these results after eight years are too little and too limited from the viewpoint of the costs of the two backcasting experiments' (which were about € 2 million each). The current niche type of development, according to this policy-maker, is too limited from the viewpoint of the system innovations towards sustainability envisioned in the mid-1990s at the STD programme.

Table 9.2 *Some guidelines for organisers of backcasting experiments*

> Give influence to committed key stakeholders
> Stimulate other types of stakeholder involvement besides 'workshop attendance', such as co-funding, substantial capacity and expertise
> Focus on a single future vision with its 'own' group of stakeholders involved
> Stimulate institutional protection at top management levels of involved stakeholders
> Stimulate high degrees of stakeholder involvement
> Involve or stimulate the emergence of (potential) vision champions that can become 'brokers' in relevant networks
> Focus strongly on follow-up of the backcasting experiment, as well as implementation and usability of its outcomes
> Do not keep several visions within a single backcasting experiment

---

However, a major purpose of the backcasting experiments at the STD programme was to contribute to greening the knowledge infrastructure and to create a 'green' knowledge base for sustainable system options for the future. Assuming that to a certain or even a considerable extent this has been achieved, the issue that then emerges is how to move from generated knowledge to innovation and implementation. This sounds a bit like the well-known Dutch innovation paradox: sufficient or excellent knowledge generation, but limited benefits in terms of competitive products and services contributing to economic growth and other societal goals.

It is possible to be more concrete on this issue than merely referring to the so-called Dutch innovation paradox. For instance, the government could develop and apply instruments that support and stimulate development of products and services, their market introduction, or related market development. Renewable energy in Germany can be taken as an illuminating example. This requires at least a change on the part of the Dutch government, which traditionally focuses on stimulating the supply of new knowledge, stimulating producers and neglecting temporarily market stimulation (or limiting this to short periods). Besides, in the environmental policy domain there seems to be a focus on developing new instruments and new initiatives and a neglect of evaluating the effects of previously developed and applied instruments and developing facilitating policies. For instance, no effect evaluations that have looked into the effects of backcasting experiments after five to ten years have been found in the policy domain.

It is also much too simple to expect system innovations towards sustainability to manage ' themselves' fully after limited budgets have been spent in backcasting experiments. Most past transitions and large-scale system innovations have involved considerable efforts by actors from all four domains in society for longer periods of time. This calls for greater financial support by government programmes, as was confirmed by the two cases with substantial follow-up and spin-off.

9.3.2 Backcasting and related approaches

Other participatory vision based approaches

Participatory backcasting belongs to larger group of future studies methods that all explicitly use normative futures or future visions. This raises the question as to how participatory backcasting relates to other approaches that use normative futures or future visions, such as Transition Management and various others. For instance, is there something that makes participatory backcasting unique in comparison with

Transition Management and other approaches in this family? I would argue that, for instance, the similarities between backcasting and Transition Management outweigh the differences. Therefore, I consider the backcasting analysis and the concept of explicit backwards looking not sufficiently unique to distinguish it completely from other normative future vision-based approaches. Nevertheless, backwards looking is a helpful concept providing a useful metaphor stimulating so-called lateral thinking outside the existing mental frameworks, while backcasting analysis methods provide useful results and can be applied in stakeholder workshops. These aspects regarding backcasting analysis and backwards looking relate to differences at the level of tools, metaphors and methods, not at the level of essentials.

One direction of further research and further development of the backcasting approach is to compare backcasting more systematically to other participatory vision-based approaches, such as Transition Management, TO3 (as proposed by In 't Veld 2001) and approaches in other countries, such as 'la prospective' in France (e.g. Godet 2000).

Visions not related to approaches

Many visions do not originate from the application of a vision-based approach or a backcasting experiment. Examples of this are the vision of bringing salmon back in the river Rhine, or closing the tidal inlets in the Netherlands before the disastrous flooding of 1953. Further research in the emergence and 'realisation' of such visions may contribute to the further development of the backcasting approach.

Transition Management

There are further distinctions between backcasting and other vision-based approaches. For instance, Transition Management assumes that transitions are controllable and manageable to a larger extent than most backcasting approaches, and also central steering or coordination is required. Transition Management also assumes that several transition visions must be articulated and taken into account in order to keep all options open and to define and stimulate selected transition paths.

By contrast, the participatory backcasting approach emphasises participatory vision development and facilitating and stimulating follow-up in line with the vision that is endorsed by stakeholders. This research has shown that better follow-up results are achieved when domestication of the vision by stakeholders and raising commitment among stakeholders takes place in single vision backcasting experiments. This research has also shown that 'strong' visions provide guidance and orientation in a decentralised way instead of through centralised coordination or management. An important condition for decentralised diffusion to other networks and domains is that there be sufficient flexibility for existing and new stakeholders to adjust the future vision in a way that it matches their own organisations and their missions better. This needs to result in what has been called nested visions, which means that the adjusted visions can be reconnected to the core of the original vision. In addition, it must be realised that backcasting methodologies show considerable variety among each other. As a consequence, comparing different backcasting experiments, or relating participatory backcasting to other normative vision based approaches always needs to be compared on a range of dimensions, instead of on a single characteristic. With regard to the comparison of participatory backcasting and transition management, transition management policies are currently being developed in the Netherlands. At the moment it is too early to compare results and effects with those of the backcasting experiments in the 1990s, for instance at the STD programme. However, some intermediate preliminary evaluation is to some extent possible, but also requires further research. Preliminary spoken, the emphasis in the backcasting experiments in the 1990s was on realising bottom-up commitment and spin-off among relevant stakeholders in emerging topics in a rather gradual way, which was called the stone in the pond. By contrast, the energy transition activities (for an analysis of the transition of the electricity system, see Hofman 2005) seem to have been organised much more top-down and with commitment at top levels in various domains, especially in the government. This may result in huge budgets for all, or selected energy transition paths; reference has been made to budgets between one and two billion euros in the Netherlands alone.

It is also possible to look upon Transition Management as a meta-approach for exploring, facilitating, managing and realising transitions demanded by society. Then backcasting and backcasting experiments may be applied in the early stages of a desired transition, as backcasting focuses on developing desirable futures and realising follow-up and spin-off. So far, backcasting has not included a philosophy on the governance or management of system innovations towards sustainability.

Strategic Niche Management

Strategic niche management (SNM) focuses on learning on functionalities of new technological artefacts, user preferences and how both influence each other. SNM takes place in niches where new radical technologies can be protected from regular market conditions and has been developed to enhance societal and user aspects in technology development. It is often connected to testing environmentally friendly technologies for which there is not a major market yet. Examples include electric cars and other alternative mobility solutions (Hoogma *et al.* 2002), demonstration projects for photovoltaic energy conversion in housing (Van Mierlo 2002) and niches for biomass based energy production (Raven 2005).

SNM and backcasting have similarities and differences. Both emphasise experimentation and learning in protected spaces. In SNM this is called a niche experiment, whereas in backcasting this is called a backcasting experiment. A major difference concerns the focus of both types of experiments. SNM focuses on experiments in real life to test new artefacts and technologies and learn about the feedback by users in order to make the new technology or artefact more socially benign and looks forward several years up to a decade. Backcasting experiments focus on developing and exploring desirable future visions as a response to complex problems that cannot be solved in the short term or through incremental change and deals with time horizons of several decades or more. In this sense, niche experiments can be part of follow-up and spin-off of backcasting experiments. However, it is also possible to broaden the niche concept in such a way that backcasting experiments can also be seen as a niche. Then, the different time horizons remain.

Foresighting at the Netherlands Scientific Council for Government Policy

In 2.1 reference has been made to the Netherlands Scientific Council for Government Policy (WRR) as an organisation that has been important to foresighting in the Netherlands, with a strong influence in the policy domain. In the 1980s, the WRR developed a foresighting approach that used political visions like socialism, conservatism and liberalism (e.g. WRR 1980, WRR 1983, WRR 1992) or different actor action perspectives (e.g. WRR 1994) to define different sets of policy goals. Using modelling and multiple-goal optimisation, different sets of policy goals were defined and translated into different policy scenarios that were agenda-setting for political and public discussions, as well as for decision-making and policy development.

Despite the use of normative political visions and normative perspectives, the WRR approach cannot be regarded as a vision-based approach focusing on normative futures, because the normative vision or value-driven action perspective is not the outcome, but an input to the process of scenario construction. Nevertheless, the WRR approach to foresighting and scenario-making is an interesting way of generating diversity in policy scenarios. It may also shed light on what type of policy instruments match particular policy scenarios, as well as the underlying political visions. In fact it may allow for the identification of more robust scenarios or policy instruments that are backed by more than one political vision.

It is interesting to look briefly into the WRR approach from the viewpoint of the backcasting approach. First of all, backcasting aims at generating and explore desirable futures in a niche setting, as well as to stimulate follow-up and spin-off supported by emerging networks. WRR scenarios describe possible futures and are used as input to political decision-making and policy-making for the long-term, but leave picking the preferred scenario to the politicians. Since a major aim is raising awareness and fuelling the debate the differences in scenarios by the WRR may be emphasised. Secondly, the impact of the WRR foresighting studies on political and public discussion, as well as on decision-making and policy development has often been very significant. From the view of participatory backcasting it would be interesting to see what is necessary to achieve such influence and if this can be used to enhance follow-up and spin-off. Thirdly, participatory backcasting using normative visions generally tries to offer opportunities to a range of stakeholders and has, at least in the Netherlands, tried to avoid to be connected to particular political visions.

But could the use of political visions be helpful in vision-based backcasting? In my opinion this could indeed be the case by using the political visions in the final steps of the approach. Whereas the political visions and normative perspectives functioned in the WRR approach as input to construct diverse scenarios, the backcasting vision could be confronted with political visions. In this way 'backcasting' visions can be evaluated, which will provide insight into their political robustness, as well as into the type of activities and policy instruments would do best in different political perspectives.

9.4 Recommendations

Based on the results of this research, recommendations can be made to various groups of actors. Here recommendations are made (1) to organisers of backcasting experiments; (2) for methodology development and learning on the backcasting approach; (3) to the government regarding follow-up and spin-off of, and; (4) for research.

Recommendations to organisers of backcasting experiments

This research has developed and tested a methodological framework for participatory backcasting that distinguishes five steps, three types of demands, different types of goals and four groups of methods. In addition, participatory backcasting has been characterised as both inter-disciplinary and trans-disciplinary. It is recommended to organisers of backcasting experiments, as well as (groups of) organisations that commission backcasting experiments to use this methodological framework. However, it is important to realise the methodological diversity that is possible. Organisers of backcasting experiments can thus use the framework to design a backcasting experiment for a particular complex problem, although it must be realised that the elements of the framework are interdependent. For instance, if realising follow-up is considered an important goal in a particular backcasting experiment, this must be reflected in the tools and methods selected, as well as in the available capacity and the competencies and skills mobilised for this goal. With respect to follow-up and spin-off, it is also recommended to organisers of backcasting experiments to apply the guidelines that are proposed in Table 9.2.

Backcasting experiments can be seen as 'protected' niches in which for the time being existing rules and institutions can be neglected. This makes it possible to develop and explore alternative solutions for the future, learn about these solutions and their underlying complex problems and try new rules and institutions. This requires not only unorthodox thinking by stakeholders involved in the backcasting experiment, it also combines design, analysis, participation, and coordination and management in a sophisticated way with high quality standards. It is strongly recommended to organisers of backcasting experiments to take these aspects into account and translate these into appropriate demands to the backcasting experiment. This recommendation not only implies that organisers should ignore core rules and institutions of existing practices and existing socio-technical systems, it also argues in favour of enacting new rules within the backcasting experiment.

Another recommendation to organisers and commissioners of backcasting experiments is that they apply backcasting to complex problems that are not framed as sustainability problems, for instance in health care or education.

Recommendations for methodology development and learning on backcasting

Recommendations can be made to both organisers of backcasting experiments and researchers with regard to methodology development, as well as to enhance learning on backcasting.

A first recommendation concerns further development of the framework, as well as of methods and tools that can be applied within the framework. It is recommended to develop new methods for the backcasting step; it is also recommended to develop design methods, as well as adjust existing design methods in such a way that they can be added to the proposed toolkit. Another recommendation is to analyse other backcasting experiments on methods that are applied and their methodological aspects. This can also contribute to further development of the framework and the toolkit.

One way of gathering researchers and professionals that have relevant expertise with regard to applying the backcasting approach and organising backcasting experiments is through a so-called Community of Practice. Such an initiative could be linked to existing activities, such as existing programmes on 'Learning for Sustainability', or to one of the initiatives that have emerged on the theme system innovations and transitions, such as the Knowledge Network for System Innovations (KSI) or the Competence Centre for Transitions. It is also recommended to include in such a Community of Practice researchers and professionals that apply related vision based approaches, such as Transition Management.

Recommendations to the government regarding spin-off and follow-up

In Chapter 8 it has been argued that all four societal groups that have been distinguished are important to bringing about system innovations towards sustainability. Stakeholders from each group may provide a necessary contribution that cannot be provide by stakeholders from the other groups. Therefore, stakeholders from all four groups need to be involved in backcasting experiments. In addition to this generic recommendation, I focus on recommendations to the government, as it appeared from the cases that government funding and resources are crucial to bringing about follow-up and spin-off, but that the current focus is strongly on knowledge development. Therefore, it is recommended not to limit government support to funding knowledge development. Instead, it is recommended to facilitate system innovations towards sustainability, as well as follow-up and spin-off of backcasting experiments by additional regulatory or market development instruments in order to stimulate next steps in a system innovation towards sustainability. As no monitoring of follow-up of backcasting experiments, or long-term effect evaluation studies have been found for the cases analysed, it is also recommended to start such monitoring or evaluation studies, as well as to connect such activities to policy-making.

Recommendations for research

Three major research recommendations can be made: (i) extending the number of evaluations of back-casting experiments and their impact; (ii) further theorising and conceptualising on mechanisms and other theoretical aspects connected to the dynamics in backcasting experiments, their follow-up and how these relate to system innovation theories; (iii) further methodology development. The latter has already been elaborated above and also involves organisers of backcasting experiment and professionals applying back-casting.

The first research recommendation allows substantiating the conclusions and findings of this research that comprised a limited number of cases. This should include longitudinal research into the impact of backcasting experiments and dynamics in its follow-up and spin-off over a longer period of time, which makes it possible to obtain a better understanding of the effects after more than ten years and how follow-up and spin-off relates to system innovation dynamics and mechanisms. Research into the application of other vision-based approaches, as well as into vision dynamics in cases where visions emerged unintended without applying special participatory vision-based approaches may also increase the understanding phe-nomena and dynamics in backcasting experiments and their follow-up. Backcasting experiments have also been applied in other countries; it is interesting to compare backcasting experiments and their follow-up and spin-off internationally.

The second recommendation aims at improving the understanding in the underlying dynamics and mechanisms in backcasting experiments and focuses on further theorising and conceptualising. It also allows further elaboration and improvement of the conceptual framework developed in this research. This should include further theorising and conceptualising with respect to higher order learning in backcasting experiments and the interaction of visions constructed in backcasting experiments with either competing alternative visions or the dominant vision. It should also include further theorising and conceptualising on what I have called the linking pin. This refers to the process by which follow-up and spin-off grow out of the backcasting experiment, and includes the question if and how it is possible to connecting multi-layered frameworks of higher order learning with dynamic models of organisational behaviour. Finally, further theo-rising on how backcasting experiments and their follow-up and spin-off relate to theories and concepts of system innovation is recommended.

EPILOGUE

An evocation of an ideal backcasting experiment

This epilogue presents an evocation of an ideal participatory backcasting experiment that is successful in terms of spin-off and broader effects, using the results of this thesis.

A backcasting experiment is initiated to explore, discuss, and elaborate possible and desirable future solutions for a persistent, complex and ambiguous (sustainability) problem. A first step is to select the scope and level of the system to focus on in the backcasting experiment; it is possible to conduct backcasting experiments at the level of countries, international regions or continents, or at a global level, and at the level of organisations. In this evocation the focus is at the level of socio-technical systems providing functions at a societal level.

Starting point is that further optimisation of the socio-technical system through incremental improvement is unlikely to provide a solution that meets long-term sustainability requirements. Resources are made available for the backcasting experiment and further preparations take place using the proposed methodological framework. The organisers of the backcasting experiment involve relevant stakeholders from all four major societal groups, including corporations, government, public and public interest groups and research or knowledge bodies. This is based on an extensive round of stakeholder identification. The backcasting experiment starts with exploring, (re)structuring and redefining the problem, using the expertise and opinions of the stakeholders involved. Furthermore, at an early stage, normative demands, boundaries and terms of reference are set for possible solutions, while also process and knowledge demands are set. The different types of demands are reflected in the goals of the backcasting experiment, which are defined interactively; consensus or congruence are sought for among the participating stakeholders. Type and degree of stakeholder involvement evolve in the course of the backcasting experiment. This is due to personal factors, changing agendas, changes in the organisation, shifts in scope and direction of the backcasting experiment, while also a particular stage can be especially interesting or less interesting to a particular stakeholder.

The stakeholders involved in the backcasting experiment interactively develop one or several future visions solving the sustainability problem and meeting requirements and demands set earlier in the process. The vision(s) are further elaborated and analysed in terms of impact and gains on relevant dimensions covering the 3Ps (Planet, People, Profit). The necessary changes and requirements for realising such a desirable future vision are analysed by reasoning backwards from the developed vision(s). This analysis includes what is necessary to achieve the desirable future vision and what possible trajectories or pathways can be envisioned when looking backwards from the desirable future vision, as well as the identification of actors that are necessary, as well as their contributions.

In addition to the problem orientation, vision development and backcasting analysis, the backcasting experiment leads to process results, as well as an agenda and concrete proposals for follow-up and implementation. Process results include stakeholder involvement, their opinions, their endorsement of outcomes or follow-up agenda, and learning by stakeholders. With regard to desired follow-up or spin-off activities it includes (follow-up) agendas and agenda building processes. The proposed follow-up activities or follow-up agenda relate to one or several domains of research, business, government, and of the wider public and public interest groups.

In the backcasting experiment, stakeholders learn from each other and from the findings and design activities in the backcasting experiment. This leads to higher order learning processes and raising awareness among stakeholders with respect to the problem, the desirable future and possible sustainable alternative solutions. It also results in congruence or shared support on the future vision and other outcomes of the backcasting experiment. Individuals in the backcasting experiment also learn about opportunities embodied in the emerging vision, how they relate to their own organisations' missions and activities, and how other participating stakeholders perceive them. These opportunities, which provide attractive behavioural alternatives to the organisation, are the starting point for mobilising colleagues and senior levels in the own organisation of these opportunities and turning the opportunity into activities. The vision and proposed activities are also used to mobilise other stakeholders that are or may become interested in further development and elaboration of the activity, which includes raising necessary resources. The interested parties partly originate from the backcasting experiment, but are also recruited from existing networks. Actors seeing opportunities in the backcasting experiment and the emerging vision shift priorities subsequently towards activities in line with these and make resources available through funds, capacity or other resources.

During this process an individual, organisation or stakeholder alliance stands up as a vision champion, having or having developed strong commitment and support to the backcasting experiment and its results. The future vision, or elements from that vision, are adjusted to specific opportunities or evolve in such a way that they are useful for making alliances and mobilising necessary resources. Several activities supported by different sets of stakeholders in different domains are established, which may function independently of one another without central coordination.

REFERENCES

Aarts HFM, de Kuijer OCH (1997a, eds.) Duurzaam landgebruik: van wensen en mogelijkheden naar voorbeeldsystemen, STD report VD4, STD office, Delft NL (in Dutch).

Aarts HFM, de Kuijer OCH (1997b, eds.) Duurzaam landgebruik: van voorbeeldsystemen naar systeemonderzoek, STD report VD5, STD office, Delft NL (in Dutch).

Aarts W (1997) Voorbeelden van duurzame technologische ontwikkeling beschreven: beschrijvingen van zoekprocessen uit het interdepartementaal onderzoekprogramma DTO 1993-1997 (Describing examples of sustainable technology development), STD report DTO-2, STD office, Delft NL (in Dutch).

Aarts W (2000) Een handreiking voor duurzame technologisch ontwikkeling (An assistance for sustainable technology development), DTO-KOV 3, STD office, Delft, NL (in Dutch).

Abernathy WJ, Clark KB (1985) Innovation: mapping the winds of creative destruction, Research Policy 14: 3-22.

Achterhuis H (1998) De erfenis van de utopie, Ambo, Amsterdam NL.

ADL (1993) Definition study Novel Protein Foods: final report, Arthur D Little, November 18, 1993.

Aiking H (2002) Profetas timeline, internal document Profetas programme, 2 pages.

Aiking H, de Boer J, Vereijken J (2006) Sustainable protein production and consumption: pigs or peas? Springer, Dordrecht NL.

Åkerman J (2005) Sustainable air transport: on track in 2050, Transportation Research D 10: 111-126.

Åkerman J, Höjer M (2006) How much transport can the climate stand? Sweden on a sustainable path in 2050, Energy Policy 34(14): 1944-1957.

Akkkerman J, van Cooten A, Noordwijn L, Sleurink D (2003) Bouwstenen voor creatief ruimtegebruik: vier jaar praktijkervaring met meervoudig duurzaam landgebruik in Winterswijk (Building bricks for creative use of space), book and cd-rom, steering goup MSL Winterswijk programme (in Dutch).

Andersen M (2006) System innovation versus innovation systems: challenges for the SCP agenda, in: M Andersen, A Tukker, Proceedings of the Workshop on 'Perspectives on radical changes to sustainable consumption', 20-21 April, Copenhagen, Denmark: 413-425.

Anderson KL (2001) Reconciling the electricity industry with sustainable development: backcasting a strategic alternative, Futures 33: 607-623.

Andringa J, Fonk G, Jansen L, Vlasman A (2001) Reflectie DTO-KOV: eindverslag, STD office, Delft NL.

Anonymous (2001a) Programmaplan meervoudig ruimtegebruik IJsselzone, gemeente Zwolle et al.

Anonymous (2001b) Programma IJsselzone: resultaten ontwerpweek september en oktober 2001, Habiforum, Gouda NL.

Anonymous (2002a) Buurtschap: visieontwikkeling, ideeën voor een gezamenlijke inrichting en gebruik, concept 6-4-2002, Habiforum, Gouda NL.

Anonymous (2002b) Plan van aanpak vervolg programma IJsselzone, internal document.

Anonymous (2005) Onderzoek vlees-imago 2005, commissioned by PVE, Zoetermeer NL.

Argyris C, Schön D (1994) Organizational learning: a theory of action perspective, Addison-Wesley, Reading MA.

Arnstein SR (1996) A ladder of citizen participation, Journal of the American Institute of Planners 35(4): 216-224.

Ashford N (1994) The industrial transformation paradigm: an innovation-based strategy for the environment, in: Finkel AM, Golding D (eds.) Worst things first? The debate over risk-based national environmental priorities, Resources for the Future, Washington DC.

Ashford N, Hafkamp W, Prakke F, Vergragt P (2001, with contributions from A Bakker, R Kemp, HW Kua, J Quist, B ter Weel, G Zwetsloot) Pathways to sustainable industrial transformations: co-optimizing competitiveness, employment and environment, Final Report.

Aurelia (2002) Bundel Congres Vleesalternatieven: de ins en outs van een veelzijdige productgroep, 23 September 2003, Den Bosch NL.

Azough (2005) Vlees noch vis, Volkskrant, 26-2-2005.

Baggerman T, Hamstra A (1995) Motieven en perspectieven voor het eten van Novel Protein Foods in plaats van vlees, STD report VN9, STD office, Delft NL (in Dutch).

Banister D, Stead D, Steen P, Åkerman J, Dreborg K, Nijkamp P, Schleicher-Tappeser R (2000a) European transport policy and sustainable mobility, Spon Press, London UK.

Banister D, KH Dreborg, L Hedberg, S Hunhammar, P Steen, Åkerman J (2000b) Transport policy scenarios for the EU in 2020: images of the future, Innovation 13: 27-45.

Beck U (1997) De wereld als risicomaatschappij: essays over de ecologische crisis en de politiek van de vooruitgang, De Balie, Amsterdam NL (in Dutch).

Beck U (2006) Reflexive governance: politics in the global risk society, in: JP Voß, D Bauknecht, R Kemp (eds.) Reflexive governance for sustainable development. Edward Elgar, Cheltenham UK: 31-56.

Beeren J, Aarts H, de Graaf H, Jansen D, de Kuijer O, Kwak, R, van der Pas A, Tiggeloven J (1997) Projectvoorstel praktijkontwikkelingsproject: duurzaam landgebruik Winterswijk, STD report, STD office, Delft NL (in Dutch).

Berkhout F (2006) Normative expectations in systems innovation, Technology Analysis & Strategic Management 18(3-4): 299-311.

Berkhout F, Hertin J, Jordan A (2002) Socio-economic futures in climate change impact assessment: using scenarios as 'learning machines', Global Environmental Change 12(2): 83-95.

Berkhout F, Smith A, Stirling A (2004) Socio-technological regimes and transition contexts, in: B Elzen, FW Geels, K Green (eds.) System innovation and the transition to sustainability, Edward Elgar, Cheltenham UK: 48-75.

Bode M (2000) Consumers' acceptance analysis of scenarios, Final Report, SusHouse Project, Lehrstuhl Markt und Konsum, Hannover University, Germany.

Boekos Food Group et al. (2000) Optimale duurzame ketenmatching: factor 10 in 2010 een duurzaamheidsprong in de agrofood productie en consumptieketen, proposal for the NIDO environmental leapfrog contest 2000 (in Dutch).

Boonstra F (2004) Laveren tussen regio's en regels: verankering van beleidsarrangementen rond plattelandsontwikkeling in Noordwest Friesland, de Graafschap en Zuidwest Salland, Van Gorcum, Assen NL.

Börjeson L, Höjer M, Dreborg KH, Ekvall T, Finnveden G (2006) Scenario types and techniques: towards a user's guide, Futures 38(7): 723-739.

Bos E, Korevaar H (2006) Multifunctioneel landgebruik biedt regio kansen, Landwerk 7(5): 10-14.

Bras-Klapwijk R (2000) Environmental assessment of scenarios, Final Report, SusHouse Project, Delft University of Technology NL.

Bras-Klapwijk RM, Knot JMC (2001) Strategic environmental assessment for sustainable households in 2050: illustrated for clothing, Sustainable Development 9(2):109-118.

Brezet H, Vergragt P, van der Horst T (2001, eds.) Vision on sustainable product innovation, Kathalys, Delft NL.

Brown N, Rappert B, Webster A (2000, eds.) Contested futures: a sociology of prospective technoscience, Aldershot, Ashgate UK.

Brown N, Michael M (2003) The sociology of expectations: retrospecting prospects and prospecting retrospects, Technology Analysis & Strategic Management 15(1): 3-18

Brown HS, Vergragt P, Green K, Berchicicci L (2003) Learning for sustainability transition through bounded social-technical experiments in personal mobility, Technology Analysis & Strategic Management 15: 291-315.

Brown HS, Vergragt PJ (2007) Bounded socio-technical experiments as agents of systemic change: the case of a zero-energy residential building, Technological Forecasting and Social Change (in press).

Buis J, van Dorst MJ, van Keeken E (2002) Leren van Lerner: duurzame ontwikkeling in Curitiba en Parana, Ecological City programme, Delft University of Technology NL.

Callon M (1986) The sociology of an actor-network: the case of the electric vehicle, in: M Callon, J Law, A Rip (eds.) Mapping the dynamics of science and technology: sociology of science in the real world, Macmillan Press Ltd, London UK: 19-34.

Callon M (1987) Society in the making: the study of technology as a tool for sociological analysis, in: W Bijker, T Hughes, T Pinch (eds.) The social construction of technological systems: new directions in the sociology and history of technology, The MIT Press, Cambridge MA: 83-103.

Callon M (1995) Technological conception and adoption network: lessons for the CTA practitioner, in: A Rip, T Misa, J Schot (eds.) Managing technology in society, Pinter, London UK: 307-330.

Callon M, Laredo P, Rabeharisoa V, Gonard T, Leray T (1992) The management and evaluation of technological programs and the dynamics of techno-economic networks: the case the AFME, Research Policy 21: 215-236.

Carlsson-Kanyama A, Dreborg KH, Engström R, Henriksson G (2003a) Possibilities for long-term changes of city life: experiences of backcasting with stakeholders, Fms-report 178, Environmental Strategies Research Group, Stockholm, Sweden, ToolSust project.

Carlsson-Kanyama A, Dreborg KH et al. (2003b) Images of everyday life in the future sustainable city: experiences of back-casting with stakeholders in five European cities, Fms-report 182, Environmental Strategies Research Group, Stockholm, Sweden, ToolSust project.

Carson R (1963) Silent spring, Hamish Hamilton Ltd, London, UK.

Chehab N, Enzing CM (1998) Sensor en Microsysteemtechnologie: Sterkte/Zwakte analyse (Sensor and microsystems technology: S/W analysis), NRLO-report 98/32, National Council for Agricultural Research (NRLO), The Hague NL (in Dutch).

Christensen CM (1997) The innovators' dilemma: when new technologies cause great firms to fail, Harvard Business School Press, Boston MA.

CLTM (1990) Het milieu: denkbeelden voor de 21e eeuw, Kerckebosch, Zeist.

Coenen L (2000) DTO illustratieprocessen, wordt vervolgd....? (STD illustration processes, to be continued....?), Report DTO-KOV-11 (in Dutch), STD office, Delft, NL.

De Bruijn H, Teisman GR, Edelenbos J, Veeneman W (2004, eds.) Meervoudig ruimtegebruik en het management van meerstemmige processen: vijftien voorbeelden in perspectief van grillig procesverloop, ruime ambities en meerstemmige sturing van meervoudige ruimtelijke ontwikkelingsprocessen, Uitgeverij LEMMA BV, Utrecht NL (in Dutch).

De Bruijn H, van der Voort H, Dicke W, de Jong M, Veeneman W (2004) Creating system innovations: how large scale transitions emerge, Balkema Publishers, Leiden NL.

De Bruijn JA, Ten Heuvelhof EF (1995) Netwerkmanagement: strategieën, instrumenten en normen, Lemma Uitgeverij, Utrecht NL.

De Bruijn JA, Ten Heuvelhof EF (2000) Process management, in: O van Heffen, W Kickert, J Thomassen (eds.) Governance in modern society: effects, changes and formation of government institutions, Kluwer Academic Publishers, Dordrecht NL: 313-328.

De Bruijn JA, Ten Heuvelhof EF, In 't Veld RJ (2002) Procesmanagement: over procesontwerp en besluitvorming (in Dutch), 2nd revised edition, Academic Service, Schoonhoven NL.

De Geus M (1996) Ecologische utopieen: ecotopia's en het milieudebat, Jan van Arkel, Utrecht NL.

De Geus M (2002) Ecotopia, sustainability and vision, Organization and Environment 15: 187-201.

De Graaf HJ, Musters CJM (1997, eds.) Ontwikkelingsperspectief duurzaam landgebruik: perspectieven voor het landelijk gebied van Winterswijk, STD report VD2, STD office, Delft NL.

De Graaf HJ, Musters CJM, ter Keurs WJ (1999) Regional opportunities for sustainable development: theory, methods and applications, Kluwer Academic Publishers, Dordrecht NL.

De Graaf HJ, Musters CJM, de Kuijer OCH, ter Keurs WJ (2003) Meervoudig duurzaam landgebruik in het landelijk gebied: hoe bereik je dat? in: FJ Dietz, Duurzaam milieugebruik: een inspiratiebron, Uitgeverij Jan van Arkel, Utrecht NL: 65-86.

De Graaf HJ, de Kuijer OCH (2004a) Gebiedsplan Winterswijk, in: H de Bruijn, GR Teisman, J Edelenbos, W Veeneman (eds.) Meervoudig ruimtegebruik en het management van meerstemmige processen, Lemma, Utrecht NL: 367-387.

De Graaf HJ, de Kuijer OCH (2004b) IJsselzone, in: H de Bruijn, GR Teisman, J Edelenbos, W Veeneman (eds.) Meervoudig ruimtegebruik en het management van meerstemmige processen, Lemma, Utrecht NL: 389-408.

De Graaf HJ, Tolkamp W (2004) Meervoudig landgebruik en dienstvelening in het landelijk gebied: reconstructie Winterswij-Oost als voorbeeld, KDO Advies, Leiden NL (in Dutch).

De Groot SJ (2002) Review of the past and present status of anadromous fish species in the Netherlands: is restocking the Rhine feasible? Hydrobiologia 478(1-3): 205-218.

De Haan H, Hermans H, de Kuijer O, Larsen I, Linsen H Quist J (1995) Kansrijke NPFs als ingrediënten voor toekomstige eiwithoudende voedingsmiddelen, STD report VN10, STD office Delft NL.

De Haan A, Quist J, Linsen BG (1996) NPF's hebben toekomst, Voeding No. 3 (March): 7-10.

De Koning R, de Kuijer O, Righarts A, Rijsdorp A (2003) Verkenning Noordwaard: denken over riververruiming, Projectbureau Benedenrivieren, RWS, Rotterdam NL.

De Kuijer O (1995a) Op weg naar een Duurzame Voeding in 2040, brochure, STD programme, Delft NL.

De Kuijer O (1995b) Illustratieproces Duurzaam Landgebruik: projectvoorstel fase A, herziene versie, STD office, Delft NL (in Dutch).

De Kuijer OCH, Aarts HFM, Beeren JTJ, de Graaf HJ, Jansen D, Kwak R, van der Pas ABWM, Tiggeloven JWB (1997a) Resultaten project Duurzaam Landgebruik (The results of the MSL project), STD report VD-7, STD office, Delft NL (in Dutch).

De Kuijer O, Beeren J, Klep L (1997b) Meervoudig Duurzaam Landgebruik: een werkend perspectief voor 2020, STD office, Delft NL (in Dutch).

De Kuijer OCH, Wielenga DK (1999) Een vergelijking van de milieubelasting van vlees en vleesalternatieven en de aantrekkelijkheid van de alternatieven voor consumenten, Publicatiereeks produktenbeleid 1999/35, Ministry of Housing, Spatial Planning & Environment (VROM), The Hague, Netherlands.

De Kuijer OCH, de Graaf HJ (2001) Gebiedsontwikkeling, hoe pakken we het aan: proceshandreiking voor meervoudig duurzaam ruimtegebruik, KDO Advies, Leiden NL.

De Kuijer O, de Graaf H, Londo M (2002) Meervoudig ruimtegebruik in het landelijk gebied, een netwerk: nieuwe concepten, gezamenlijk leren en uitwisselen van ervaringen in een Community of Practie, projectvoorstel 2e fase, KDO Advies, Leiden (in Dutch).

De Laat B (1996) Scripts for the future: technology foresight, strategic evaluation and socio-technical networks, the confrontation of script-based scenarios, PhD thesis, University of Amsterdam NL.

De Laat B, Larédo P (1998) Foresight for research and technology policies: from innovation studies to scenario confrontation, in: R Coombs, K Green, A Richards, V Walsh (eds.) Technological change and organization, Edward Elgar, Cheltenham UK: 150-179.

De Vries J, te Riele H (2005) Spelen met hyena's, consumption sustained, commissioned by the Foundation for Nature & Environment (SNM), Utrecht NL.

De Wilde R (2000) De Voorspellers: een kritiek op de toekomstindustrie, De Balie, Amsterdam NL (in Dutch).

De Wolf P, de Buck A, Grin J (2006) Werken aan de toekomst: beschrijving en evaluatie van het toekomstbeeldenproject in de plantaardige sectoren, report 351, WUR-PPO, Wageningen NL.

Dekker C (1999) Consumer acceptance of future scenarios in food consumption: an explorative research, MSc-Thesis, Dept. of Household and Consumer Studies, Wageningen Agricultural University, Wageningen NL.

Dekker C, Quist J (2000) Consumer acceptance of shopping, cooking and eating scenarios in the Netherlands, Background Report, SusHouse Project, Delft University of Technology NL.

Diepenmaat HB, te Riele H (2001) Boven het klaver bloeien de margrieten: een maatschappelijk netwerk voor innovaties richting duurzaamheid, in opdracht van het ministerie van VROM, Actors Procesmanagement, Zeist NL/Storrm CS, The Hague NL.

Dierkes M, Hoffmann U, Marz L (1992) Leitbild und Technik: zur Entstehung und Steuerung technischer Innovationen, Wissenschaftszentrum Berlin fur Socialforschung, Rainer Bohn Verlag, Berlin.

Dierkes M, Hoffmann U, Marz L (1996) Visions of technology: social and institutional factors shaping the development of new technologies. Campus Verlag/St.Martin's Press, Frankfurt/New York.

DLG (2001) Concept gebiedsplan WCL Winterswijk/Winterswijk-Oost, WCL Winterswijk & Dienst Landelijk Gebied (in Dutch).

Dortmans PJ (2005) Forecasting, backcasting, migration landscapes and strategic planning maps, Futures 37: 273-285.

Dosi G (1982) Technological paradigms and technological trajectories: a suggested interpretation of the determinants and directions of technical change, Research Policy 11: 147-162.

Dreborg KH (1996) Essence of backcasting, Futures 28 (9): 813-828.

Dreborg KH, Hunhammar S, Kemp-Benedict E, Raskin P (1999) Scenarios for the Baltic Sea region: a vision of sustainability, International Journal of Sustainable Development and World Ecology 6: 34-44.

Driessen P, Glasbergen P, Huigen P, Hijmans van den Bergh F (1995) Vernieuwing van het landelijk gebied: een verkenning van de strategieën voor een gebiedsgerichte aanpak, VUGA, Den Haag NL.

Dunn WN (1994) Public policy analysis: an introduction, 2nd edition, Prentice Hall, Englewood Cliffs NJ.

EBS (2005) Single Cell Protein (SCP) als alternatief voor soja: een haalbaarheidsstudie, Easthouse Business Solutions, commissioned and published by INGRA, Utrecht NL.

Elzen B, Geels FW, Hofman PS (2002) Sociotechnical Scenarios (STSc): development and evaluation of a new methodlogy to explore transitions towards a sustainable energy supply, report for NWO/NOVEM, University of Twente, Centre for Studies of STS, Enschede NL.

Elzen B, Geels F, Hofman PS, Green K (2004) Socio-technical scenarios as a tool for transition policy: an example from the traffic and transport domain, in: B Elzen, FW Geels, K Green (eds.) System innovation and the transition to sustainability, Edward Elgar, Cheltenham UK: 251-281.

Elzen B, Geels FW, Green K (2004) System innovation and the transition to sustainability, Edward Elgar, Cheltenham UK.

Enserink B (2000) Building scenarios for the university, Int Transactions Operational Research 7: 569-583.

Enzing C, Gijsbers G, Vullings W (2005) Een blik vooruit verrijkt de toekomst, Innovatienetwerk Groene Ruimte en Agrocluster, Utrecht NL.

Falkena HJ, Moll HC, Kok R, Eenkhoorn R (2003) Duurzaam perspectief op lange termijn verandering van het dagelijks leven in Groningen: ervaringen met het gebruik van de backcasting methodologie in twee workshops met stakeholders in Groningen, ToolSust project, Centre for Energy & Environment, University of Groningen NL.

Falkenmark M (1998) Dilemma when entering 21st century: rapid change but lack of sense of urgency, Water Policy 1(4): 421-436.

Fischer F (1980) Politics, values and public policy: the problem of methodology, Westview Press, Boulder CO.

Fischer F (1995) Evaluating public policy, Nelson-Hall, Chicago.

Fishbein M, Ajzen I (1975) Belief, attitude, intention and behavior: an introduction to theory and research, Addison-Wesley, Reading MA.

Fonk G (1994) Een constructieve rol van de consument in technologieontwikkeling (A constructive contribution of consumers in technology development), PhD thesis, University of Twente NL.

Fonk G, Hamstra A (1996) Toekomstbeelden voor consumenten van Novel Protein Foods, (Future Images for consumers of Novel Protein Foods), DTO-report VN12, STD office, Delft NL.

Freeman RE (1984) Strategic management: a stakeholder approach, Pitman, Boston MA.

Freeman C (1987) Technology policy and economic performance: lessons from Japan, Pinter, London UK.

Freeman C, Perez C (1988) Structural crises of adjustment, business cycles and investment behaviour, in: G Dosi, C Freeman, R Nelson, G Silverberg, L Soete (eds.) Technical change and economic theory, Pinter Publishers, London/New York: 38-66.

Freeman C, Louca F (2001) As time goes by: from the industrial revolutions to the information revolution, Oxford University Press, Oxford UK.

Freeman RE, McVea J (2005) A stakeholder approach to strategic management, in: MA Hitt, RE Freeman, JS Harrison (eds.) The Blackwell handbook of strategic management, Blackwell Publishing, Malden MA/Oxford UK: 189-207.

Fresco LO (1998) Schauwdenkers en lichtzoekers, Huizinga-lezing 1998, Bert Bakker, Amsterdam NL.

Garud R (1997) On the distinction between know-how, know-why, and know-what, Advances in Strategic Management 14: 81-101.

Geels FW (2002) Technological transitions as evolutionary reconfiguration processes: a multi-level perspective and a case-study, Research Policy 31: 1257-1274.

Geels FW (2005) Technological transitions and system innovations: a co-evolutionary and socio-technical analysis, Edward Elgar, Cheltenham UK/Northampton MA.

Geels FW, Smit WA (2000) Failed technology futures: pitfalls and lessons from a historical survey, Futures 32: 867-885.

Geurs K, van Wee B (2000) Backcasting as a tool to develop a sustainable transport scenario assuming emission reductions of 80-90%, Innovation 13 (1): 47-62.

Geurs K, van Wee B (2004) Backcasting as a tool for sustainable transport policy making: the environmentally sustainable transport study in the Netherlands, EJTIR 4: 47-69.

Giddens A (1984) The constitution of society: outline of the theory of structuration, Polity Press, Cambridge UK.

Godet M (2000) The art of scenarios and strategic planning: tools and pitfalls, Technological Forecasting and Social Change 65: 3-22.

Goekoop AC (2000) Workshop Novel Protein Foods: een kans voor nieuwe productinnovaties, Kathalys report 938, TNO Industry, Delft NL.

Goldemberg J, Johansson TB, Reddy AKN, Williams RH (1985) An end-use oriented global energy strategy, Annual Review of Energy 10: 613-688.

Green K (1998) Shopping, cooking and eating function format, internal document, SusHouse project.

Green K, Young W (2000) The shopping, cooking and eating function, Final Report, SusHouse Project, Manchester School of Management, UMIST, Manchester UK.

Green K, Vergragt P (2002) Towards sustainable households: a methodology for developing sustainable technological and social innovations, Futures 34: 381-400.

Green K, Harvey M, McMeeking A (2003) Transformations in food consumption and production systems, Journal of Environmental Policy & Planning 5: 145–163.

Green K, Foster C (2005) Give peas a chance: transformations in food consumption and production systems, Technological Forecasting and Social Change 72: 663-679.

Grin J (2000) Vision assessment to support shaping 21st century society? Technology Assessment as a tool for political judgement, in: Grin J, Grunwald A (eds.) Vision assessment, shaping technology in the 21st century: towards a repertoire for Technology Assessment, Springer Verlag, Berlin: 9-30.

Grin J (2004) De politiek van omwenteling met beleid, inaugural speech (in Dutch), University of Amsterdam, Vossiuspers, Amsterdam.

Grin J (2006) Reflexive modernization as a governance issue or: designing and shaping Re-structuration, in: JP Voß, D Bauknecht, R Kemp (eds.) Reflexive governance for sustainable development, Edward Elgar, Cheltenham UK: 57-81.

Grin J, Felix F, Bos B, Spoelstra S (2004) Practices for reflexive design: lessons from a Dutch programme on sustainable agriculture, International Journal of Foresight and Innovation Policy 1: 126-149.

Grin J, Grunwald A (2000) Vision assessment: shaping technology in the 21st century towards a repertoire for Technology Assessment, Springer Verlag, Berlin Germany.

Grin J, van de Graaf H (1996a) Technology assessment as learning, Science, Technology & Human Values 20 (1): 72-99.

Grin J, van de Graaf H (1996b) Implementation as communicative action: an interpretive understanding of interactions between policy actors and target groups, Policy Sciences 29: 291-319.

Grin J, van de Graaf H, Hoppe R (1997) Technology assessment through interaction, Rathenau Institute, The Hague NL.

Håkansson H (1987, ed.) Industrial technological development: a network approach, Routledge London UK.

Håkansson H (1989) Corporate technological behaviour: cooperation and networks, Routledge London UK.

Hall (1993) Policy paradigms, social learning, and the state, Comparative Politics 25(3): 275-296.

Heidemij, ATO-DLO, LUW, TNO-Voeding (1997) Een verkenning van noodzakelijke trendbreuken in de voedsel-voorziening, STD report V0, STD office, Delft NL.

Hekkert M, Suurs R, Negro S, Kuhlman, S, Smits R (2007) Functions of innovation systems: a new approach for analysing technological change, Technological Forecasting and Social Change (in press).

Hennicke P (2004) Scenarios for a robust policy mix: the final report of the German study commission on sustainable energy supply, Energy Policy 32(15): 1673-1678.

Hoevenagel R, van Rijn U, Steg L, de Wit H (1996) Milieurelevant consumentengedrag: ontwikkeling conceptueel model, Sociaal en Cultureel Planbureau, The Hague NL.

Hofman P (2005) Innovation and institutional change: the transition to a sustainable electricity system, PhD thesis, University of Twente, Enschede NL.

Hofman PS, Elzen BE, Geels FW (2004) Sociotechnical scenarios as a new policy tool to explore system innovations: co-evolution of technology and society in the Netherlands' electricity domain, Innovation: Management, Policy & Practice 6(2): 344-360.

Höjer M (1998) Transport telematics in urban systems – a backcasting Delphi study, Transportation Research D 3(6): 445-463.

Höjer M, Mattsson L-G.(2000) Determinism and backcasting in future studies, Futures 32: 613-634.

Holmberg J (1998) Backcasting: a natural step in operationalising sustainable development, Greener Management International 23: 30-51.

Holmberg J, Robèrt KH (2000) Backcasting: a framework for strategic planning, Int Journal of Sustainable Development and World Ecology 7: 291-308.

Hoogma R, R Kemp, J Schot, B Truffer (2002) Experimenting for sustainable transport: the approach of strategic niche management, Spon Press, London UK.

Hoogma R, Schot J (2001) How innovative are users? A critique of learning-by-doing and –using, in: R Coombs, K Green, A Richards, V Walsh, Technology and the market - demand, users and innovation, Edward Elgar, Cheltenham UK: 216-233.

Hughes (1983) Networks of power: electrification in western society 1880-1930, The John Hopkins University Press, Baltimore.

Hughes (1987) The evolution of large technical systems, in: W Bijker, T Hughes, T Pinch (eds.) The social construction of technological systems: new directions in the sociology and history of technology, The MIT Press, Cambridge MA: 51-82.

In 't Veld RJ (2001, red.) Eerherstel voor Cassandra: een methodologische beschouwing over toekomstonderzoek voor omgevingsbeleid (rehabilitation for Cassandra: a methodological reflection on future studies for spatial planning policy), Lemma, Utrecht NL (in Dutch).

Isaksen SG (2000) Facilitative leadership: making a difference with creative problem solving, Kendall/Hunt Publishing Company, Dubuque Iowa: 1-23.

Jansen JLA (1993) Discussienota illustratieproces vleesachtige producten, (discussion paper on the illustrative process meat-like products), STD office Delft, 9 April 1993, 4 pages.

Jansen L (2003) The challenge of sustainable development, Journal of Cleaner Production 11: 231-245.

Jansen L (2005) Transition management: from vision to action, in: P. Lens, P Westermann, M Haberbauer, A Moreno (eds.) Biofuels for fuel cells: renewable energy from biomass fermentation, IWA Publishing Alliance, London: 5-15.

Jansen JLA, Vergragt PJ (1992) Sustainable Development: a challenge to technology: proposal for the inter-departmental research programme Sustainable Technology Development, Ministry of the Environment, Leidschendam NL, 10 June 1992.

Jansen JLA, Vergragt PJ (1993) Naar duurzame ontwikkeling met technologie: uitdaging in programmatisch perspectief, Milieu 1993(5): 179-183.

Jantsch E (1967) Technological forecasting in perspective, OECD, Paris, France.

Jégou F, Joore P (2004) Food delivery solutions: cases of solution oriented partnership, Cranfield University, UK.

Johansson TB, Steen P (1980) Solar versus nuclear: choosing energy futures, Pergamon Press, Oxford UK.

Jongen WMF, Meulenberg MTG (1998, eds.) Innovation of food production systems: product quality and consumer acceptance, Wageningen Press NL.

Kamp L (2002) Learning in wind turbine development: a comparison between the Netherlands and Denmark, PhD thesis, Utrecht University NL.

Kemp R (1994) Technology and the transition to environmental sustainability: the problem of technological regime shifts, Futures 26(10): 1023-1046.

Kemp R, Schot J, Hoogma R (1998) Regime shifts to sustainability through processes of niche formation: the approach of strategic niche management, Technology Analysis & Strategic Management 10 (2): 175–195

Kemp R, Rip A, Schot J (2001) Constructing transition paths through the management of niches, in: Garud R, Karnøe P, Path dependence and creation, LEA Publishers, Mahwah NJ/London UK: 269-299.

Kemp R, Rotmans J (2004) Managing the transition to sustainable mobility, in: B Elzen, FW Geels, K Green (eds.) System innovation and the transition to sustainability, Edward Elgar, Cheltenham UK: 137-167.

Kemp R, Rotmans J (2005) The management of the co-evolution of technical, environmental social systems, in: M Weber, J Hemmelskamp (eds.) Towards environmental innovation systems, Springer, Heidelberg: 33-56.

Kemp R, Parto S, Gibson RB (2005) Governance for sustainable development: moving from theory to practice, Int Journal of Sustainable Development 8: 12-30.

Kern S (2000) Dutch innovation policies for the networked economy: a new approach?, TNO-STB, paper, TNO, Delft NL.

Keune H, Goorden L (2002) Interactieve backcasting: een duurzame energietoekomst voor Belgie (Interactive backcasting: a sustainable energy future for Belgium), Centre for Technology, Energy and Environment (STEM), University of Antwerp, Belgium.

Keune H, Goorden L (2003) A future for nuclear power in Belgium?, paper ENER Forum 5: Technological, market reform and climate policies, Bucharest, Romania, 16-17 October.

Kok K, Rothman DS, Patel M (2006a) Multi-scale narratives from an IA perspective: Part I, European and Mediterranean scenario development, Futures 38: 261-284.

Kok K, Patel M, Rothman DS, Quaranta G (2006b) Multi-scale narratives from an IA perspective: Part II, participatory local scenario development, Futures 38: 285-311.

Korevaar H (2006) Perspectieven van veranderend landgebruik: extensivering kan lonend zijn, brochure, 16 pages, PRI-WUR.

Korevaar, H, van der Werf A, Oomes MJM (1999) Meervoudig Duurzaam Landgebruik: van visie naar realisatie, proceedings programme symposium 25-6-99, AB-DLO, Wageningen (in Dutch).

Korevaar H, van Loenen P (2002) Meervoudig Duurzaam Landgebruik Winterswijk, kennisoverdracht en publiciteit: publicaties, rapporten, nieuwsbrieven en artikelen over MDL Winterswijk in 1998-2002, report 60, PRI Wageningen NL.

Kramer KJ (2000) Food matters: on reducing energy use and greenhouse gas emissions from household food consumption, PhD thesis, University of Groningen NL.

Kramer KJ, Moll HC (1995) Energie voed(t), nadere analyses van het indirecte energieverbruik van voeding, IVEM research report 77, Rijksuniversity Groningen NL.

Kruijsen J (1999) Photovoltaic technology diffusion, PhD thesis, Delft University of Technology NL.

Kuusi O, Meyer M (2002) Technological generalizations and leitbilder: the anticipation of technological opportunities, Technological Forecasting and Social Change 69: 625-639.

Lagendijk A, Wissenhof J (1999) Geef ruimte de kennis geef kennis de ruimte!, deel 1: Verkenning van de kennisinfrastructuur voor meervoudig ruimtegebruik, RMNO-report 136, RMNO, Rijswijk NL.

LeBlansch K, van der Heijden N, Smit W (2003) Burgerparticipatie in het milieubeleid: ervaringen met methoden en instrumenten in het stimuleringsprogramma Burger en Milieubeleid, QA+, The Hague NL, commisioned by the Ministry of the Environment.

LEI/DUT (2000) Workshopverslag implementatie duurzaam voeden in het huishouden (Report Workshop Implementation sustainable nutrition in the household), LEI/DUT, The Hague/Delft.

Lindeijer E (2000) Achtergronddocument domeinverkenning voeden: resultaten milieu-analyses, IVAM-ER/Schuttelaar & Partners, Amsterdam/Den Haag (in opdracht van VROM).

List D (2004) Multiple pasts, converging presents and alternative futures, Futures 36(1): 23-43.

LNV (2000) Nota voedsel en groen: het Nederlandse agro-foodcomplex in perspectief, Ministry of Agriculture, Nature Management and Fisheries, The Hague NL (in Dutch).

Loeber A (1997) Novel Protein Foods illustratieproces: evaluatie gebruikte methoden en behaalde resultaten (Novel Protein Foods project: evaluation of applied methods and obtained results), STD (DTO) report VN23, STD office, Delft (www.dto-kov.nl).

Loeber A (2004) Practical wisdom in the risk society: methods and practice of interpretive analysis on questions of sustainable development, PhD thesis University of Amsterdam NL.

Londo HM, de Kuijer OCH (2002) Novel Protein Foods in de catering, verkenning van de mogelijkheden: interim report for Novem, KDO Advies, Amsterdam, April 2002.

Londo M, de Kuijer O, de Graaf H (2002) Community of Practice landelijk gebied: onderlinge kennismaking en gezamnlijke verkenning van de aanpak in de periode december 2001 – juni 2002, KDO Advies, Leiden NL.

Londo M, de Graaf H, de Kuijer O (2005) Publiek vergoeden, privaat belonen: een publiek-private handelsonderneming als spil voor een veelzijdig kleurrijk landelijk gebied, INGRA, Utrecht NL.

Lönnroth M, Johansson TB, Steen P (1980) Sweden beyond oil: nuclear commitments and solar options, Science 208: 557-563.

Loorbach D, Rotmans J (2006) Managing transitions for sustainable development, in: X Olsthoorn, A Wieckzorek (eds.) Understanding industrial transformation: views from different disciplines, Springer, Dordrecht NL: 187-206.

Lovins A (1976) Energy strategy: the road not taken? Foreign Affairs 55: 63-96.

Lovins AB (1977) Soft Energy Paths: toward a durable peace, Friends of the Earth/Ballinger Publishing Company, Cambridge MA.

Luiten E, van Lente H, Blok K (2006) Slow technologies and government intervention: energy efficiency in industrial process technologies, Technovation 26: 1029-1044.

Lundvall BA (1992) National systems of innovation: towards a theory of innovation and interactive learning, Pinter Publishers, London UK.

MacDonald JP (2005) Strategic sustainable development using the ISO 14001 standard, Journal of Cleaner Production 13(6): 631-643.

MacFarlane A (2001) Design of a sustainable energy strategy for a planned resort development in India: a case study utilizing a backcasting approach, MSc thesis, Lund University, Sweden.

Mak G (1996) Hoe God verdween uit Jorwerd: een Nederlands dorp in de 20e eeuw (How God disappeared from Jorwerd: a Dutch village in the 20th century), Atlas, Amsterdam NL (in Dutch).

Malerba F (2002) Sectoral systems of innovation, Research Policy 31(2): 247-264.

Mambrey P, Pateau M, Tepper A (1994) Controlling visions and metaphors, in: Duncan K, Krueger K (eds.) Proceedings 13th World Computer Congress 94, Volume III, Elsevier: 223-228.

Mambrey P, Tepper A (2000) Technology assessment as metaphor assessment: visions guiding the development of information and communications, in: Grin J, Grunwald A (eds.) Vision assessment, shaping technology in the 21st century: towards a repertoire for Technology Assessment, Springer Verlag, Berlin: 33-52.

Manzini E, Jégou F (1998) Scenarios for sustainable households, in: E Brand, T de Bruijn, J Schot (eds.) Partnerships and leadership: building alliances for a sustainable future, Proceedings Greening of Industry Network Conference, Rome, Italy, November 15-18.

Manzini E, Jégou F (2000) The construction of Design Orienting Scenarios, Final report, SusHouse project, Politecnico di Milano, Dept of Industrial Design, Milan, Italy.

Marchau V, van der Heijden R (2003) Innovative methodologies for exploring the future of automated vehicle guidance, Journal of Forecasting 22: 257-276.

Mayer I (1997) Debating technologies: a methodological contribution to the design and evaluation of participatory policy analysis, PhD thesis Tilburg University, Tilburg NL.

McDowall W, Eames M (2006) Forecasts, scenarios, visions, backcasts and roadmaps to the hydrogen economy: a review of the hydrogen futures literature, Energy Policy 34: 1236-1250.

Meadows DH, Meadows DL, Randers J (1992) De grenzen voorbij: een wereldwijde catastrofe of een duurzame wereld, Uitgeverij Spectrum/Aula, Utrecht NL.

Meadows DL (1972) De grenzen aan de groei: rapport van de Club van Rome, Uitgeverij Spectrum, Utrecht NL.

Meurs P (1998) Curitiba: een integraal stedelijk proces, in: P Meurs, E Agricola (red.) Brazilië: laboratorium van architectuur en stedenbouw, NAi Uitgevers, Rotterdam NL: 194-211.

Milieu Centraal (2001) Factsheets indirecte en directe energie van voeding & verliezen en verspilling van voeding, internal documents, Milieu Centraal, Utrecht NL.

Minister of Agriculture (2006) Reactie op onderzoek Profetas, Letter to the Second Chamber of the Dutch Parliament, dd 16-10-2006, The Hague NL.

Moll H, Groot-Marcus A (2002) Households past, present and opportunities for change, in: M Kok, W Vermeulen, A Faaij, D de Jager (eds.) Global warming and social innovation: the challenge of a climate neutral society, Earthscan, London UK: 83-106.

Moors E (2000) Metal making in motion: technology choices for sustainable metals production, PhD thesis, Delft University of Technology NL.

Mulder K (1992) Choosing the corporate future: technology networks and choice concerning the creation of high performance fiber technology, PhD thesis, University of Groningen NL.

Mulder H (1995) Back to our future: physical constraints on sustainable development paths in an energy-based backcasting approach, PhD thesis, University of Groningen NL.

Mulder H, Biesiot W (1998) Transition to a sustainable society: a backcasting approach to modelling energy and ecology, Edward Elgar, Cheltenham UK.

Mulder KF (2005) Innovation by disaster: the ozone catastrophe as experiment of forced innovation, International Journal of Environment & Sustainable development 4(1): 88-103.

Nattras B, Altomare M (1999) The natural step for business: wealth, ecology and the evolutionary corporation, New Society Publishers, Canada.

Nelson R, Winter S (1977) In search for a useful theory of innovation, Research Policy 6: 36-76.

Nelson R, Winter S (1982) An evolutionary theory of economic change, Harvard University Press, Cambridge MA.

Noorman KJ, Schoot Uiterkamp T (1998, eds.) Green households? Domestic consumers, environment and sustainability, Earthscan, London UK.

NRLO (1999a) Sensor- en microsysteemtechnologie: speerpunten voor actie (Sensor and microsystem technology), NRLO report 99/2, NRLO, The Hague NL (in Dutch).

NRLO (1999b) Beleidsprocessen en ICT in de groene ruimte: speerpunten voor actie (Policy processes and ICT in 'green' space, NRLO report 99/11, NRLO, The Hague NL (in Dutch).

Oerlemans L (1996) De ingebedde onderneming: innoveren in industriële netwerken, PhD thesis (in Dutch) Tilburg University Press, Tilburg NL.

Oliver C (1996) Strategic responses to institutional processes, Organization Studies 13(4): 563-588.

Olsthoorn X, Wieckzorek A (2006) Understanding industrial transformation: views from different disciplines, Springer, Dordrecht NL.

Ostrom E (1999) Institutional rational choice: an analysis of the institutional analysis and development framework, in: Sabatier PA (ed.) Theories of the policy process, Westview Press, Boulder CO/Oxford UK: 35-72.

Oudshoff BC (1996) Energieverbruik voor buitenshuis geconsumeerde maaltijden, IVEM-doctoraalverslag no 42, Rijksuniversiteit Groningen NL.

Partidario PJ (2002) 'What-If', from path dependency to path creation in a coatings chain: a methodology for strategies towards sustainable innovation, PhD thesis, Delft University of Technology NL.

Partidario PJ, Vergragt P (2002) Planning of strategic innovation aimed at environmental sustainability: actor-networks, scenario acceptance and backcasting analysis within a polymeric coating chain, Futures 34: 841-861.

Peek GJ (2007) Locatiesynergie. Een participatieve start van de herontwikkeling van binnenstedelijke stationslocaties, PhD thesis, Eburon, Delft NL.

Poot E (2004) Back from the future: towards sustainable glasshouse horticulture in the Netherlands, Acta Hortic 655: 253-258

Porter M (1990) The competitive advantage of nations, Free Press, New York.

Porter ME (2001) Innovation and competitiveness: findings on the Netherlands, Innovation Lecture 2001, Ministry of Economic Affairs, The Hague NL.

Profetas (1998) Protein Foods, Environment, Technology and Society: excerpt from the 1998 programme proposal, internal document Profetas programme, 17 pages.

Profetas (2004), Abstracts and programme of the symposium 'Transition towards sustainable protein supply chains', Wageningen, 29 October 2004.

Quist J (1998) Inputdocument SusHouse workshop 'duurzame vervulling huishoudfunctie voeden', 28 november 1998, Oegstgeest, Delft University of Technology NL.

Quist J (1999a) Report of the Dutch stakeholder creativity workshop on sustainable shopping, cooking and eating function, November 28 1998, internal document, SusHouse project, Delft University of Technology, Technology Assessment Group.

Quist J (2000a) Towards sustainable shopping, cooking and eating in the Netherlands, Final Report, SusHouse Project, Delft University of Technology, Technology Assessment Group, Delft NL.

Quist J (2000b) Report of the Dutch implementation & strategy workshop on sustainable shopping, cooking and eating (Oegstgeest, 13 January 2000), internal document, SusHouse project, Delft University of Technology, Technology Assessment Group, draft, March 2000.

Quist J (2001a) Eindrapport duurzaam voeden in het huishouden, SusHouse project, Delft University of Technology NL.

Quist J (2001b) Towards sustainable shopping, cooking and eating in the Netherlands: results and follow-up towards implementation, proceedings 9th GIN Conference on 'Sustainability at the millennium: globalization, competitiveness, and the public trust', Bangkok, 21-25 January.

Quist J (2003) Regime shifts and sustainability: the STD programme revisited, Conference on 'Knowledge and economic and social change: new challenges to innovation studies', Manchester April 7-9, 2003, http://les1.man.ac.uk/cric/2003conf/programme.htm.

Quist J (2006) Backcasting in the Netherlands: the STD programme and beyond, in: K Mulder (ed.) Sustainability made in Delft, Eburon, Delft NL: 93-104.

Quist J, de Haan A, Linsen H, de Kuijer O, Hermans H, Larsen I (1996a) Ontwikkelingstraject Novel Protein Foods (Novel Protein Foods development trajectory, DTO-werkdocument VN22, (in Dutch), STD office, Delft.

Quist J, de Kuijer O, De Haan A, Linsen H, Hermans H, Larsen I (1996b) Restructuring meat consumption: novel protein foods in 2035, paper 5th GIN Conference, 24-27 November 1996, Heidelberg, Germany.

Quist J, Linsen B, de Haan A, de Kuijer O (1996c) Minder vlees en toch lekker eten in een duurzame toekomst, in: Duurzame consumptie: voorbeelden en kosten, Reeks Achtergrondstudies P96-11, Raad voor het Milieubeheer, Den Haag, 50-63.

Quist J, Maas T (1999) SusHouse workshop verslag: verduurzaming huishoudfunctie Voeden in Nederland, document SusHouse project, Delft University of Technology NL.

Quist J, Pacchi C, Van der Wel M (2000) Workshop organisation and stakeholder management, Final Report, SusHouse Project, Delft University of Technology NL/Avanzi, Milano Italy.

Quist J, Vergragt P (2000) System innovations towards sustainability using stakeholder workshops and Scenarios, paper at POSTI Conference on 'Policy agendas for sustainable technological innovation', London, UK, December 1-3 2000, http://www.esst.uio.no/posti/workshops/programme.html.

Quist J, Knot M, Young W, Green K, Vergragt P (2001a), Strategies towards sustainable households using stakeholder workshops and scenarios, Int Journal of Sustainable Development 4: 75-89.

Quist J, Pacchi C, Van der Wel M (2001b) Stakeholder workshops and management applied to sustainable households, proceedings 9th GIN Conference, 'Sustainability at the millennium: globalization, competitiveness, and the public trust', Bangkok, Thailand, January 21-25.

Quist J, Silvester S (2001) Duurzame horeca door gedragsverandering, nieuwe technologie en diensinnnovatie, programma-opdracht tbv het Nationaal Initiatief Duurzame Ontwikkeling (NIDO), Delft University of Technology, September 2001.

Quist J, Vergragt P (2001) Multiple sustainable land use: towards a new sustainable socio-economic perspective for rural areas, Case Study Report Pathways Project, Delft University of Technology, NL.

Quist J, Green K, Szita Toth K, Young W (2002a) Stakeholder involvement and alliances for sustainable shopping, cooking and eating, in: T de Bruijn, A Tukker (eds.) Partnerships and leadership: building alliances for a sustainable future, Kluwer Academic Publishers: 273-294.

Quist J, Young W, Bras-Klapwijk R, Tóth K (2002b) Assessing sustainable scenarios on their environmental gain, economic implications and consumer attractiveness: the case of shopping, cooking and eating, proceedings IAIA-2002 Conference, The Hague NL, June 16-22.

Quist J, Silvester S, van der Horst H (2003) Towards sustainable eating out through innovation, behavioural changes and system changes, in: M Charter (ed.) proceedings TSPD8 Conference on 'Creating sustainable products, services and product-service systems', Stockholm, Sweden, 27-28 October 2003.

Quist J, Vergragt P (2004) Backcasting for Industrial transformations and system innovations towards sustainability: relevance for governance? in: K Jacob, M Binder, A Wieczorek (eds.) Governance for industrial transformation. Proceedings of the 2003 Berlin Conference, Environmental Policy Research Centre, FFU-Report 3-04: 409-437.

Quist J, Vergragt P, Thissen W (2005) The impact of backcasting: what is the relevance for sustainable system innovations and transition management? in: Proceedings SWOME/GaMON Market Day 2005, S van den Burg, G Spaargaren, H Waaijers (eds.): 115-120.

Quist J, Rammelt C, Overschie M, de Werk G (2006) Backcasting for sustainability in engineering education: the case of Delft University of Technology, Journal of Cleaner Production 14: 868-876.

Quist J, Vergragt P (2006) Past and future of backcasting: the shift to stakeholder participation and a proposal for a methodological framework, Futures 38(9): 1027-1045.

Rabinovitch J (1996) Innovative land use and public transport policy: the case of Curitiba, Brazil, Land Use Policy 13(1): 51-67.

Raskin P, Banuri T, Gallopin G, Gutman P, Hammond A, Kates R, Swart R (2002) Great transition: the promise and lure of the times ahead, a report of the Global Scenario Group, Tellus Institute, Boston.

Raven R (2005) Strategic niche management for biomass, PhD thesis, TUE, Eindhoven NL.

Reijnders L (1998) The factor X debate: setting targets for eco-efficiency, Journal of Industrial Ecology 2: 13-22.

Remmers J (2004) People, Planet, Profit in de supermarkt: insprerende voorbeelden uit binnen- en buitenland van verantwoord ondernemen in supermarkten, SNM (foundation for Nature & Environment), Utrecht NL.

Rip A, Kemp R (1998) Technological change, in: S Rayner, EL Malone (eds.) Human choice and climate change, Volume 2 Resources and technology, Battelle Press, Columbus, Ohio: 327-399.

Rip A, Misa T, Schot J (1995, eds.) Managing technology in society: the approach of constructive technology assessment, Harold Pinter, London UK.

RMB (1996) Advies duurzame consumptie: een reëel perspectief (Advice on sustainable consumption: a reasonable perspective, Advice 96-07, Council for Environmental Management, The Hague NL.

RMC (1996) Novel Protein Foods: ontwikkeling strategisch aktieplan, unpublished document, Rienstra Marketing Consultancy/STD office Delft.

Robert M (2005) Backcasting and econometrics for sustainable planning: information technology and individual preferences of travel, Journal of Cleaner Production 13: 841-851.

Robinson J (1982) Energy backcasting: a proposed method of policy analysis, Energy Policy 10: 337-344.

Robinson J (1990) Futures under glass: a recipe for people who hate to predict, Futures 22: 820-843.

Robinson J (2003) Future subjunctive: backcasting as social learning, Futures 35: 839-856.

Robinson JB (1988) Unlearning and backcasting: rethinking some of the questions we ask about the future, Technological Forecasting and Social Change 33: 325-338.

Rogers EM (1983) Diffusion of innovations, 3rd edition, The Free Press, New York

Rogers EM (1995) Diffusion of innovations, 4th edition, The Free Press, New York.

Rood T, van Wijk J, van der Knoop J (2002) Zonder actoren geen transitie: een denkraam met vegetarisch voedsel als voorbeeld, Arena (June/July, no 4): 61-64.

Rooijers FJ, Van Soest JP (1998) Ontwikkelingen in wetenschap en technologie: energietechnieken in landelijke gebieden (Developments in science and technology: energy technology in rural areas), NRLO report 98/16 (in Dutch), NRLO, The Hague NL.

Roth A, Kaberger T (2002) Making transport systems sustainable, Journal of Cleaner Production 10: 361-371.

Rotmans J (2003) Transitie management, sleutel voor een duurzame samenleving, van Gorcum, Assen NL.

Rotmans J, Kemp R, van Asselt M (2001) More evolution than revolution: transition management in public policy, Foresight 3(1): 15-31.

Rotmans J, Kemp R, van Asselt M, Geels F, Verbong G, Molendijk K (2000) Transities en transitiemanagement: the casus van een emissiearme energievoorziening (Transitions and transition management: the case of a low-emission energy supply), background report for the Dutch 4th National Environmental Policy Plan (NEPP-4), ICIS & MERIT, Maastricht NL.

Rotmans J, van Asselt M, Anastasi C, Greeuw S, Mellors J, Peters S, Rothman D, Rijkens N (2000) Visions for a sustainable Europe, Futures 32: 809-831.

Ruijgh-van der Ploeg T, Verhallen A (2002) Envisioning the future of trans-boundary river basins: case-studies from the Scheldt river basin, Delft University of Technology NL & Wageningen University NL.

Sabatier PA (1999, ed.) Theories of the policy process, Oxford Westview Press, Boulder, Colorado.

Sabatier PA, Jenkins-Smith HC (1999) The advocacy coalition framework: an assessment, in: PA Sabatier (ed.) Theories of the policy process, Oxford Westview Press, Boulder, Colorado: 117-166.

Schmidt T, Postma A (1999) Minder energiegebruik door een andere leefstijl: Eindrapportage Project Perspectief, ministerie van VROM, Den Haag NL.

Schmidt-Bleek F (1994) Wieviel Umwelt braucht der Mensch? (How much environment does man need?), Birkhauser Verlag (in German), Berlin.

Scholtens (2005) Vleesvervanger zonder problematisch mondgevoel, in: Volkskrant, 4-2-2005.

Schot J (2001) Towards new forms of participatory technology development, Technology Analysis & Strategic Management 13(1): 39-52.

Schot J, Rip A (1996) The past and future of constructive technology assessment, Technology Forecasting and Social Change 54: 251-268.

Scott WR (1995) Institutions and organizations, Sage Publications, Thousand Oaks CA.

Scott WR (2001) Institutions and organizations, 2nd edition, Sage Publications. Thousand Oaks CA.

Senge PM (1990) The fifth discipline: the art and practice of the learning organization, Doubleday, New York.

SER (2003) Duurzaamheid vraagt om openheid: op weg naar een duurzame consumptie, The Social and Economic Council of the Netherlands (SER), publication 03/02, The Hague NL.

Slob AFL, Bouman MJ, de Haan M, Bok K, Vringer K (1996) Trendanalyse consumptie en milieu (a trend analysis on consumption and environment), VROM, The Hague NL.

Slocum N (2003) Participatory methods toolkit, a practitioner's manual, joint publication of ViWTA and King Baudouin Foundation, Belgium.

Smith A, Stirling A, Berkhout F (2005) The governance of sustainable socio-technological transitions, Research Policy 34: 1491-1510.

Smits M (2002a) Monsterbezwering: de culturele domesticatie van nieuwe technologie, Boom, Amsterdam NL.

Smits R (2002b) Innovation studies in the 21st century: questions from a user's perspective, Technological Forecasting and Social Change 69: 861-883.

Smits R, J Leijten and P den Hertog (1995) Technology assessment and technology policy in Europe: new concepts, new goals, new infrastructures, Policy Sciences 28: 271-299.

Spoelstra SF, Waninge A, Menting B, Van der Woude M (2003, eds.), Wending naar duurzaamheid: programma nieuwe veehouderijsystemen, WUR.

STD (1994a) Stakeholder analysis for Novel Protein Foods, concept of January 1994, unpublished document (in Dutch).

STD (1994b) Novel Protein Foods: a study on high-quality protein food products, a report on the period April 1993 – April 1994, unpublished document, 9 pages, STD office, Delft.

STD (1994c) Input documents and minutes NPF kick-off meeting, 23-6-94, unpublished documents.

STD (1994d) Handleiding B-fase (Project guidelines phase B), unpublished document, 15-11-94.

STD (1994e) Documentatiemap behoefteveldanalyse voeden, Heidemij Advies, ATO-DLO, LUW, TNO-Voeding.

STD (1995a) Handleiding C-fase: deel 1 NPF-analyse, unpublished document, 13-7-95.

STD (1995b) Handleiding C-fase: deel 2 Inbedding & Evaluatie, unpublished document, 6-11-95.

STD (1995c) Concept stakeholder analyse C-fase, 20-12-95, unpublished document.

STD (1996a) Anders eten in een duurzame toekomst: Novel Protein Foods in 2035, Final Report NPF project, STD office, (in Dutch with a summary in English).

STD (1996b) Procesbeschrijving NPF (Process description NPF), internal document NPF project, March 1996.

STD (1996c) ideaaltypisch DTO-proces (afgeleid van het illustratieproces Novel Protein Foods), unpublished document, 3-5-1996.

STD (1996d) Duurzaam landgebruik: resultaten definitiestudie (Sustainable land-use: results of the feasibility study), STD office Delft NL.

STD (1996e) Duurzaam landgebruik: resultaten fase A & werkplannen fase B/C, STD report VD-3, STD office, Delft NL (in Dutch).

STD (1997) Visie en sleutels tot een duurzame welvaart (Vision and keys to sustainable prosperity), STD office, Delft, The Netherlands.

Street P (1997) Scenario workshops: a participatory approach to sustainable urban living? Futures 29: 139-158.

Stuiver M, Wiskerke JSC (2004) The VEL and VANLA environmental co-operatives as a niche for sustainable development, in: in: Wiskerke JSC, van der Ploeg JD (eds.), Seeds of transition: essays on novelty production, niches and regimes in agriculture, Van Gorcum, Assen NL: 119-148.

Stuurgroep MDL (1999) Meervoudig duurzaam landgebruik: programmaplan 1999-2003, MSL Winterswijk Steering group.

Stuurgroep T&O (2001) Handorakel voor toekomstonderzoek: kennisvereisten en procesvereisten aan toekomstonderzoek voor het strategisch omgevinsbeleid, stuurgroep toekomstonderzoek en strategisch omgevingsbeleid (WRR/COS/Innovatienetwerk/RMNO), Lemma, Utrecht NL.

Suurs R, Hekkert M, Meeus M, Nieuwlaar E (2004) Assessing transition trajectories towards a sustainable energy system: a case study on the Dutch transition to climate-neutral transport fuel chains, Innovation: Management, Policy & Practice 6(2): 269-285.

Tansey J, Carmichael J, Van Wynsberghe R, Robinson J (2002) The future is not what it used to be: participatory integrated assessment in the Georgia Basin, Global Environmental Change 12: 97-104.

Tassoul M (1998) Making sense with backcasting: the future perfect, Creativity and Innovation Management 7(1): 32-45.

Tazelaar RJ (1997) Nieuwjaarstoespraak door de voorzitter van het productschap Vee, Vlees en Eieren, unpublished document, Rijswijk.

Teisman GR, Edelenbos J (2004) Getting through the 'twilight zone': managing transitions through process-based, horizontal and interactive governance, in: B Elzen, FW Geels, K Green (eds.), System innovation and the transition to sustainability, Edward Elgar, Cheltenham UK: 168-190.

Tempelman E, Joore P, van der Horst T, Lindeijer E, Luiten H, Rampino L, van Schie M (2004) Product-service systems in the need area food, Suspronet report, www.suspronet.org.

Thissen W (1999) A scenario-approach for identification of research topics, in: M Weijnen, E ten Heuvelhof (eds.) The infrastructure playing field in 2030, Delft University Press: 5-10.

Tóth KS, Tóth L, Szekeres Z, Szuts L, Galbács Z, Fenyvessy J (2000) Shopping, cooking and eating in Hungary, Final report, SusHouse project, College of Food Industry, Szeged University, Hungary.

Tukker A, Tischner U (2006, eds.) New business for old Europe: product-service development, competitiveness and sustainability, Greenleaf Publishers, Sheffield UK.

Uitdenbogerd DE, Brouwer NM, Groot-Marcus JP (1998) Domestic energy saving potentials for food and textiles – an empirical study, Wageningen Agricultural University, Wageningen NL.

Vaessen P (2003) Samenwerking in het programma meervoudig duurzaam landgebruik Winterswijk: een procesevaluatie met accent op de projecten Stortelersbeek en Bioraffinage, Nijmengen Business School, Nijmegen University NL (in Dutch).

Van Asselt M, Rotmans J, Rothman D (2005) Scenario innovation: experiences from a European experimental garden, Taylor & Francis, Leiden NL.

Van Calmthout M (2004) Wrongelvlees, in: Volkskrant 27-11-2004.

Van Dam-Mieras R (2003) Midterm evaluation of the Profetas integration process, Open University of the Netherlands, Heerlen NL.

Van de Kerkhof M (2004) Debating climate change: a study of stakeholder participation in an integrated assessment of long-term climate policy in the Netherlands, PhD thesis, Free University, Amsterdam, Lemma Publishers, Utrecht NL.

Van de Kerkhof M, Hisschemoller M, Spanjersberg M (2003) Shaping diversity in participatory foresight studies: experiences with interactive backcasting on long-term climate policy in the Netherlands, Greener Management International 37: 85-99.

Van de Kerkhof M, Wieczorek A (2005) Learning and stakeholder participation in transition processes towards sustainability: methodological considerations, Technological Forecasting and Social Change 72 (2005): 733-747.

Van de Poel I (1998) Changing technologies: a comparative study of eight processes of transformation of technological regimes, Twente University Press, Enschede NL.

Van de Poel I (2000) On the role of outsiders in technical development, Technology Analysis & Strategic Management 12(3): 383-397

Van de Poel I (2003) The transformation of technological regimes, Research Policy 32: 49-68.

Van den Ende J, Kemp R (1999) Technological transformations in history: how the computer regime grew out of existing computing regimes, Research Policy 28: 833-851.

Van den Hoed R (2004) Driving fuel cell vehicles: how established industries react to radical technologies, PhD thesis, Delft University of Technology NL.

Van der Cammen H, De Lange MA (1998) Ontwikkelingen in wetenschap en technologie: sturingstheorieen en landelijke gebieden, NRLO report 98/11, NRLO, The Hague NL.

Van der Heijden K (1996) Scenarios, the art of strategic conversation, John Wiley Publishers.

Van der Horst H (2002) Duurzaam uit eten in de horeca: ontwikkeling van de drive-in, een mobiele koelunit voor de levering van dagvoorraden aan de horeca- en cateringbranche, MSc thesis (in Dutch), Design for Sustainbility Group, Delft University of Technology NL.

Van der Knijff A, Westerman E, Geerling-Eiff F, Verstegen J (2004) Maatschappelijke ontwikkelingen en verkenningen nieuwe eiwitten: research guidance, rode daad bij systeeminnovaties, WUR, Nota 300.

Van der Lans IA (2001) Het imago en de evaluatie van vers vlees en vleesvervangers in 2000 en 2001 (The image and the evaluation of fresh meat and meat substitutes in 2000 and 2001), (in Dutch) Product Boards for Livestock, Meat and Eggs (PVE), Rijswijk NL.

Van der Meer HG, Spiertz JHJ (1992) Stikstofstromen in agro-ecosystemen (Nitrogen flows in agro-ecosystems), Agrobiologische Thema`s 6, Wageningen, 115 pp. (in Dutch).

Van der Pijl S, Krutwagen B (2000, with contributions from K Bemelmans, J van 't Bosch, E Lindeijer) Domeinverkenning voeden: ingredienten voor een gezond milieu (Exploration of the nutrition domain: ingredients for a healthy environment), Schuttelaar & Partners/IVAM Environmental Research, The Hague/Amsterdam NL (in Dutch).

Van der Vleuten E (2000) Twee decennia van onderzoek naar grote technische systemen: thema's, afbakening en kritiek, NEHA jaarboek voor economische, bedrijfs- en techniekgeschiedenis: 328-364.

Van Eeten M (1999) Dialogues of the deaf: defining new agendas for environmental deadlocks, PhD thesis, Delft University of Technology, Eburon, Delft NL.

Van Est R (1999) Winds of change: a comparative study of the politics of wind energy innovation in California and Denmark, PhD thesis, University of Amsterdam, International Books, Utrecht NL.

Van Gaasbeek AF, Meeusen-van Onna MJG, Wiersma G, Kamminga KJ, Moll HC (2000) Eindrapport consument, voeding en milieu, programmabureau DTO/NRLO, Delft/Den Haag.

Van Gaasbeek ÀF, Wiersma G, Meeusen MJG, Kamminga KJ, Wilemse-van den Bergh AC, Moll HC, (1999) Consument, voeding en milieu: probleemanalyse van de aan voeding gerelateerde milieubelasting in de consumentenfase, LEI/IVEM, Den Haag/Groningen.

Van Hemel C (1998) Eco-design empirically explored: design for environment in Dutch small and medium-sized enterprises, PhD thesis, Delft University of Technology NL.

Van Kasteren J (2001) Transities in de praktijk: ervaringen met duurzame technologische ontwikkeling, kennisoverdracht en verankering, STD office, Delft NL (in Dutch).

Van Lente H (1993) Promising technology: the dynamics of expectations in technological developments, PhD thesis, University of Twente, Enschede NL.

Van Merkerk RO, van Lente H (2005) Tracing emerging irreversibilties in emerging technologies: the case of nanotubes, Technological Forecasting and Social Change 72: 1094-1111.

Van Mierlo B (2002) Kiem van maatschappelijke verandering: verspreiding van zonnecelsystemen in de woningbouw met behulp van pilotprojecten, PhD thesis University of Amsterdam, Aksant Publishers, Amsterdam NL.

Van Notten PWF, Rotmans J, van Asselt MBA, Rothman DS (2003) An updated scenario typology, Futures 35: 423-443.

Van Otterloo A (1990) Eten en eetlust in Nederland (1840-1990): een historisch-sociologische studie, Bert Bakker, Amsterdam NL.

Van Otterloo AH (2000) Voeding, in: A van Otterloo (ed.) Techniek in Nederland in de twintigste eeuw, deel III Landbouw en voeding, stiching Historie der Techniek, Walburg Pers, Zutphen NL: 235-374.

Van Stuijvenberg JH (1969) Margarine: an economic, social and scientific history 1869-1969 (Honderd jaar margarine 1869-1969), Liverpool University Press UK.

Van Twist, ten Heuvelhof E, Edelenbos J (1998) ICT: mogelijkheden voor sturing en ontwerp in landelijke gebieden, NRLO report 98/12, National Council for Agricultural Research, The Hague NL.

Van Wijk JJ, Rood GA (2002) Besluitvormingsmodellen in het transitieproces: toegepast op vegetarisch voedsel, RIVM-report, National Institute for Public Health and the Environment (RIVM), Bilthoven NL.

Veldhoen L, van den Ende J (2003) Technische mislukkingen: deel 1 (Technical failures: part 1), Ad Donker, Rotterdam, 3rd edition (first published in 1995).

Vellinga P, Herb N (1999, eds.), Industrial transformation science plan, IHDP-report 12, International Human Dimensions Programme on Global Environmental Change.

Vereijken PH (2002) Transition to multifunctional land use including agriculture, Netherlands Journal of Agricultural Science 50(2): 171-179.

Vergragt P et al. (no date) Strategies towards the sustainable household, technical annex & workprogramme submitted to the EU, internal document SusHouse project.

Vergragt P (2000) Strategies towards the sustainable household, Final Report, SusHouse project, Delft University of Technology NL.

Vergragt P, Diehl JC, Soumitri GV (2001) Creating an ecodesign network in Delhi, paper 9th Int GIN Conference on 'Sustainability at the millennium: globalization, competitiveness, and the public trust', Bangkok, January 21-25.

Vergragt P, Mulder K, Rip A, van Lente H (1989) De matrijs van verwachtingen ingevuld voor de polymeren Tenax en Twaron, Werkdocument W12, NOTA, The Hague NL.

Vergragt P, Van der Wel M (1998) Back-casting: an example of sustainable washing, in: N Roome (ed.) Sustainable strategies for industry, Island Press, Washington DC: 171-184.

Vergragt P, Van Noort D (1996) Sustainable technology development: the mobile hydrogen fuel cell, Business Strategy and the Environment 5: 168-177.

Vergragt PJ (2005) Back-casting for environmental sustainability: from STD and SusHouse towards implementation, in: M Weber, J Hemmelskamp (eds.), Towards environmental innovation systems, Springer, Heidelberg Germany: 301-318.

Vergragt PJ, Jansen L (1993) Sustainable technological development: the making of a long-term oriented technology programme, Project Appraisal 8: 134-140.

Verheul H, Vergragt PJ (1995) Social experiments in the development of environmental technology: a bottom-up perspective, Technology Analysis & Strategic Management 7 (3): 315-326

Vig NJ, and Paschen H (2000) Parliaments and technology: the development of technology assessment in Europe, Chapter 1: Technology assessment in comparative perspective, State University of New York Press: 3-35.

Von Weizsäcker E, Lovins AB, Lovins HL (1997) Factor Four: doubling wealth - halving resource use, the new report to the Club of Rome, Earthscan, London UK.

Voß JP, Bauknecht D, Kemp R (2006) Reflexive governance for sustainable development, Edward Elgar, Cheltenham UK.

Vringer K, Aalbers T, Drissen E, Hoevenagel R, Bertens C, Rood G, Ros J, Annema J (2001) Nederlandse consumptie en energiegebruik in 2040: een verkenning op basis van twee lange termijn scenario's, RIVM report, National Institute of Public Health and the Environment (RIVM), Bilthoven NL.

VROM (1998) Nationaal Milieubeleidsplan 3 (National Environmental Policy Plan 3), Ministry of Housing, Spatial Planning and the Environment, The Hague NL.

VROM (2001) Nationaal Milieubeleidsplan 4, Een wereld en een wil: werken aan duurzaamheid, (National Environmental Policy Plan 4), Ministry of Housing, Spatial Planning and the Environment, SDU publishers, The Hague NL.

VROM (2005) Hoofdlijnennotitie toekomstagenda milieu, 14-9-05, The Hague.

WCED (1987) Our common future, World Commission on Environment and Development, Oxford University Press, Oxford, UK.

Weaver P, Jansen L, Van Grootveld G, van Spiegel E, Vergragt P (2000) Sustainable technology development, Greenleaf Publishers, Sheffield UK.

Westerman F (1999) De graanrepubliek (The corn republic), Atlas, Amsterdam NL (in Dutch).

Weterings RAPM, Opschoor JB (1992) The environmental capacity as a challenge to technology development, Advisory Council for Research on Nature & Environment (RMNO), Rijswijk NL.

Wiek A, Binder C, Scholz RW (2006) Functions of scenarios in transition processes, Futures 38(7): 740-766.

Wiersma G, Van Gaasbeek T (1999) Discussienota consument, voeding, milieu voor de workshop 'Implementatie Duurzaam Voeden in het Huishouden', 12 januari 2000, Oegstgeest.

WRR (1980) Beleidsgerichte toekomstverkenning, deel 1: een poging tot uitlokking, Scientific Council for Government Policy, SDU Publishers, The Hague NL.

WRR (1983) Beleidsgerichte toekomstverkenning, deel 2: een verruiming van het perspectief, Scientific Council for Government Policy, SDU Publishers, The Hague NL.

WRR (1992) Grond voor keuzen: vier perspectieven voor de landelijke gebieden in de Europese Gemeenschap, Report 42, Scientific Council for Government Policy, SDU Publishers, The Hague NL.

WRR (1994) Duurzame risico's: een blijvend gegeven (Sustained risks: a lasting phenomenon), Report 44, Scientific Council for Government Policy, SDU Publishers, The Hague NL.

Wynne B (1995) Technology assessment and reflexive social learning: observations from the risk field, in: A Rip, T Misa, J Schot (eds.), Managing technology in society: the approach of constructive technology assessment, Harold Pinter, London: 19-36.

Yin RK (1994) Case study research design and methods, 2nd edition, Sage Publications.

Young W (2000) Shopping, cooking and eating in the United Kingdom, Final Report, SusHouse Project, UMIST, Manchester UK.

Young W, Quist J, Tóth K, Anderson, K Green K (2001) Exploring sustainable futures through design orienting scenarios: the case of shopping, cooking and eating, Journal of Sustainable Product Design 1(2): 117-129.

Young W, Simms J (2000) Economic analysis of scenarios, Final Report, SusHouse Project, UMIST, Manchester UK.

Zimmermann KL, Kramer KJ, Klein Essink G, Koelemeijer K, Londo M, Guinée J (2006) Ketenproject vervanging vleesproducten door plantaardige eiwitproducten in bedrijfsrestaurants, AKK rapport.

Zwart KB (1997) Onderzoekprogramma multifunctioneel landgebruik: oplossen van knelpunten in voorbeeldsystemen duurzaam landgebruik, STD report VD-7, STD office, Delft (in Dutch).

www.akk.nl, visited November 2004.

www.bedr-horeca.nl, visited June 2005.

www.energietransitie.nl, visited on December 12, 2006.

www.epz.novem.nl, visited 10-02-2005.

www.habiforum.nl, visited in September 2005 & October 2006.

www.hetlankheet.nl, visited October 2006.

www.hicsproject.org, visited June 2005.

www.ijsselzone.nl, visited October 2006.

www.ipcc.org, visited October 2006.

www.milieucentraal.nl, visited February 2005.

www.nederlandmooi.nl, visited October 2006.

www.planetgreen.nl, visited on November 29, 2004 & August 2, 2006.

www.profetas.nl, visited on November 13, 2003.

www.scheppenvanruimte.nl, visited November 2004.

www.senternovem.nl/eet, visited 2003.

www.suspronet.org, visited June 2005.

www.syscope.nl, visited November 2005 & October 2006.

www.thenaturalstep.org, visited on November 15, 2006.

APPENDIX A

Interview checklist

A1. General and introduction
- Background, affiliation and position of respondent.
- Involvement in and familiarity with backcasting experiment or and follow-up and spin-off.
- Opinions on why involvement was attractive to the organisation.
- Problem formulation with regard to the problem addressed in the backcasting experiment.
- Alternative solutions and visions dealing with similar problems.

A2. The backcasting experiment
- Major successes and strengths of the backcasting experiment and its approach.
- Opinions on what was less successful and should be improved (when repeating the backcasting experiment).
- Opinions on stakeholder involvement (quality, heterogeneity, influence, relevance, level).
- Opinions on methods within the backcasting experiment, especially about design methods.
- Views on how the future vision was generated and on the content of the vision.
- Views on what has been learnt (personally as well as at the level of the organisation).
- Views on what other stakeholders could have learnt from the backcasting experiment.
- Impact of the backcasting experiment on the organisation.

A3. Follow-up activities and spin-off of the backcasting experiment
- Views on activities that are the follow-up of, or considerably influenced by the backcasting experiment (in research domain, business domain, government domain and the public domain).
- 'History' of these activities, involved (key) actors, mobilised resources, crucial contributions and conducted actions by these actors.
- Views on the role of the future vision and how it has evolved.
- Results of follow-up and spin-off activities and how these were perceived by stakeholders.
- Views on broader spin-off and effects, such as entrenchment and resistance.

A4. Context developments and external factors
- Relevant developments and major events in the socio-technical system in which the backcasting experiment is conducted, as well as in its context.
- Possible negative or positive influence of external factors on the follow-up and spin-off of the backcasting experiment.

A5. Other issues discussed
- Expectations with regard to further development of the envisaged system innovation.
- Possible opportunities and threats for further development.
- Next steps and by whom they should be taken.

APPENDIX B

List of interviews and contacts

B.1 Novel Protein Foods case (Chapter 5)

List of interviews

Dr H. Aiking, Institute for Environmental Studies (IVM), Free University, Amsterdam, interview December 1, 2004.

Mr C. Dutilh, Unilever Netherlands, Rotterdam, interview November 26, 2004.

Mr W. Engels, Aurelia Marketing & Planet Green, Amersfoort, interview November 19, 2004.

Professor J.L.A. Jansen, former chair STD programme & member Profetas programme committee, interview November 10, 2004.

Mrs L. Kemper, Technology Foundation STW, Utrecht, interview December 14, 2004.

Dr B.G. Linsen, chair Profetas programme committee, former project manager of the NPF project, interview November 11, 2004.

Professor T. van Boekel, Agrotechnology & Food Sciences, WUR, Wageningen, interview January 4, 2005.

Mr P. van Egmond & Mr A. Sein, DSM Food Specialties, Delft, interview November 26, 2004.

Mr B. Timmerman, Netherlands Vegetarian Society (NVB), Hilversum, interview December 20, 2004.

Dr J. Vereijken, Agrotechnology & Food Innovations, Wageningen, interview November 25, 2004.

Mr H. Waaijers, Social Science Research Council (NWO-MaG), The Hague, interview December 1, 2004.

Mr A. Zwijgers, Unilever Health Institute, Vlaardingen, interview December 21, 2004.

Additional contacts

Additional information has been provided by Mr D. Brand (VROM), Dr H. Huizing (Innovation Network), Mr J Remmers (NM), Mr H. Kroft (Heinz), Mr J. Cornelese (formerly LNV), Mrs M. de Groot (LNV), Mrs A Kuipers (LNV).

B.2 Household Nutrition case (Chapter 6)

List of interviews

Dr G. Fonk, Innovation Network for Green Space and Agrocluster, The Hague, telephonic interview May 25, 2005.

Professor K. Green, Manchester Business School, University of Manchester, telephonic interview April 26, 2005.

Dr K.J. Kramer, LEI, The Hague, telephonic interview April 25, 2005.

Mrs H. Luiten, TNO, Delft, telephonic interview April 25, 2005.

Mr H. te Riele, Storrm Consultancy, The Hague, telephonic interview April 15, 2005.

Dr K. Szita Tóth, University of Miskolc, Miskolc, Hungary, written correspondence, December 2005.

Dr S. Silvester, TU Delft, section Design for Sustainability, interview April 21, 2005.

Professor P.J. Vergragt, Tellus Institute, Boston, USA, interview March 29, 2005.

Additional contacts

Additional information has been obtained from Mr D. Brand (Ministry of the Environment), Dr G. Beers (LEI Agricultural Economic Institute), Professor G. Spaargaren (Wageningen University), Dr A. Tukker (TNO) and Professor R. Roy (Open University UK).

B.3 Multiple Sustainable Land-use case (Chapter 7)

List of interviews

Dr H.J. de Graaf, KDO Advies & CML, Leiden University, interview October 24, 2005.

Dr J.H.A. Hillebrand, Innovation Network for Rural Areas and Agricultural Systems, interview October 13, 2005.

Mr G.M. Kiljan & Mr T.S.P. Moolenaar, Province of Gelderland, interview September 27, 2005.

Dr H. Korevaar, Plant Research International (PRI), WUR, interview September 29, 2005.

Mr O.C.H. de Kuijer, KDO Advies, interview August 23, 2005.

Mr P.J.A.M. Smeets, Alterra, WUR, interview October 10, 2005.

Professor J.H.J. Spiertz, Crop and Weed Ecology Group, WUR, interview November 15, 2005.

Dr S. Spoelstra, WUR Animal Sciences, interview November 11, 2005.

Mr J. Tiggeloven, WCL Winterswijk, interview September 27, 2005.

Mr G.J. Verkade, Habiforum, interview September 20, 2005.

Additional contacts

Additional information has been provided by Mrs S. van der Pas (Province of Utrecht), Mr E. Koldewey (LTO Noord, previously called GLTO), Mr A. Spekschoor (formerly GLTO, currently DLG), Mr J. Leemkuil (farmer Winterswijk), Mr J. Wytema (Winterswijk), Dr J. Ketelaars and Dr P. Vereijken (WUR-PRI), Mr A. Oosterbaan (Alterra), Dr J. Vogelezang (WUR-Plant Sciences), Mr E. Poot (WUR-PPO), Mr G. van der Veen and Mr E. Kok (Water Board Rhine and IJssel), Dr H Hetsen (formerly NRLO), Mr J. Eisen (Nutreco), Professor R. Rabbinge (WUR).

ABBREVIATIONS

3P	People, Planet, Profit
AFI	Agrotechnology and Food Innovation
AKK	Agri Chain Competence (Agro Keten Kennis)
ANWB	Netherlands Association of Motorists (Algemene Nederlandse Wielrijdersbond)
BHC	Dutch Board for the Hotel and Catering industry (Bedrijfschap Horeca & Catering)
BSE	Bovine Spongiform Encephalopathy (mad cow disease)
BSTE	Bounded Socio-Technical Experiment
CoP	Community of Practice
COOL	Climate Options for the Long-term
CPS	Creative Problem Solving
CTA	Constructive Technology Assessment
DLG	Division for Rural Areas (Dienst Landelijk Gebied)
DLO	Agricultural Research Division (Dienst Landbouwkundig Onderzoek)
DOS	Design-Oriented Scenario
DTO-KOV	Sustainable Technology Development – Knowledge Transfer and Embedding (Duurzame Technologische Ontwikkeling – Kennisoverdracht en Verankering)
DUT	Delft University of Technology
EET	Ecology, Economy, Technology
EHS	National Ecological Network (Ecologische Hoofd Structuur)
EU	European Union
FAO	Food and Agricultural Organisation
GDP	Gross Domestic Product
GLTO	Farmer Union in the province of Gelderland (Gelderse Land- en Tuinbouw Organisatie)
GMO	Genetic Modified Organisms
HiCS	Highly Customised Solutions
ICS	Intelligent Cooking and Storing
ICT	Information and Communication Technology
IHDP	International Human Dimensions Programme on Global Environmental Change
INGRA	Innovation Network for Green Space and Agro-cluster (Innovatie Netwerk Groene Ruimte & Agrocluster)
IOP	Innovative Research Programme (Innovatief Onderzoeksprogramma)
ITA	Interactive Technology Assessment
IVM	Institute for Environmental Studies (Instituut voor Milieuwetenschappen)
KSI	Knowledge Network on System Innovations (Kennisnetwerk Systeem Innovaties)
LCA	Life Cycle Analysis
LEI	Agricultural Economic Institute (Landbouw-Economisch Instituut)
LG	Local and Green
LNV	Ministry of Agriculture, Nature Management and Food Quality (Ministerie van Landbouw, Natuurbeheer en Voedselkwaliteit)
LTO	National Farmer Union

LTS	Large Technological System
MaG	Social Science Research Council (Maatschappij- en Gedragswetenschappen)
MF	Multifunctional
MFZH	Environmental Federation of South-Holland (Milieufederatie Zuid-Holland)
MLP	Multi-Level Perspective
MNP	Netherlands Environmental Assessment Agency (Milieu en Natuur Planbureau)
MSL	Multiple Sustainable Land-use (Meervoudig Duurzaam Landgebruik)
NBPV	Dutch Association of Country Women (Nederlandse Bond van Plattelandsvrouwen)
NIDO	National Initiative for Sustainable Development (Nationaal Initiatief voor Duurzame Ontwikkeling)
NM	Foundation for Nature and Environment (stichting Natuur en Milieu)
NPF	Novel Protein Foods
NRLO	National Council for Agricultural Research (Nationale Raad voor Landbouwkundig Onderzoek)
NSI	National System(s) of Innovation
NSW	National Foundation for Valuable Country-estates (Nationale Stichting Waardevolle Landgoederen)
NVB	Netherlands Vegetarian Society (Nederlandse Vegetariersbond)
NWO	Netherlands Organisation for Scientific Research (Nederlandse Organisatie voor Wetenschappelijk Onderzoek),
OVO	Research, Transfer/Dissemination, Education (Onderzoek, Voorlichting, Onderwijs)
PIA	Participatory Integrated Assessment
PRI	Plant Research International
Profetas	Protein Foods, Environment, Technology and Society
PTA	Participatory Technology Assessment
PVE	Product Boards for Livestock, Meat and Eggs (Productschappen voor Vee, Vlees en Eieren)
R&D	Research and Development
RIVM	National Institute for Public Health en the Environment (Rijkstinstituut voor Volksgezondheid en Milieuhygiene)
RIZA	Institute for Inland Water Management and Waste Water Treatment (Rijksinstituut voor Integraal Zoetwaterbeheer en Afvalwaterbehandeling).
RMNO	Research Council for Spatial Planning, Nature and the Environment (Raad voor Ruimtelijk, Milieu en Natuur Onderzoek)
RTD	Research and Technology Development
RWS	Directorate-General of Water Management, Ministry of Public Works and Water Mangement (Rijkswaterstaat van het Ministerie van Verkeer en Waterstaat)
SBB	State Forest Management (Staatsbosbeheer)
SCB	Society for Consumers and Biotechnology (Stichting Consument & Biotechnologie)
SCE	Shopping, Cooking, Eating
SCP	Single Cell Protein
SENSE	Research school for Socio-Economic and Natural Sciences of the Environment
SER	Social and Economic Council of the Netherlands (Sociaal-Economische Raad)
SME	Small and Medium-sized Enterprise
SNM	Strategic Niche Management
SR	Super-Rant

STD	Sustainable Technology Development
STW	Technology Foundation STW (Stichting Technische Wetenschappen)
SusProNet	Sustainable Product Service Systems Network
TM	Transition Management
TNS	The Natural Step
ToR	Terms of Reference
TTI	Technological Top Institute (Technologisch Top Instituut)
TvC	Future Images for Consumers (Toekomstbeelden voor Consumenten)
TVP	Textur(is)ed Vegetable Proteins
VLAG	Research school for Nutrition, Food Technology, Agro-biotechnology and Health Sciences (Onderzoeksschool Voeding, Levensmiddelen en Agrotechnologie)
VNM	Association for Nature Conservation (Vereniging Natuur Monumenten)
VROM	Ministry of Housing, Spatial Planning and the Environment (Ministerie van Volkshuisvesting, Ruimtelijke Ordening en Milieu)
VU	Free University Amsterdam
WCFS	Wageningen Center for Food Sciences
WCL	Valuable Man-made Landscape (Waardevol Cultuur Landschap)
WRR	Scientific Council for Government Policy (Wetenschappelijke Raad voor het Regeringsbeleid)
WTO	World Trade Organisation
WUR	Wageningen University and Research Centre

SUMMARY

Backcasting for a sustainable future: the impact after 10 years

<u>System innovations towards sustainability and participatory backcasting</u>

Contemporary societies face the challenge of realising sustainable development and have to deal with underlying complex sustainability problems. Sustainable development demands that the needs of the present generation be fulfilled in such a way that future generations will also be able to meet their needs. This requires interrelated technological, cultural, organisational and institutional changes at the level of socio-technical systems, such as the system of food production and consumption, industrial sectors, and house-hold consumption domains like mobility and shelter. A system innovation towards sustainability is defined in this research as the transformation from one socio-technical system towards another, sustainable one. System innovations take several decades and require the involvement of many actors. These actors possess relevant expertise and resources to contribute to system innovations towards sustainability. They are also needed because they can provide legitimacy and support with regard to the changes that are associated with the system innovation.

Participatory backcasting is a promising approach to explore system innovations towards sustainability, as well as to define and initiate first steps towards an envisaged system innovation. Backcasting literally means looking back from the future. In the backcasting approach the desirable future is envisaged first, before looking back to how that future may be achieved, and defining what steps need to be taken to bring about the envisaged future. Backcasting is particularly useful in the case of complex problems, where there is a need for major change, where dominant trends are part of the problem, where there are side-effects or externalities that cannot be satisfactorily solved in markets, and where long time horizons allow for future alternatives that need time to develop. In this research the term backcasting experiments is used, which are defined as studies or processes in which the backcasting approach is applied and a broad range of stakeholders is involved.

<u>Research topic and research questions</u>

Since the early 1990s, several dozens of backcasting experiments have been conducted in the Nether-lands. There are considerable differences in the degree to which the various experiments have led to follow-up and spin-off after a few years, as well as with regard to other characteristics. In general, the available reports focus on (1) the way the backcasting approach has been applied and what content results have been generated, or on (2) the stakeholder learning process and the social dynamics among stakeholders in the backcasting experiment. By contrast, very little is known about the impact of backcasting experiments after five to ten years, as well as how follow-up and spin-off relate to (system) innovation theories. As yet, there has been no systematic evaluation and comparison of the impact of backcasting experiments after five to ten years, while conceptual and analytical frameworks to analyse the follow-up and spin-off are also lacking.

To improve our insight into the impact of backcasting experiments, two main questions have been de-veloped to guide this research into backcasting experiments and their impact after five to ten years:

A. *What factors determine the impact of backcasting experiments after five to ten years?*
B. *How should participatory backcasting be applied for exploring and shaping system innovations towards sustainability?*

The main questions have been dealt with in eight steps:

STEP 1 Elaboration of research topic, research approach and research questions (Chapter 1).

STEP 2 Exploration of backcasting and related approaches (Chapter 2).

STEP 3 Theoretical explorations of visions, learning, stakeholder participation, system innovation theories and network theories (Chapter 3).

STEP 4 Development of conceptual framework, propositions and multiple case study research methodology (Chapter 4).

STEP 5 Three case studies (Chapters 5-7).

STEP 6 Comparison of cases and evaluation of propositions (Chapter 8).

STEP 7 Conclusions, reflections and recommendations (Chapter 9).

Conceptual framework and research methodology

In this research I have used an ex post multiple case study approach. As part of this a conceptual framework has been developed that comprises both the backcasting experiment phase and the follow-up and spin-off phase after five to ten years, as well as a set of propositions and a research methodology. The conceptual framework uses the theoretical exploration on future visions, stakeholder participation theories, higher order learning and various network theories and system innovation theories, as well as the developed methodological framework for backcasting. To do so, literature from innovation studies, policy sciences, technology assessment and future studies have been studied. As a result, theories and concepts have been selected to develop the conceptual framework, which have been adjusted when it was necessary for the purpose of this research. The theoretical exploration has also shown that network theories offer better possibilities to map and analyse the follow-up and spin-off of backcasting experiments after five to ten years, than system innovation theories.

The conceptual framework comprises two phases: the backcasting experiment and the follow-up and spin-off after five to ten years. The backcasting experiment phase consists of four building blocks: (1) stakeholder participation, (2) future visions, (3) learning, and (4) settings and methodological aspects. The building blocks are based on various theories about actor and stakeholder participation, the *Leitbild* concept from German sociology of technology, theories about higher order learning and the methodological framework for backcasting, respectively. The building block **participation** includes various aspects derived from a number of actor and stakeholder participation theories, including stakeholder heterogeneity, stakeholder influence and the degree of involvement. The building block **future visions** includes the aspects 'guidance' (where to go) and 'orientation' (what to do), which are derived from the *Leitbild* concept, while the aspect 'competing visions' has been added. The building block **learning** emphasises higher order learning by actors and includes shifts in problem definitions, perceived solutions and principal approaches to dealing with the problems at the level of individual actors, as well as joint and congruent learning at the level of groups of stakeholders. Whereas joint learning refers to consensus and joint opinions, congruent reflects non-conflicting issues. The building block **settings and methodological aspects** comprises various aspects reflecting how the participatory backcasting approach has been applied, as well as various settings.

The follow-up and spin-off phase consists of three building blocks: (1) network formation, (2) future visions (3) institutionalisation. The building block **network formation** is based on industrial network theory and contains the aspects 'activities', ' actors' and 'resources'. Like in the phase of the backcasting experiment the building block **future visions** comprises the aspects 'guidance', 'orientation' and 'competing visions'. The building block **institutionalisation** is based on institutional theory. In this building block the aspect 'institutionalisation' reflects changes in institutions and rules, whereas the aspect 'institutional resistance' reflects the resistance from vested interests and institutions and the actors backing them.

The conceptual framework also proposes **internal factors** and **external factors** that both exert influence on the emergence of follow-up and spin-off. Internal factors are characteristics of the backcasting experiment. External factors are exerted by the socio-technical system and its context, which surround the backcasting experiment and its follow-up and spin-off. Internal factors and external factors can have both a positive (enabling) and a negative (constraining) influence on follow-up and spin-off. The backcasting experiment and its follow-up and spin-off both occur within a socio-technical system. Four domains are distinguished in which follow-up and spin-off may occur: (1) research, (2) business, (3) government, and (4) the public domain that includes public interest groups as well as the wider public.

Based on the conceptual framework a set of propositions has been developed that focuses on the relationships between aspects of the backcasting experiment and the degree of follow-up and spin-off. A research methodology has been developed using indicators for each building block of the conceptual framework, which has allowed to analyse the cases and to evaluate the propositions. A number of sources was used to conduct the case studies, which included in-depth interviews, other personal communication, and a range of printed and internet sources.

Three case studies and their results

Three case studies have been conducted, each consisting of a completed backcasting experiment and its follow-up and spin-off after five to ten years. The first was the Novel Protein Foods (NPF) case, which focused on sustainable and attractive meat alternatives and which envisioned a system innovation in which a substantial share of meat and meat products is replaced by protein foods from non-animal sources. The second was the Sustainable Household Nutrition (SHN) case. The third was the Multiple Sustainable Land-use (MSL) case, which dealt with function integration in rural areas involving agriculture and other functions related to landscape, nature, recreation, water production and water management. The NPF case and the MSL cases were based on backcasting experiments that were conducted from 1993 to 1997 at the Dutch governmental Sustainable Technology Programme (STD). The SHN backcasting experiment was conducted between 1998 and 2000 as part of the international project on sustainable households.

While all cases relate to the food production and consumption system in the Netherlands, each case focuses on different socio-technical systems with different characteristics. The NPF case focuses on a production and consumption system of protein foods, which includes meat alternatives, meat and meat products, and in which food companies are the central players. The SHN case views food consumption and production from the viewpoint of households and consumers. Finally, the MSL case focuses on a regional system in which the agricultural function was integrated with other functions and as a consequence spatial planning aspects are important.

All three backcasting experiments studied involved a wide range of stakeholders, developed one or several desirable future visions, proposed follow-up activities and action agendas and induced higher order learning among participating stakeholders. Higher order learning occurred on the topics under study in the backcasting experiments, as well as on the backcasting approach itself. Follow-up agendas included R&D-activities, strategy development, policy recommendations and short-term proposals.

Despite these similarities, the three backcasting experiments varied considerably in terms of stakeholder influence, the degree of stakeholder involvement, whether a vision champion and institutional protection emerged, whether other types of participation than the capacity to participate were mobilised (e.g. co-funding), and the degree to which the future visions have provided guidance and orientation. There also differences in joint and congruent learning at the level of groups of actors.

The case studies showed strongly varying extents of follow-up and spin-off after five to ten years. The SHN case showed very limited follow-up and spin-off. By contrast, the MSL case and the NPF case showed considerable follow-up and spin-off across the four societal domains distinguished, as well as instances of institutionalisation. The emphasis of follow-up and spin-off was in the research domain, while in the NPF case considerable follow-up also occurred in the business domain. Both cases showed instances of initial institutionalisation.

Nearly all activities in the MSL and NPF cases involved often actors from more than one domain. It is possible to cluster follow-up and spin-off activities into groups of activities that relate to shared adjustment in the future vision. All clusters include actors from the backcasting experiment, as well as newly mobilised actors. The main share of financial resources at follow-up and spin-off activities in all domains involved government funding and budgets. In the NPF case a second major source of mobilised resources involved investments by companies for R&D, product development and market introduction.

Empirical conclusions

The **first** conclusion is that backcasting experiments involving various stakeholders from different societal domains can result in the development, exploration and analysis of desirable visions of the future that provide guidance (where to go) and orientation (what to do) to involved stakeholders; backcasting experiments can also lead to instances and processes of higher order learning among participating stakeholders and in the formulation of follow-up agendas.

The **second** conclusion is that this does not automatically lead to follow-up, spin-off and implementation in line with the vision and the follow-up agenda, but that this depends on various internal and external factors that can be both enabling and constraining.

Important enabling **internal factors**, relating to the backcasting experiment, are institutional protection by top level of participating stakeholders, the emergence of vision champions, a high degree of stakeholder involvement, other types of participation in addition to the capacity to participate like co-funding or substantial 'free' capacity, a strong focus on follow-up and spin-off, a single vision backcasting experiment, and high degrees of influence to key stakeholders. Constraining internal factors related to the backcasting experiment include a multiple vision backcasting experiment and a strong focus on academic achievements. **External factors** evolving in the surrounding socio-technical system or its context can be both constraining and enabling, but these are case-specific.

The **third** conclusion is that the vision of the desirable future is relevant to follow-up and spin-off activities and provides high degrees of guidance (where to go) and orientation (what to do). Follow-up and spin-off activities are constituted by networks of actors that have been successful in mobilising sufficient resources for establishing the activities. Visions of the future at play in follow-up and spin-off show both stability and flexibility; visions co-evolve with networks in the sense that networks and actors are influenced and inspired by the visions, while networks and actors involved in follow-up and spin-off influence and adjust the vision too.

The **fourth** conclusion is that when substantial follow-up and spin-off occur after five to ten years, they still take place at the level of niche activities, or concern a set of niches in the four distinguished domains of research, business, government, and of public interest groups and the general public. This follow-up and spin-off comes along with first instances of broader impacts and institutionalisation. The niches have 'grown out' of the backcasting experiments and can be seen as first steps or stepping stones towards system innovations towards sustainability.

Selected theoretical results and reflections

The conceptual framework as a whole, as well as the set of propositions have proved a relevant and useful guide during the research and helped to obtain the results and conclusions described in this thesis. Another theoretical conclusion is that industrial network theory can be adjusted to analyse follow-up and spin-off activities of backcasting experiments in various societal domains. A third conclusion is that the Leitbild concept is not only applicable to emerging technologies and radical innovations, but after some refinements also to system innovations towards sustainability using visions with an explicitly normative component.

Clusters of activities as found in this research can be seen as niches, which can be found in all four societal domains distinguished. This finding contains various suggestions for refining the existing niche concept in innovation studies. Firstly, a niche is not necessarily a simple phenomenon like a market niche or a technological niche, but can comprise various types of niche activities in different societal domains that may have an implicit decentralised kind of coordination provided by a future vision.

One element of the conceptual framework that certainly deserves further theorising and conceptualisation is the process through which the follow-up and spin-off 'grow' out of the backcasting experiment. So far, the backcasting experiment and its impact have been conceptualised separately, but the process that connects the two phases has not been conceptualised. This is certainly a direction in which further theorising is possible by integrating elements from theories on higher order learning as well as on organisational behaviour. Eventually this could result in connecting higher order learning among groups of stakeholders to organisational behaviour and behavioural alternatives. This could allow for relating actor learning to behaviour by actors, as well as to take into account emerging opportunities and context factors.

Higher order learning at the level of individual actors was not difficult to achieve, whereas higher order learning at the level of groups of actors was. This finding points to (1) making the applied learning concept more dynamic and assuming and (2) to distinguish better between higher learning by individual actors and joint or congruent learning among groups of actors.

Vision development and network formation are mutually interdependent: a vision exerts influence on networks around activities (through attraction, mobilisation, learning and internalisation), and the networks and the actors making up the network influence the vision leading to adjustment or emphasis on specific elements. Existing theories focus on either visions or on networks, but not on the interaction and dependencies. This calls for further theory development, as well as to take into account flexibility and stability of future visions.

Methodical results and reflections

Backcasting originates from the 1970s and was originally developed as an alternative for traditional forecasting and planning. The original focus was on policy analysis for energy planning, and shifted later to exploring sustainable futures and solutions. Stakeholder participation and achieving implementation became important in the 1990s in which the Netherlands was a front runner. Backcasting can be applied at the level of organisations, regions, industrial sectors, socio-technical systems, countries as well as on a global scale. While the approach is named after backcasting or backwards looking from the desirable future, the backcasting step itself is at the same time the least developed and formalised part of the approach. Backcasting is not the only approach using normative or desirable future visions; participatory backcasting approaches are part of a family of approaches, all combining desirable future visions or normative scenarios with stakeholder participation.

A methodological framework for participatory backcasting has been developed, which consists of five steps and the outline of a toolkit containing four groups of methods and tools: (1) design tools, (2) participatory tools, (3) analytical tools and (4) management, coordination and communication tools. This framework also distinguishes three types of demands, as well as different types of goals. A distinction has been drawn between normative demands, process demands and knowledge demands.

The way that the three backcasting experiments were conducted matched well with the developed methodological framework for participatory backcasting. The five steps could be distinguished in the cases, but were not applied in the proposed order in all three cases; iteration of steps took place, especially with respect to the future vision and backcasting. Especially design activities and the backcasting analysis deserve further methodological elaboration. Finally, embedding of outcomes and follow-up action agenda, as well as raising commitment among stakeholders was part of stakeholder communication throughout the three backcasting experiments.

Recommendations

It is recommended to **organisers** of backcasting experiments, as well as organisations that commission backcasting experiments, to use the developed methodological framework for participatory backcasting to design and conduct backcasting experiments. It is recommended to **developers** of backcasting to extend and elaborate the four groups of methods that can be distinguished in backcasting. This should be not only be done based on research, but also based on the experience of organisers and appliers of backcasting. This could for instance be done through establishing a Community of Practice on backcasting or include backcasting in a Community of Practice on foresighting and future studies for sustainability.

To **initiators** and **commissioners** of backcasting experiments, especially the government, it is recommended not to limit support and facilitation to the backcasting experiment, immediate follow-up and knowledge development, but to extend support and facilitation to follow-up and spin-off after more than ten years. The latter allows further facilitation of desirable system innovations towards sustainability. This could include support of market development, new regulation and adjustment of existing regulation.

Three major recommendations to **researchers** can be made: (i) extending the number of evaluations of backcasting experiments and their impact studies; (ii) further theorising and conceptualising on mechanisms and other theoretical aspects connected to the dynamics in backcasting experiments, their follow-up and how these relate to system innovation theories; (iii) further methodology development.

Jaco Quist, March 2007

SAMENVATTING

Backcasting voor een duurzame toekomst: het effect na 10 jaar

Duurzame ontwikkeling, systeeminnovaties en backcasting

Duurzame ontwikkeling (DO) is een ontwikkeling waarin huidige generaties zodanig in hun behoeften voorzien dat toekomstige generaties ook in hun behoeften kunnen voorzien. DO gaat over het bereiken van een betere balans tussen de ecologische, de sociale en de economische dimensie van ontwikkeling. Het gaat onder meer om een eerlijkere verdeling van welvaart tussen landen, het behoud van biodiversiteit en het sterk verminderen van de milieubelasting.

DO vereist structurele veranderingen op het niveau van socio-technische systemen, zoals industriële sectoren, de landbouw, en consumptiedomeinen als mobiliteit, voeding, wonen. Voor het complexe veranderingsproces waarin een bestaand socio-technische systeem wordt omgevormd tot een duurzaam socio-technisch systeem wordt hier de term systeeminnovatie voor DO gebruikt. Hoewel duurzaam(heid) niet exact gedefinieerd kan worden en mede beïnvloed wordt door de opvattingen van actoren en belanghebbenden in een specifiek socio-technisch systeem, gaan systeeminnovaties voor DO altijd over verbeteringen in de ecologische dimensie en het verbeteren van de balans tussen de ecologische, sociale en economische dimensies. Bij systeeminnovaties voor DO is de participatie van belanghebbenden en actoren van verschillende maatschappelijke domeinen belangrijk. Deze partijen hebben relevante kennis en hulpbronnen en zijn nodig om aan activiteiten bij te dragen. Ook zijn ze nodig voor legitimatie, steun en draagvlak voor de veranderingen waarmee een systeeminnovatie gepaard gaat.

Backcasting is een veelbelovende aanpak om systeeminnovaties voor DO te verkennen, de eerste stappen hiervoor te definiëren en tot uitvoering te brengen. De backcasting-aanpak staat (1) voor het ontwikkelen van een wenselijk toekomstbeeld en hieruit terug te kijken naar mogelijke ontwikkelingstrajecten, gevolgd door (2) het ontwerpen van een traject van het heden naar het toekomstbeeld toe en dit in gang zetten waarvoor een vervolgagenda en vervolgacties gedefinieerd worden. Voor participatieve backcasting-studies wordt hier de term backcasting-experiment gebruikt, waarbij experiment verwijst naar sociaal experiment.

Probleemstelling en onderzoeksvragen

In de afgelopen vijftien jaar hebben enige tientallen participatieve backcasting-experimenten plaatsgevonden in Nederland. De rapporten en artikelen hierover gaan vooral in op de aanpak, de inhoudelijke resultaten en soms op proces- en leerresultaten kort na afloop. Studies naar de effecten en doorwerking van backcasting-experimenten na vijf tot tien jaar ontbreken echter. Wel is bekend dat backcasting-experimenten aanzienlijk verschillen in hoe participatie van actoren en belanghebbenden is georganiseerd, de wijze waarop het toekomstbeeld is ontwikkeld en de mate waarin vervolgactiviteiten en andersoortige spin-off is gerealiseerd. Ook zijn er verschillen in de methoden die gebruikt zijn en in wat en hoeveel betrokken actoren hebben geleerd. Empirisch ex-post onderzoek naar de doorwerking van backcasting-experimenten en de factoren die dit beïnvloeden is nodig om beter inzicht te krijgen, maar ontbreekt tot nu toe. Ook ontbreekt het aan conceptuele en analytische kaders om doorwerking en effecten van backcasting experimenten in kaart te brengen. Onderzoek hiernaar kan bijdragen aan de verdere ontwikkeling van de backcasting-aanpak en tot beter inzicht in de relatie tot (systeem)innovatietheorie. Ook kan het leiden tot aanbevelingen voor partijen die deelnamen aan backcasting-experimenten of deze initiëren, zoals de overheid. Dit heeft geleid tot de volgende hoofdvragen voor dit onderzoek:

A. *Welke factoren beïnvloeden de doorwerking en de effecten van backcasting-experimenten?*
B. *Hoe kan participatieve backcasting verbeterd worden voor het verkennen en vormgeven van systeeminnovaties voor DO?*

De hoofdvragen zijn onderzocht in acht stappen:

STAP 1: Uitwerking probleemstelling, onderzoeksvragen en onderzoeksaanpak (H1).

STAP 2: Studie van de backcasting-aanpak en verwante methoden (H2).

STAP 3: Verkenning van de theorie over toekomstbeelden, stakeholder participatie, leren van actoren, systeeminnovatie en netwerken (H3).

STAP 4: Uitwerking conceptueel kader, proposities en onderzoeksmethodologie (H4).

STAP 5: Onderzoek naar drie geselecteerde casussen (H5-H7).

STAP 6: Vergelijking van de bevindingen uit de casussen en toetsing van de proposities (H8).

STAP 7: Conclusies, reflecties en aanbevelingen (H9).

Onderzoeksopzet, conceptualisering en onderzoeksmethodologie

Het onderzoek is opgezet als een empirische ex-post studie met meerdere casussen, waarvoor een conceptueel kader, proposities en een onderzoeksmethodologie zijn ontwikkeld. Daarvoor is eerst een methodisch kader voor participatieve backcasting gedefinieerd. Ook is het nodig het backcasting experiment en de doorwerking na vijf tot tien jaar theoretisch te funderen. Daartoe beschrijft en vergelijkt de theoretische verkenning een aantal theoretische concepten over toekomstbeelden, participatie van actoren, leren door actoren, en diverse systeeminnovatie- en netwerktheorieën. Hiervoor gebruik is gemaakt van literatuur uit de innovatiewetenschappen, de beleidswetenschappen, de bestuurskunde en technology assessment. Er is onder meer gebleken dat netwerktheorieën beter geschikt zijn voor het beschrijven van de doorwerking van backcasting-experimenten dan systeeminnovatietheorieën. Op basis van de theoretische analyse zijn concepten geselecteerd voor de ontwikkeling van het conceptuele kader.

Het conceptuele kader onderscheidt twee fases. Het backcasting-experiment vormt de eerste fase. Vervolgactiviteiten en spin-off na vijf tot tien jaar vormen de tweede fase. Elke fase is opgebouwd uit onderdelen (bouwstenen) die weer bestaan uit verschillende aspecten. Het backcasting-experiment bestaat uit vier onderdelen: (1) toekomstbeelden; (2) participatie; (3) leren; (4) methodische en organisatorische aspecten. Voor de ontwikkeling van de vier onderdelen is respectievelijk gebruik gemaakt van het Leitbild concept uit de Duitse technieksociologie, diverse theorieën over de participatie van actoren, theorie over hogere-orde leren van actoren en het ontwikkelde methodische kader voor backcasting. Voor de invulling van **toekomstbeelden** zijn van het *Leitbild* concept de aspecten 'richting' (waar naartoe) en 'oriëntatie' (wat te doen) afgeleid, terwijl het aspect 'concurrerende toekomstbeelden' is toegevoegd. **Participatie** van actoren omvat diverse aspecten gebaseerd op verschillende participatietheorieën, waaronder 'stakeholderheterogeniteit', de 'invloed van stakeholders en het 'niveau van betrokkenheid'. Bij **leren** door actoren ligt de nadruk op het zogeheten hogere-orde leren van actoren. Deze vorm van leren, gaat over veranderde inzichten van actoren over probleemdefinities, oplossingen voor deze problemen en hoe deze aangepakt dienen te worden. Hogere-orde leren gaat ook over de mate waarin veranderde inzichten door meerdere actoren gedeeld of geaccepteerd worden. Het onderdeel **methodische en organisatorische aspecten** bevat aspecten over hoe de participatieve backcasting-aanpak is toegepast en georganiseerd.

De fase van vervolg en spin-off is geconceptualiseerd met drie onderdelen: (1) toekomstbeelden; (2)

netwerkvorming; (3) institutionalisering. Net als in het backcasting-experiment bestaat toekomstbeelden uit de aspecten 'richting', 'oriëntatie' en 'concurrerende toekomstbeelden'. Bij **netwerkvorming** worden 'activiteiten', 'actoren' en 'hulpbronnen' onderscheiden als relevante aspecten. Netwerkvorming is afgeleid uit de industriële netwerktheorie. Voor **institutionalisering** is gebruik gemaakt van institutionele theorie; dit onderdeel gaat enerzijds over veranderingen in instituties en regels en anderzijds over de institutionele weerstand hiertegen vanuit gevestigde belangen en bestaande regels.

Twee groepen factoren worden onderscheiden die invloed uitoefenen op de vorming en mate van vervolgactiviteiten en spin-off. Eigenschappen van het backcasting-experiment vormen een eerste groep van **interne factoren**. Een tweede groep bestaat uit **externe factoren**. Externe factoren zijn in dit onderzoek externe gebeurtenissen en ontwikkelingen die invloed uitoefenen op de totstandkoming van vervolgactiviteiten en andere spin-off.

Zowel het backcasting-experiment als vervolgactiviteiten en spin-off vindt plaats in een socio-technisch systeem. Vier maatschappelijke domeinen worden onderscheiden, waarin vervolgactiviteiten en spin-off kunnen plaatsvinden: (1) onderzoek; (2) bedrijven; (3) overheid; en (4) het publieke domein van maatschappelijke organisaties, burgers en algemene media.

Op basis van het conceptuele kader zijn proposities en een onderzoeksmethodologie ontwikkeld. De proposities richten zich op de mogelijke samenhang tussen kenmerken van een backcasting-experiment en de mate van vervolg en spin-off. Om casussen te analyseren en de proposities te toetsen zijn indicatoren ontwikkeld voor de aspecten waaruit het backcasting experiment is opgebouwd. Ook zijn indicatoren ontwikkeld voor de aspecten waaruit de doorwerking is opgebouwd. Voor elke casus is gebruik gemaakt van diepte-interviews met betrokkenen in de twee fasen en experts. Daarnaast zijn verschillende soorten documenten gebruikt.

Drie casussen en analyseresultaten

Voor de drie casussen heeft een beschrijving en een analyse plaatsgevonden van (1) het backcasting-experiment en de directe resultaten daarvan, en van (2) de gerelateerde vervolgactiviteiten en spin-off vijf tot tien jaar later. De eerste casus gaat over Novel Protein Foods (NPF) en vlees vervangers. De tweede casus gaat over Duurzaam Voedselgebruik in het Huishouden (DVH). De derde casus gaat over Meervoudig Duurzaam Landgebruik (MDL). De eerste en derde casus betreffen backcasting-experimenten die uitgevoerd zijn bij het programma Duurzame Technologische Ontwikkeling tussen 1993 en 1997. De tweede (DVH) casus betreft een backcasting-experiment dat tussen 1998 en 2000 is uitgevoerd als onderdeel van een internationaal project over duurzame huishoudens.

Alle casussen hebben betrekking op delen van het productie- en consumptiesysteem voor landbouw en voeding. Wel het gaat om verschillende delen (of subsystemen) met verschillende eigenschappen. De NPF casus gaat over het productie- en consumptiesysteem van vlees en vleesvervangers met bedrijfsactiviteiten en consumptieactiviteiten als kern. De DVH casus gaat over voeding en voedselproductie vanuit huishoudelijk perspectief met huishoudens en huishoudelijke consumptie als kern. De MDL casus gaat over regionale ruimtelijke systemen waarin meerdere ruimtelijke functies zoals landbouw, recreatie, natuur, landschap en diverse waterbeheerfuncties gecombineerd worden op hetzelfde grondoppervlak. Behalve de vertegenwoordigers van de verschillende functies zijn hier verschillende overheden belangrijke actoren.

Alle onderzochte backcasting-experimenten worden gekenmerkt door participatie van een heterogene groep actoren en belanghebbenden, de ontwikkeling van tenminste één toekomstbeeld, het definiëren van vervolgactiviteiten en een vervolgagenda in lijn met het toekomstbeeld en hogere-orde leren door individuele actoren in het backcasting-experiment. Leren omvat veranderingen in de opvattingen over (1) probleemdefinities en oplossingen voor de (duurzaamheids)problemen in het socio-technische systeem en (2) de aanpak om deze problemen op te lossen en de oplossingen te realiseren.

Er zijn ook verschillen tussen de backcasting-experimenten over MDL en NPF enerzijds, en het DVH back-casting-experiment anderzijds. Deze verschillen hebben onder meer betrekking op de invloed van actoren en belanghebbenden op de inhoud van het toekomstbeeld; de wijze waarop participatie georganiseerd is; het niveau van betrokkenheid; de aanwezigheid van 'visiekampioenen' of 'visietrekkers'; institutionele be-scherming en de mate waarin toekomstbeelden richting en oriëntatie hebben gegeven aan actoren. Ook is er verschil in de mate waarin actoren gezamenlijk leren over problemen, oplossingen, aanpak en prioriteiten.

De drie casussen verschillen aanzienlijk in de mate waarin vervolgactiviteiten en andere spin-off gevon-den zijn. De DVH casus kent nauwelijks vervolgactiviteiten en spin-off. De NPF casus en de MDL casus tonen beiden een aanzienlijke mate van vervolgactiviteiten en spin-off verspreid over de vier domeinen, maar met het zwaartepunt in het onderzoeksdomein. In de NPF casus ligt een tweede zwaartepunt in het bedrijfsdo-mein. Zowel de NPF casus als de MDL casus bevat voorbeelden van beginnende institutionalisering.

Bij vrijwel alle activiteiten waren actoren uit meerdere domeinen betrokken. Het is mogelijk vervolg-activiteiten en spin-off te groeperen in clusters van activiteiten met gemeenschappelijke veranderingen in het toekomstbeeld. Deze clusters bevatten meestal zowel nieuwe actoren, als actoren uit het back-casting-experiment. Financiële middelen van verschillende overheden bleken essentieel te zijn voor de totstandkoming van vervolgactiviteiten en andere spin-off in zowel de MDL casus als de NPF casus. De NPF casus toont ook substantiële investeringen van bedrijven in productontwikkeling, productiefaciliteiten en marktintroducties.

Empirische conclusies

De **eerste** conclusie is dat backcasting-experimenten, waarin stakeholders van verschillende maatschap-pelijke domeinen participeren, kunnen leiden tot het ontwikkelen, verkennen en onderzoeken van wenselijke toekomstbeelden die voorzien in richting (waar naartoe) en oriëntatie (wat te doen). Backcasting-experi-menten resulteren ook in hogere-orde leren bij betrokken actoren en in het formuleren van agenda's voor vervolg en implementatie.

De **tweede** conclusie is dat backcasting-experimenten met dergelijke resultaten niet altijd leiden tot vervolgactiviteiten en spin-off in lijn met het wenselijke (duurzame) toekomstbeeld en de geformuleerde vervolgagenda. De mate van vervolgactiviteiten en spin-off blijkt zowel afhankelijk te zijn van verschillende interne factoren als van externe factoren.

Belangrijke **interne factoren** met een positieve invloed zijn: (i) institutionele bescherming door de top van betrokken actoren; (ii) de aanwezigheid van een of meerdere 'visiekampioenen' (individuen van naam en faam die hun netwerk inzetten om het backcasting-experiment te laten slagen en vervolgactiviteiten en spin-off te verwezenlijken); (iii) een hoge mate van betrokkenheid van stakeholders; (iv) andere vormen van participatie dan alleen deelname aan bijeenkomsten, zoals cofinanciering of substantiële extra capaciteit; (v) focus op het realiseren van vervolgactiviteiten en spin-off en (vi) het werken met één toekomstbeeld in het backcasting-experiment. **Interne factoren** met een negatieve invloed op de doorwerking zijn (i) meer-dere toekomstbeelden in een backcasting-experiment en (ii) focus op andere doelen dan vervolg en spin-off, bijvoorbeeld wetenschappelijke resultaten. Daarnaast zijn in alle casussen **externe factoren** gevonden die een negatieve of positieve invloed uitoefenen op de totstandkoming van vervolg en spin-off.

De **derde** conclusie is dat de in de backcasting-experimenten ontwikkelde toekomstbeelden belangrijk zijn voor vervolgactiviteiten en spin-off doordat ze richting (waar naartoe) en oriëntatie (wat te doen) geven aan het denken en handelen van actoren en belanghebbenden. Vervolgactiviteiten en spin-off zijn zichtbaar als netwerken van actoren die succesvol zijn in het mobiliseren van voldoende heterogene hulpbronnen om de activiteiten te realiseren. Toekomstbeeld en netwerken van actoren rond vervolgactiviteiten beïnvloeden elkaar wederzijds. Toekomstbeelden vertonen qua inhoud zowel flexibiliteit als stabiliteit. Zowel flexibiliteit als stabiliteit zijn belangrijk voor de totstandkoming van vervolg en spin-off.

De **vierde** conclusie is dat vervolgactiviteiten en spin-off na vijf tot tien jaar nog niet gezien kunnen worden als een systeeminnovatie, maar gezien moeten worden als een niche of een aantal gerelateerde niches in verschillende maatschappelijke domeinen. De niches zijn voortgekomen uit de backcasting-experimenten en kunnen gezien worden als mogelijke kiemen voor systeeminnovaties voor DO.

Voornaamste theoretische resultaten en conclusies

Het conceptuele kader en de proposities en hun theoretische basis hebben dit onderzoek op een goede wijze ondersteund. Een andere theoretische conclusie is dat industriële netwerktheorie na aanpassing ook toepasbaar is voor het beschrijven en analyseren van vervolgactiviteiten en spin-off van backcasting-experimenten. Hierbij maken netwerken van heterogene actoren activiteiten mogelijk als ze voldoende hulpbronnen weten te mobiliseren. Het Leitbild concept is ontwikkeld voor emergente nieuwe technologieën en technologische artefacten. Na aanpassing is dit concept goed toepasbaar voor systeeminnovaties voor DO met een expliciete normatieve component.

Ook is gebleken dat het nicheconcept uit de innovatietheorie niet beperkt hoeft te worden tot marktniches of technologische niches waarin over nieuwe technologische innovaties en gebruikersaspecten geleerd kan worden. Vervolg en spin-off van de onderzochte backcasting-experimenten kunnen getypeerd worden als een grotere niche die vertakkingen heeft in verschillende maatschappelijke domeinen, of als een aantal kleinere niches in verschillende maatschappelijke domeinen die verbonden zijn door een toekomstbeeld. Dit inzicht maakt het mogelijk beleidsinitiatieven, beleidsprogramma's, onderzoeksprogramma's en sociale experimenten ook als niche(-onderdeel) te zien. Dit vormt een aanvulling op de bestaande theoretische inzichten over niches in de innovatietheorie.

Het gebruik van het conceptuele kader in dit onderzoek levert aanknopingspunten voor verdere conceptualisering en theorieontwikkeling. Als eerste kan genoemd worden dat het proces van backcasting-experiment tot niche niet is uitgewerkt in dit onderzoek. Een dergelijke uitbreiding biedt mogelijkheden om de leereffecten bij actoren niet alleen te koppelen aan gedragsalternatieven voor actoren, maar ook aan daadwerkelijk gedrag van actoren. Hierbij dienen omgevingsfactoren en alliantiemogelijkheden die gedragsalternatieven vergemakkelijken of bemoeilijken meegenomen te worden.

Daarnaast is uit dit onderzoek gebleken dat in elk van de onderzochte casussen hogere-orde leren op het niveau van individuele actoren voorkomt. Hogere-orde leren door groepen actoren leidend tot gedeelde en congruente (niet conflicterende) inzichten en het omzetten hiervan in gedragsverandering komt voor, maar is veel moeilijker te bereiken. Dit biedt aanknopingspunten om theoretisch en conceptueel (1) beter onderscheid te maken tussen hogere-orde leren door individuele actoren en door groepen actoren en (2) aandacht te schenken aan de tijdsdimensie, omdat leren door individuele actoren meestal vooraf gaat aan gezamenlijk of congruent leren.

Een ander punt vormt de interactie tussen wenselijk toekomstbeeld en netwerken die elkaar wederzijds beïnvloeden. Bestaande theorieën richten zich hetzij op toekomstbeelden, hetzij op netwerken, maar niet op het combineren van beiden. Dit vergt theorieontwikkeling over de samenhang en wisselwerking tussen netwerken en toekomstbeelden en hoe zich dit verhoudt tot flexibiliteit en stabiliteit van toekomstbeelden, evenals over de empirisch gevonden geneste karakter van toekomstbeelden.

Het is niet duidelijk geworden in dit onderzoek welke mate van invloed op de inhoud door actoren en belanghebbenden nodig is om tot een hoge mate van vervolg en spin-off te komen. In de casussen met substantiële doorwerking is een gevarieerd beeld gevonden waarin verschillende groepen actoren en belanghebbenden in verschillende mate invloed op de inhoud uitoefenen. Wel is empirisch een strategisch niveau (consultatie van sleutelactoren) en een operationeel niveau (met uitwerking van samenwerkende onderzoekers) gevonden. Ook hebben de aard (meer dan alleen tijd) en de mate van participatie (in tijd en geld) een positieve invloed op doorwerking. Op basis van deze uitkomsten is het mogelijke de bestaande theorievorming over participatie van actoren en belanghebbenden en het sturingsconcept netwerkmanagement aan te vullen.

Enkele methodische resultaten en conclusies

Op basis van een viertal backcasting-methoden uit de wetenschappelijke literatuur is een methodisch kader voor participatieve backcasting ontwikkeld. Dit methodische kader bestaat uit vijf stappen en onderscheidt vier groepen methoden die binnen het kader toegepast kunnen worden: (1) methoden voor interactie met actoren en belanghebbenden; (2) ontwerpmethoden; (3) analysemethoden en (4) methoden voor communicatie, coördinatie en management. Daarnaast worden verschillende soorten vereisten onderscheiden: kennisvereisten, procesvereisten en normatieve vereisten.

In dit onderzoek blijkt dat de wijze waarop de backcasting-aanpak is uitgevoerd in de casussen in lijn is met het ontwikkelde methodische kader. De vijf gedefinieerde stappen zijn duidelijk onderscheidbaar, maar zijn meestal niet in de voorgestelde volgorde uitgevoerd. In plaats daarvan vindt iteratie van verschillende stappen plaats. Een toekomstbeeld kan bijvoorbeeld gradueel ontwikkeld worden, waarna weer een vorm van backcasting-analyse wordt uitgevoerd. Ook is communicatie met actoren buiten het backcasting-experiment en het verwerven van de steun van actoren voor uitkomsten en vervolgactiviteiten niet een afgebakende activiteit aan het eind van het backcasting-experiment, maar een doorlopende activiteit die geïntensiveerd wordt naar het einde toe.

Backcasting behoort tot een grotere groep van methoden voor toekomstverkenningen waarin wenselijke normatieve toekomstbeelden en toekomstvisies gecombineerd worden met participatie van stakeholders. Backcasting is vernoemd naar het terugkijken vanuit wenselijke toekomstbeelden. Toch is de backcasting-stap zelf methodisch het minst uitgewerkte deel van de aanpak. Participatieve backcasting is ook geschikt voor complexe problemen waarin de duurzaamheidscomponent minder of niet belangrijk is.

Aanbevelingen

Toepassers van participatieve backcasting wordt aanbevolen het ontwikkelde methodische kader als startpunt voor het opzetten van backcasting-experimenten te nemen en gebruik te maken van de opgestelde richtlijnen hiervoor. Ook wordt aanbevolen de selectie van methoden af te stemmen op de prioriteit van verschillende typen doelen. **Ontwikkelaars** van backcasting wordt aanbevolen verder invulling te geven aan elk van de vier groepen methoden op basis van zowel wetenschappelijk onderzoek als praktijkervaringen van toepassers. Het verdient aanbeveling om hiervoor een Community of Practice op te richten of dit onderdeel te maken van een Community of Practice over participatieve toekomstverkenningen en DO.

Aan **opdrachtgevers** en **initiatiefnemers** van backcasting-experimenten rond complexe duurzaamheidsproblemen, met name de overheid, wordt aanbevolen niet alleen het backcasting-experiment, onmiddellijke vervolgactiviteiten en wetenschappelijke kennisontwikkeling te stimuleren en te faciliteren. De **overheid** wordt aanbevolen om ook die doorwerking na vijf tot tien jaar te faciliteren welke verdergaande systeemveranderingen mogelijk maken. Het kan bijvoorbeeld gaan om het ondersteunen van (markt)ontwikkeling, het faciliteren via nieuwe regulering of het aanpassen van bestaande regulering.

Met betrekking tot **onderzoek** wordt aanbevolen vervolgstudies naar de doorwerking van backcasting-experimenten uit te voeren. Ook kan onderzoek bijdragen aan (1) methodologieontwikkeling over backcasting en (2) theorieontwikkeling voor de doorwerking van backcasting. Daarnaast wordt aanbevolen vergelijkende studies te doen naar backcasting-experimenten in andere landen en naar andere participatieve toekomstverkenningsmethoden waarin wenselijke toekomstbeelden ontwikkeld worden. Nader onderzoek is ook gewenst naar (1) het relatieve belang en onderlinge afhankelijkheid van de verschillende interne factoren, en (2) de gewenste verhouding tussen stabiliteit en flexibiliteit in toekomstbeelden om tot effectieve sturing te komen.

Jaco Quist, maart 2007

CURRICULUM VITAE

Jaco Quist (1966) is an assistant professor at the Technology Dynamics & Sustainable Development Group of the Faculty of Technology, Policy and Management, Delft University of Technology (DUT).

Between mid-2003 and 2007 he combined teaching with writing a PhD thesis on the follow-up and spin-off of participatory backcasting. His research interests include system innovations towards sustainability, foresighting and constructive technology assessment. His teaching includes technology policy, technology assessment, sustainable innovation and sustainable entrepreneurship. Occasionally, he facilitates workshop sessions and trainings on visioning, backcasting and scenario-building, for instance at the bi-annual Rathenau Technology Assessment Summer School.

From 1994 to 1997 Jaco Quist was involved in the Sustainable Technology Development (STD) programme, where he was a project coordinator in several projects on sustainable technologies in the field of food production and nutrition. In 1998 he started working at the Technology Assessment Group of DUT as a researcher for the international project 'Strategies for the Sustainable Household (SusHouse)'. From 2000 to 2002 he also coordinated the annual four weeks introduction programme for international MSc students at DUT.

Jaco Quist holds an MSc in Molecular Sciences from Wageningen University, where he graduated in 1992 on photo-acoustic laser spectroscopy applied to trace detection in water and the atmosphere, and NMR flow imaging of plant model systems. Part of his research training was conducted in 1990 at the University of Malaysia, Kuala Lumpur and the Gadjah Mada University in Yogyakarta, Indonesia.